Extended Producer Responsibility

UPDATED GUIDANCE FOR EFFICIENT WASTE MANAGEMENT

OECD
BETTER POLICIES FOR BETTER LIVES

Please cite this publication as:
OECD (2016), *Extended Producer Responsibility: Updated Guidance for Efficient Waste Management*, OECD Publishing, Paris.
http://dx.doi.org/10.1787/9789264256385-en

ISBN 978-92-82-25629-3 (print)
ISBN 978-92-82-25638-5 (PDF)
ISBN 978-92-64-25637-8 (e-pub)

Photo credits: © Alexmillos/Dreamstime.com.

Corrigenda to OECD publications may be found on line at: *www.oecd.org/publishing/corrigenda*.

Foreword

In a world in which sustained demographic and economic growth are exerting increasing pressures on natural resources, establishing a resource efficient economy is central to greening growth. The total volume of material extracted or harvested worldwide reached nearly 72 billion metric tonnes in 2010, doubling since 1980 and an estimated ten-fold increase over the last century. Curbing these trends requires policies that improve resource productivity and that ensure sustainable materials management, building on the principle of the 3Rs – reduce, reuse, recycle – and encouraging more sustainable consumption patterns.

In this context, extended producer responsibility (EPR), an environmental policy approach in which a producer's responsibility for a product is extended to the post-consumer stage of a product's life cycle have become common. Currently about 400 such schemes are in operation across the world, most of them in the OECD and some in emerging market economies. While extended producer responsibility systems have contributed to significantly increasing material recovery rates from certain waste streams, as well as providing a reliable source of funding for the collection and recycling of waste, the economic and environmental performance of these systems appears to be highly disparate. As a consequence, there is a sustained debate about the way EPR systems should be designed.

This report contributes to this debate by providing updated guidance on the design of EPR. This work updates the OECD Guidance Manual for Governments, published in 2001, by drawing on recent experience, deepening the analysis in selected areas and adding a discussion on the particular challenges faced by developing countries.

Chapters 1 and 2 provide an introduction and compile updated guidance and recommendations on EPRs. These chapters constitute Part 1 of the Guidance. The remaining chapters constitute Part 2, and provide more in-depth analyses of a number of key elements of EPR schemes that underlie the updated guidance. Chapter 2 discusses governance, Chapter 3 competition, Chapter 4 design for the environment, and Chapter 5 on the informal sector. The annex provides summary versions of country case studies that were provided by member countries and partners as an input to the Global Forum on Environment focusing on Extended Producer Responsibility that took place in June 2014 in Tokyo.

This review was overseen by the OECD Working Party on Resource Productivity and Waste (WPRPW) and benefited from the documents prepared for, and the discussions at, the Global Forum on EPR that took place in Tokyo and the support provided by an ad hoc expert group.

At a time when many governments are grappling with the challenge of developing a policy mix that can support the transition towards a more resource efficient, circular economy, this report provides some valuable guidance concerning one of the policy approaches that can assist.

Simon Upton
Director, OECD Environment Directorate

Acknowledgements

This report has been developed under the responsibility of the Working Party on Resource Productivity and Waste (WPRPW). Within the OECD Secretariat, the work was managed by Peter Börkey under the overall responsibility of Shardul Agrawala, the Head of the Environment and Economy Integration Division. The work was further supported by Shunta Yamaguchi, Andrew Prag, Maria Dahl and Maroussia Klep, Elisabeth Corbett, Pascale Rossignol and Katjusha Boffa provided secretarial support to the project.

The report brings together a number of papers that were prepared by several experts. Chapter 1, which provides an introduction and overview, was drafted by Brendan Gillespie (Consultant) and Peter Börkey (OECD), drawing on the other chapters. Chapter 2 on the governance of extended producer responsibility (EPR) was prepared by Reid Lifset and Gonzalo Urbina from Yale University; Chapter 3 on EPR and competition was developed by Sally van Siclen (Consultant); Chapter 4 on EPR and design for environment was drafted by Maarten Dubois and Jeff Peters from KU Leuven; Chapter 5 that focuses on the role of the informal sector was prepared by Anne Scheinberg, Brendan Gillespie (both consultants) and Shunta Yamaguchi (OECD). The report further incorporates a survey of EPR and literature review that was prepared by Daniel Kaffine from University of Colorado Boulder. The country case studies, of which summary versions are included in this report, were prepared by: Japan: Yasuhiko Hotta and Atsushi Santo (IGES), Tomohiro Tasaki (NIES), Hajime Yamakawa (Kyoto Prefectural University); Canada: Jacinthe Séguin (Environment Canada); People's Republic of China: Liu Chunlong, (MEP); France: Baptiste Legay (MEDDE); Belgium: Katleen Dierick and Roeland Bracke (OVAM); United States of America: Scott Cassel (Product Stewardship Institute); Australia: Bruce Edwards and Declan O'Connor-Cox (Department of Environment). Case studies for Chile and Colombia were developed with the help of the authorities and Christoph Vanderstricht (Ernst&Young). Maroussia Klep from the OECD Secretariat developed the summary versions of these.

We would also like to thank the numerous experts, many of which represented in an expert group that was set-up specifically to support the development of this work, for their tireless and high-quality inputs to the project. They include Garth Hickle (Minnesota Pollution Control Agency), Joachim Quoden (EXPRA), Alexandra Lange (Eco-Emballages), Arne Campen (Sony), Michel Sponar and Olivier de Clercq (European Commission), Jacinthe Séguin (Environment Canada), Bruce Edwards and (Australia), Yasuhiko Hotta (IGES), Tomohiro Tasaki (NIES), Stéphane Arditi (European Environment Bureau), Daniel Kaffine (University of Colorado Boulder), Mathieu Hestin (Bio IS Deloitte), Ross Bartley (BIR), Shruti Rai (Ministry of Environment India), Scott Cassel (Product Stewardship Institute), Marina Franke, Linda Godfrey (CSIR South Africa), Elisabet Kock (SEPA), Baptiste Legay and Loic Lejay (Ministry of Environment France), Beata Klopotek (Ministry of Environment Poland), Koji Yamada (Ministry of Environment Japan), Jiro Hiratsuka (Permanent Delegation of Japan), Joost Meijer (Ministry of Environment Chile), Erika Mink (EUROPEN),

Kamal Sharma (Confederation of Indian Industries), Christoph Vanderstricht (Ernst&Young), Yang Zheng (Ministry of Environmental Protection China), Valentina Bolognesi (Digitaleurope), Virginia Janssens (EUROPEN), Simon Webb (Procter&Gamble).

In addition, we would like to thank all participants who provided useful insights to this work at the Global Forum on Environment: Promoting Sustainable Materials Management through Extended Producer Responsibility (EPR) that took place in Tokyo, 17-19 June 2014.

Finally, this work would not have been possible without the generous financial support from the European Commission, Switzerland, Finland and Japan.

Table of contents

Tables

Figures

Abbreviations and acronyms

ADF	Advance disposal fee
CNY	Renminbi
CPR	Collective producer responsibility
DfE	Design for environment
DSD	Duales System Deutschland
ELV	End-of-life vehicles
EoL	End-of-life
EPR	Extended producer responsibility
IPR	Individual producer responsibility
LCA	Life-cycle assessment
MRF	Materials recovery facility
NGO	Non-governmental organisation
PET	Polyethylene terephthalate
PRN	Packaging waste recovery note
PRO	Producer responsibility organisation
PS	Product stewardship
PVC	Polyvinyl cloride
R&D	Research and development
RFID	Radio frequency identification
UCTS	Upstream combination tax/subsidy
VAT	Value added tax
WEEE	Waste of electric and electronic equipment
WPRPW	Working Party on Resource Productivity and Waste

Executive summary

Since the late 1980s, the concept of "Extended Producer Responsibility" (EPR) has become an established principle of environmental policy in an increasing range of countries. It aims to make producers responsible for the environmental impacts of their products throughout the product chain, from design to the post-consumer phase. It was hoped that this would relieve the burden on municipalities and taxpayers for managing end-of-life products, reduce the amount of waste destined for final disposal, and increase rates of recycling.

OECD provided a platform for countries to exchange experience, and, in 2001, produced a Guidance Manual to support the development of EPR systems. Since 2001, the number and variety of EPR systems have increased significantly. Thus, a review of recent experience is timely, particularly in view of the support that EPR could provide to enhancing resource productivity and the circular economy, issues that are now high on the environmental policy agendas of many countries. In its first part, this report provides updated Guidance on EPRs, building on the 2001 Manual and in view of the developments and lessons learnt since then. In the second part, the report brings together four selected challenges within EPR and examines them in greater detail.

Evolution and impacts of EPRs

A recent survey identified about 400 EPR systems currently in operation. Nearly three-quarters were established since 2001. Legislation has been a major driver, and most EPRs appear to be mandatory rather than voluntary. Small consumer electronic equipment accounts for more than one-third of EPR systems, followed by packaging and tyres (each 17%), end-of-life vehicles, lead-acid batteries and a range of other products. Various forms of take-back requirements are the most commonly used instrument, accounting for nearly three-quarters of those surveyed. Advance disposal fees (ADF) and deposit/refund account for most of the rest. While in some cases individual firms have established their own systems, in most cases, producers have established collective EPR systems managed by Producer Responsibility Organisations (PROs).

Assessing the impacts of EPR systems is difficult for several reasons: a considerable lack of data, analytical difficulties in distinguishing the impact of EPR systems from other factors, and the wide variety of EPR systems which limits comparison among them. Bearing in mind these caveats, there is evidence that in some countries, EPRs have helped to shift some of the financial burden for waste management from municipalities and taxpayers to producers, and to reduce the public costs of waste management. In addition, it seems likely that EPR systems have contributed to the decreased share of waste destined for final disposal and to the increased rates of recycling recorded in many OECD countries. However, progress in these areas varies very widely among countries, suggesting that there is scope in many countries to improve their performance by emulating the best performers.

On the other hand, the consensus appears to be that while EPR systems have contributed to waste prevention (e.g. eco-design) in some countries and some sectors, they are seldom sufficient to serve as the triggering factor.

Guidance and recommendations

Many of the recommendations regarding the good governance of EPR systems identified in the 2001 OECD Guidance Manual are still relevant and should be applied more systematically. For instance, the 2001 guiding principles for EPRs state that these systems should provide producers with incentives to change product designs, stimulate innovation, take a life-cycle approach, clearly define responsibilities and chose flexible policy instruments adapted to the particular product and waste stream.

Similarly, the key factors in designing EPRs already provided by the 2001 Guidance Manual are still as relevant today as they were 15 years ago. Among other things, these recommendations point out that the objectives and the scope of EPR should be clearly defined and the producers of products subject to EPR clearly identified; mechanisms for reporting and monitoring should be established and appropriate enforcement mechanisms and sanctions developed. There are also opportunities to make EPRs more effective including by: increasing their level of ambition, broadening the scope of products covered, better internalising environmental costs. Arguably, the single most important challenge is to make EPR systems more transparent. EPRs should be required to make available the information needed to assess their performance and to identify ways in which they can be made more efficient and effective.

The 2001 Guidance Manual also provided more specific recommendations on a range of issues, the key elements of which have been integrated with the more recent experience.

Design and governance of EPR

The design and governance of EPR are crucial to their performance. The issues range from target setting and monitoring & enforcement, to free-riding and financing.

- The targets of EPR policies should be periodically reviewed and adjusted, taking account of changes in market conditions and technology.
- In mandatory systems, governments should establish consistent and credible means for enforcing EPR obligations, including registers of producers, official accreditation of producer responsibility organisations (PROs) and appropriate sanctions.
- Adequately resourced monitoring systems need to be established; the performance of EPR operations should be regularly audited, preferably independently. In the same jurisdiction, EPR systems should be harmonised to the extent possible, and a means for checking the quality and comparability of data established.
- Free-riding, which still is a challenge to many EPR systems, should be addressed through peer pressure and strict enforcement.
- Governments should identify ways in which EPR systems can be financed in a sustainable manner. This should include analysis of how risks such as price volatility and leakage could be managed.

Promote the integration of competition policy and EPR

As the recycling and waste management industries have grown and become more concentrated, the potential financial gains for producers, as well as the additional costs to society that result from collusion among producers and other forms of anti-competitive behaviour, have become more significant. Since 2001, some competition authorities and courts have reviewed alleged anti-competitive behaviour within EPR systems.

- Most attention should be placed on competition issues in product markets, where the welfare effects are potentially largest, followed by collection and sorting markets, recovery and disposal markets, and the market of producer responsibility services.
- Concerns persist about collusion among producers and about the potential abuse of vertical agreements between PROs and companies involved in downstream operations. An important means for minimising anti-competitive behaviour is to consult competition authorities when EPR systems are being established.
- Services such as waste collection, sorting, as well as material recovery and disposal should be procured by transparent, non-discriminatory and competitive tenders.
- EPR schemes should allow single PROs only when it can be demonstrated that the benefits (e.g. the capacity to manage the waste would otherwise not be built) outweigh the costs of less competition.

Incentives for design for environment

Better internalisation of end-of-life costs and stricter enforcement would also strengthen incentives for improving the eco-design of products and packaging. Setting fees at a level where they recover the full cost of the end-of-life management of the products covered by the EPR is therefore a key measure.

- Ideally, producer responsibility would be implemented at the level of individual producers, but due to the significant economies of scale and scope that are often available, most EPR systems apply collective producer responsibility, which dilutes incentives for eco-design.
- Where possible, producers' fees should therefore be more closely linked to the actual end-of life treatment costs of their products, for instance through the use of variable (e.g. weight-based) rather than fixed (e.g., unit-based) fees, and/or modulated fees that differ according to specific design features that make products more easily recyclable.
- In the case of globally-traded products, better eco-design incentives could also be achieved by harmonising environmentally-sensitive design.

Integrating informal workers in EPRs in emerging and developing countries

Since 2001, EPR systems have been established in many developing and emerging economies. In contrast to the most developed OECD countries, there are large numbers of informal waste workers engaged in recycling; an estimated 20 million globally. Waste picking is often hard, dangerous and socially precarious. While there are serious concerns about downstream informal dismantling and recycling which can generate negative economic and environmental impacts, the potentially positive contribution of informal waste collection and sorting activities is increasingly recognised.

- EPR systems need to find ways for informal operators to work with rather than against formal waste management systems, unless there is a risk that they will be undermined by them.
- However, this is not always easy or possible, and it will be important to draw lessons from current initiatives to guide further policy development in this area.

PART I

Overview and updated guidance

Extended Producer Responsibility
Updated Guidance for Efficient Waste Management

PART I

Chapter 1

Extended producer responsibility – an overview

This chapter provides an introduction to extended producer responsibility (EPR) by discussing the policy rationale behind the approach, the main instruments as well as the most important trends. It finds that there has been a significant increase in the use of EPR in the past 15 years, with about 400 systems now being in use around the globe, most of them in the OECD region. This has led to important achievements, such as an increase in material recovery rates from different waste streams and the generation of significant financial resources from producers that now contribute to a market that is worth about 300 billion EUR globally. A number of areas where EPRs need to be strengthened are also identified.

1.1. Introduction

Extended producer responsibility (EPR) for the end-of-life management of products emerged in a number of OECD countries in the late 1980s. It was a response to the challenges that many municipalities were facing in managing waste that was growing in volume and complexity, and in a context where the siting of waste management facilities was often opposed by the public. EPR policy sought to shift the burden of managing certain end-of-life products from municipalities and taxpayers to producers. It was hoped that this redefinition of responsibilities, and the incentives it provided to producers to redesign products and packaging, would reduce the share of waste destined for final disposal and increase recycling.

OECD provided a platform to exchange good practices and to analyse common challenges related to EPRs. Following an extensive phase of research and policy dialogue, the OECD published a Guidance Manual on EPRs in 2001 to support Member countries to implement EPR policies (hereafter the 2001 Manual). Since then, the number and variety of EPR systems have increased significantly, not only in OECD countries but also in emerging economies. In many countries EPRs have helped to reduce the share of waste destined for final disposal and increased material and energy recovery, thereby enhancing the resource productivity of those economies. At the same time, EPRs have contributed to the development of a multi-billion dollar recycling industry.

Part one of this report provides updated Guidance on EPRs, building on the 2001 Manual and in view of the developments and lessons learnt since then. While many of the original recommendations remain valid, recent experience gained suggests that additional guidance could help to enhance the environmental effectiveness and economic efficiency of EPR systems.

The first chapter begins by providing a brief summary of the policy rationale and main instruments for implementing EPR. Some of the main recent trends in EPR systems are then described. The following chapter aims to integrate the main elements of the 2001 Guidance Document with the findings and recommendations emerging from the most recent analysis of EPRs. A concluding section examines some remaining challenges.

To support the development of more up-to-date and policy-relevant guidance, four issues were examined in more depth. These are presented in the second part of the report: design and governance of EPR systems (Chapter 3); the anti-competitive behaviour that has been observed in EPR systems, a concern that has increased with the growth and increased concentration of the waste and recycling sectors (Chapter 4); the role of EPRs in promoting more environmentally friendly design of products (Chapter 5); and the operation of EPR systems in emerging economies, particularly the important role played by the informal waste sector (Chapter 6).

1.2. EPR policy rationale and instruments

1.2.1. *Definition and policy rationale*

The OECD defined EPR as an environmental policy approach in which a producer's responsibility for a product is extended to the post-consumer stage of a product's life cycle. In practice, EPR involves producers taking responsibility for collecting end-of-life products, and for sorting them before their final treatment, ideally, recycling. EPR schemes can allow producers to exercise their responsibility either by providing the financial resources required and/or by taking over the operational and organisational aspects of the process from municipalities. They can do so individually or collectively.

EPR was not entirely a new concept; recycling markets existed well before the 1980s, particularly for end-of-life products with a market value. However, these markets were limited in scope and operated with many imperfections. As a result, they did not generate a socially optimal level of recycling and the burden of treating the residual waste fell on municipalities. By the late 1980s, the volume and complexity of waste generated exceeded the management capacity of municipalities in the most developed economies. The task of municipalities was further complicated by public opposition to the siting of landfills and incinerators: the NIMBY (Not In My Back Yard) concept. The dominant role of the public sector also meant that the opportunity to mobilise the technical and managerial skills of the private sector in managing waste products was not being realised. EPR aimed to address these challenges by shifting the financial burden of managing end-of-life products from municipalities and taxpayers to producers. It was hoped that this would reduce the volume of waste going for final disposal, increase rates of recycling, and provide incentives for waste prevention and reduction at source.

EPR policy is consistent with the Polluter-Pays Principle in so far as financial responsibility for treating end-of-life products is shifted from taxpayers and municipalities to producers and, ultimately, consumers. However, EPR policy alone does not aim to achieve a full internalisation of environmental costs; the task of establishing an environmental price for a wide range of environmentally diverse waste streams makes this impractical. EPR policy nevertheless aims to provide producers with incentives to internalise environmental costs throughout the product life-cycle, including at the design stage. EPRs seek to provide incentives to producers to (re)design products and packaging to facilitate their end-of-life management, and to avoid using materials that may pose risks to human health or the environment. Without this, some products can require significant amount of resources before they can be recycled.

1.2.2. *Policy instruments*

Four broad categories of EPR instruments exist, even though they are sometimes used in combinations, (also see Figure 1.1):

- *Product take-back requirements* involve assigning responsibility, for example to producers or retailers, for the end-of-life management of products. This type of requirement is often achieved by establishing recycling and collection targets for a product or material. The targets may be either mandatory or voluntary. A further approach is to provide incentives for consumers to return the used product to a specified location such as the selling point.
- *Economic and market-based instruments* provide a financial incentive to implement EPR policy. They come in several forms, including:

Figure 1.1. **EPR policy instruments in the product cycle**

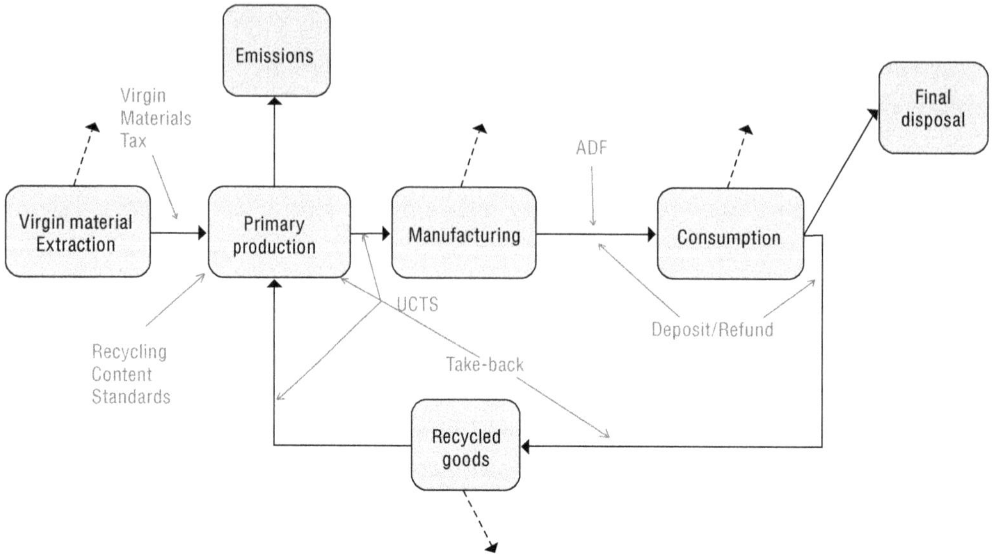

Note: ADF > Advance disposal fee; UCTS > Upstream combination tax/subsidy
Source: OECD (2013), *What have we learned about extended producer responsibility in the past decade? – A survey of the recent EPR economic literature*, Paris

- *Deposit-refund*: an initial payment (deposit) is made at purchase and is fully or partially refunded when the product is returned to a specified location.
- *Advanced Disposal Fees (ADF)*: fees levied on certain products at purchase based on the estimated costs of collection and treatment. The fees may be collected by public or private entities and used to finance post-consumer treatment of the designated products. Unused fees may be returned to consumers.
- *Material taxes*: involve taxing virgin materials (or materials that are difficult to recycle, contain toxic properties, etc.) so as to create incentives to use secondary (recycled) or less toxic materials. Ideally, the tax should be set at a level where the marginal costs of the tax equal the marginal treatment costs. The tax should be earmarked and used for the collection, sorting, and treatment of post-consumer products.
- *Upstream combination tax/subsidy (UCTS)*: a tax paid by producers subsequently used to subsidise waste treatment. It provides producers with incentives to alter their material inputs and product design and provides a financing mechanism to support recycling and treatment.

- *Regulations and performance standards* such as minimum recycled content can encourage the take back of end-of-life products. When used in combination with a tax, such standards can strengthen incentives for the redesign of products. Standards can be mandatory or applied by industries themselves through voluntary programmes.
- *Information-based instruments* aim to indirectly support EPR programmes by raising public awareness. Measures can include reporting requirements, labelling of products and components, communicating to consumers about producer responsibility and waste separation, and informing recyclers about the materials used in products

The 2001 Manual noted that most of the EPR systems that had been established at that time included targets or quotas. These maybe quantitative or qualitative and could be expressed in various ways depending on the ultimate policy objective; for example, in terms of reuse or recycling rates, the volume of waste collected or going to final disposal, performance quotas or environmental quality objectives.

1.3. Main trends and achievements of EPRs

This section examines some of the main trends and achievements in EPRs since the 2001 Guidance Document was published. It shows that the number of EPR systems in operation has increased significantly. Most have been established in the EU and US, at national and sub-national levels, often in response to legislative requirements. The established systems manage a range of end-of-life products and employ a limited number of instruments. Governance arrangements have continued to evolve and to reflect the widely differing contexts in which the EPR systems were established and the specific goals they were intended to achieve.

Assessing the impacts of EPR systems is hindered by a considerable lack of data, methodological difficulties in distinguishing the impacts of EPRs from other factors, and the wide variations in EPR systems which limits comparison. Nevertheless, there is some evidence that EPR systems have helped to decrease the volume of waste destined for final disposal, increased rates of recycling, and, as a result, relieve pressures on public budgets. EPR systems have also contributed to the development of a multi-billion dollar waste and recycling industry. On the other hand, the consensus appears to be that while EPR systems have contributed to eco-design in some countries and some sectors, they are seldom sufficient to serve as the triggering factor.

1.3.1. Main trends

Trends in the adoption of EPRs

There has been a significant increase in the adoption of EPRs since 2001, in line with an increased emphasis on waste management policies in many countries. A recent study (OECD, 2013) reviewed 384 EPR policies. Of these, more than 70% were implemented since 2001 (Figure 1.2).

Figure 1.2. **Cumulative EPR policy adoption globally, 1970-2015**

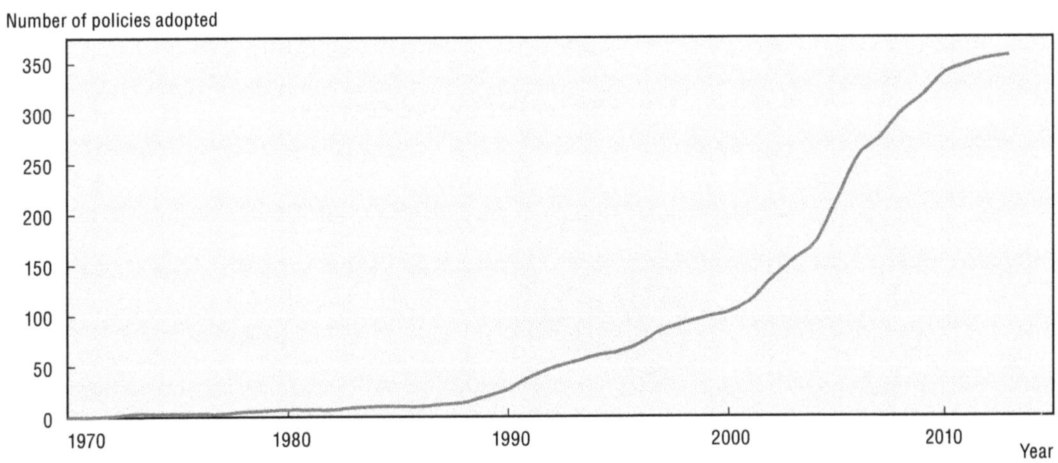

Source: OECD (2013), *What have we learned about extended producer responsibility in the past decade? – A survey of the recent EPR economic literature*, Paris

In terms of products covered, small consumer electronics appear to be the most prevalent (see Figure 1.3). When mobile phones, renewable batteries, thermostats and auto switches are included, this category accounts for 35% of EPR policies globally. Packaging (including beverage containers) and tyres each account for 17%. End of life vehicles (ELVs) (7%) and lead-acid batteries (4%) are the next largest groups of products covered. The remaining 20% of polices cover less common products including used oil, paint, chemicals, large appliances, and florescent light bulbs. Thus it appears that products with potentially high costs of disposal and relatively high levels of consumption have been the main focus of attention in EPRs, reflecting both policy and market drivers. EPRs have been used less for products with relatively low levels of consumption.

Figure 1.3. **EPR by product type, worldwide**

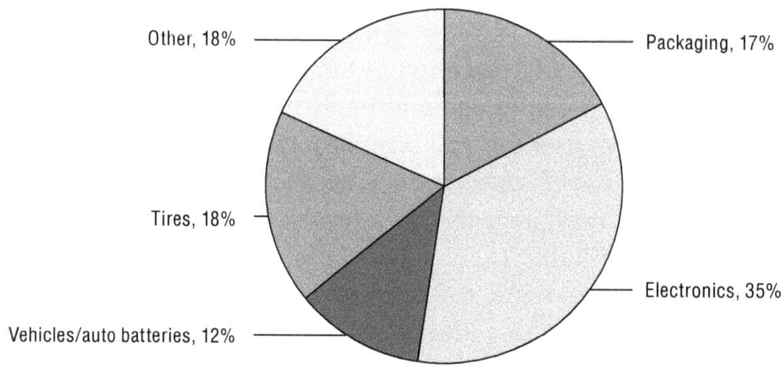

Source: OECD (2013), *What have we learned about extended producer responsibility in the past decade? – A survey of the recent EPR economic literature*, Paris.

Regarding the policy instruments employed in EPRs (Figure 1.4), various forms of take-back requirements are the most commonly used (72% globally), sometimes in combination with advances disposal fees (ADF). These instruments are used for a wide range of products. Advance disposal fees are the next most frequently used instrument (16%), and they have also been applied to many different products. Deposit/refund instruments (11%) are concentrated in the used beverage container and lead-acid battery markets, sometimes in combination with take-back requirements. The other possible EPR policy instruments identified in the 2001 Manual – upstream combined tax/subsidy, recycling content standards, and virgin material taxes – appear to be used infrequently, if at all.

Figure 1.4. **EPR by Policy, worldwide**

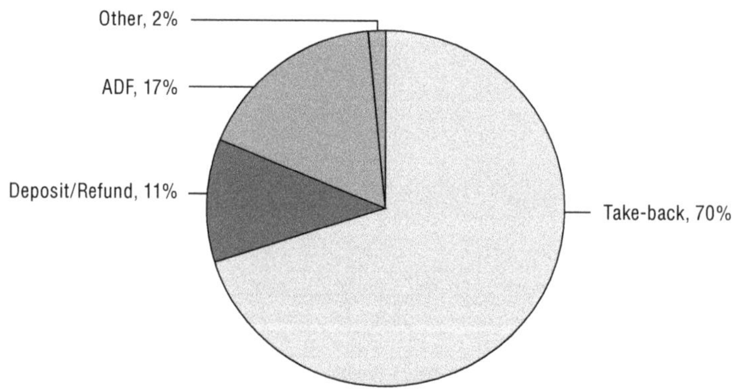

Source: OECD (2013), *What have we learned about extended producer responsibility in the past decade? – A survey of the recent EPR economic literature*, Paris

Comparing the regional distribution of EPRs (Figure 1.5), 90% of the EPRs systems have been implemented in the EU and North America. EPR systems in the United States appear slightly more inclined to use instruments such as deposit/refund and ADF. These instruments were used in a bit less than half of US state-level policies compared with 21% for the rest of the world. In terms of products covered, there is some variation across regions. Within the EU, 34% of policies cover electronics, 18% packaging, 14% tyres, and 20% vehicles/auto batteries. In the US, 50% of policies cover electronics, 8% cover packaging, 24% cover tyres, and 7% cover vehicles/auto batteries.

Figure 1.5. **Regional Distribution of EPRs**

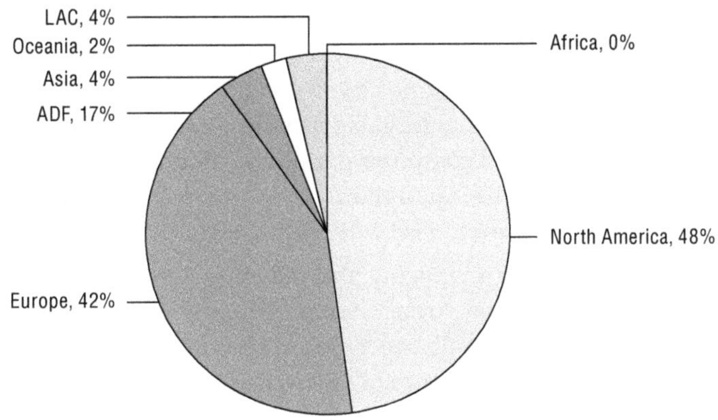

Source: OECD (2013), *What have we learned about extended producer responsibility in the past decade? – A survey of the recent EPR economic literature*, Paris

Another recent survey (Tasaki et al., 2015) examined how various stakeholders in different countries perceived the concept of EPR.[1] Respondents were asked what they thought EPRs should achieve. Out of 16 responses, the top three were: to increase the possibility of disassembling or recycling a product; to reduce a product's environmental impact; and to promote recycling or recovery. The bottom three were: to reduce the costs of waste management; to promote recovery; and to promote new business models. Responses varied among regions and stakeholders. While it is difficult to draw any general conclusion, the results of the survey help to underline stakeholders' and countries' different expectations in regard to EPRs, as well as the diversity of EPRs.

Legal and policy drivers of EPRs

Within the European Union (EU), all Member States have established take back EPR systems for the four waste streams identified in EU Directives: packaging, batteries, ELVs and Waste Electrical and Electronic Equipment (WEEE). Some of the recycling targets are currently under review as part of a broader review of EU waste management policy. While European waste legislation provides the enabling framework, national legislation by Member States specifies the operational aspects of EPR systems. As a result, EPR policies have been designed and implemented in a very heterogeneous manner across EU members. Some EU Member States have also put in place EPR systems for products not directly addressed in EU legislation e.g. for tyres, graphic paper, oil and medical waste.

In North America, EPR programmes in the United States and Canada cover a wide array of products and are primarily designed and implemented at sub-national level, by states and provinces. The 2009 Canada-wide action plan for EPR, emphasises a

Box 1.1. Legal frameworks for Extended Producer Responsibility in Australia and Canada

Australia's National Waste Policy sets a comprehensive agenda for national and co-ordinated action on waste and marks a fundamental shift in the national approach to waste management and waste resource recovery in that it aims to provide a common approach across the country, where responsibility for waste management is essentially located at the state level. The National Waste Policy was endorsed by the Australian Government, and state, territory and local governments in 2009. The policy identifies key areas of focus for all governments to pursue. Product stewardship was identified as one of these areas which lead to the development of the Product Stewardship Act 2011 (the Act), with end-of-life televisions and computers identified as the first products to be regulated under the Act. The Act provides a national framework to support voluntary, co-regulatory and mandatory product stewardship schemes. The scheme commenced in May 2012 with industry funded recycling services gradually becoming available around Australia. The scheme accepts all televisions, computers, printers and computer peripheral products (such as keyboards, mice and hard drives) for recycling. The Australian Government led the development of the scheme across all levels of government and with industry stakeholders. (See Annex A)

In Canada, the responsibility for managing and reducing waste is shared among federal, provincial, territorial, and municipal governments. EPR programs are regulated under the jurisdiction of provincial and territorial authorities, each using varying approaches to reach common EPR objectives. A national picture of EPR in Canada, therefore, accounts for these provincial variations. In 2009, the Canadian Council of Ministers of the Environment (CCME), a body that brings the environmental ministers of federal provincial, and territorial governments together, developed the Canada-wide Action Plan (CAP) for Extended Producer Responsibility (EPR). Through the CAP, the CCME and its member jurisdictions committed to working towards the development and implementation of EPR programs to provide guidance on how to strengthen the use of EPR and promote the harmonization and consistency of programs across the country. The goal of the CAP is to increase diversion and recycling of municipal solid waste through the harmonization of provincial EPR programs. Phase 1 of the CAP calls for a number of commitments from provinces and territories, including the implementation of EPR programs for electrical and electronic products by 2015. The jurisdiction for EPR programs rests with the provinces. (See Annex C)

Source: Case studies prepared for OECD Global Forum on Extended Producer Responsibility, 17-19 June 2014, *www.oecd.org/env/waste/gfenv-extendedproducerresponsibility-june2014.htm/*.

harmonised, outcome-driven model that has mostly resulted in systems where producers discharge their responsibilities collectively with oversight by provincial governments. In the US, there is no federal law regarding EPR; individual states have developed and implemented their own policies, reflecting local conditions and each state's specific political dynamics. Between 1991 and 2011, US states enacted more than 70 EPR laws generally requiring manufacturers to implement EPR programmes, but without specifying recycling targets. In parallel, producers themselves have implemented voluntary and stewardships programmes for the collection and recycling of their products.

In Latin America and the Caribbean (LAC), several countries including OECD members Chile and Mexico, but also Brazil, Argentina and Colombia, have recently taken steps towards implementing their first EPR systems. Their main focus is the large markets for potentially hazardous electronic waste (e-waste). In 2013, Chile submitted draft legislation on EPR to the Congress in response to a recommendation made in the context of its accession to the OECD to strengthen its waste management policies. Legislation was approved by the Chamber of Deputies in 2015 and is now discussed in the Senate. Most EPR policies in the LAC region are at an early stage of implementation and require further efforts to be fully operational. In some cases they are complemented by voluntary initiatives by the private sector.

The current landscape of EPR in Asia varies significantly across countries and between OECD and non-OECD members. OECD economies like Japan and Korea have well-established, legally-based EPR systems, supported by a solid monitoring and enforcement. Some rapidly emerging economies, such as India and Indonesia have started to develop EPR programmes, though they are generally not yet fully operational. Malaysia and Thailand are working towards EPRs for e-waste, although these initiatives generally rely on the voluntary participation of producers. China put in place an EPR for e-waste in 2012, which is now beginning to show first results.

In Africa, EPR, and waste management policies in general, remain at a less advanced stage. E-waste is a growing concern on the whole continent. In some countries, informal recyclers play an important role, mostly for valuable waste fractions. Concerns have been raised about the health and environmental impacts of their operations (see below). In South Africa, a broad waste management act was introduced in 2009 which empowers the environment minister to require EPR measures on a product-by-product-basis. Although EPR initiatives in South Africa have been mostly initiated by industry, the government has enacted regulations to ensure enforcement of some of these initiatives, for example for an industry-led tyres recycling initiative.

Developments in governance arrangements

The governance arrangements that are being used for EPR schemes have evolved and converged to some extent of the past decade.

Many EPR systems are mandatory, while the use of voluntary schemes remains limited. There is little systematic, quantitative information available on the proportion of end-of-life products managed through voluntary or mandatory EPRs. However, the evidence that is available suggests that voluntary programmes are confined to a few, specific products or product categories where firms have an incentive to take back products because it is profitable to do so (see Chapter 2). For example, voluntary EPR programs, often known as "product stewardship" schemes, for consumer electronics, rechargeable batteries, mercury thermostats and auto switches were reviewed in the US. Despite a potentially large number of potential product categories that could be covered, the review indicated that voluntary EPR systems were limited to a range of products with high visibility, such as TVs and computers (see Chapter 4). There is also some evidence that voluntary programs have tended to achieve low collection rates (see Chapter 4). However, business-led voluntary schemes have had some effect in developing countries where mandatory EPR is not yet on the policy horizon (see Chapter 5).

There is a sizeable literature on voluntary environmental initiatives, though EPRs have not been a particular focus (OECD, 2003). Doubts have been raised about the effectiveness of such initiatives for various reasons including failure to progress beyond business-as-usual, regulatory capture, lack of transparency, poor monitoring and free riding. Their effectiveness appears to be contingent on a credible threat of regulation and on the establishment of a robust governance system.

Most EPR systems are organised collectively rather than individually. EPR schemes can be designed so that producers may discharge their EPR obligations individually or collectively. However, in some countries, there is no clear distinction between the two, and individual and collective responsibilities are designed to be mutually reinforcing (e.g. in Japan's EPR for home appliances). Individual responsibility systems have been established for some waste streams examined in a recent study for the European Commission, particularly when the product market is concentrated and it is viable for producers to operate a take-back system (e.g. German car producers [European Commission, 2014]). However, in a great number of cases, producers have established collective systems managed by Producer Responsibility Organisations (PROs, see Box 1.2; and Chapter 2).

Box 1.2. **Producer responsibility organisations**

Under take-back programmes, it could be impractical and not particularly economically feasible for each producer to take back its own products. Therefore, third party organisations are often formed allowing producers to collectively manage the take-back (and most often arrange for the treatment) of products. These organisations are often referred to as Producer Responsibility Organisations (PRO) and can be an effective structure for managing and collecting post-consumer products. The need to create a PRO depends on the policy instrument selected and other factors such as the product group, number of producers and importers, and secondary materials to be collected. The advantages of a PRO as a means to implement the EPR programme should be examined in the design stage. Most PROs in operation to date collect a fee directly from the producers based on a specific fee structure and the revenue is used to pay for the costs of waste collection, sorting and treatment.

Source: OECD, 2001, *Extended Producer Responsibility: A Guidance Manual for Governments*, OECD Publishing, Paris, *http://dx.doi.org/10.1787/9789264189867-en*.

Several reasons have been advanced to explain the greater use of collective systems: they generate economies of scale (or density) and hence reduce costs for participants; they help to share risk among participants, particularly in the difficult start-up phase; they can help to reduce free-riding through collective action by producers and peer pressure; they can simplify operations and reduce administrative burdens for consumers, retailers and municipalities; and they provide a means for governments to manage waste generated by orphan products.

Collective EPR systems may have one or more PROs. EPR schemes can be designed so that product markets may be served by individual or multiple PROs. Since the publication of the 2001 Manual, concerns about anti-competitive behaviour by monopolistic PROs has fostered the development of more multiple-PRO EPRs. Among the 36 EPR systems in the European Union that were reviewed (European Commission, 2014), all of the EPR systems for e-waste were managed by multiple PROs. Batteries were also frequently managed by multiple PROs. In contrast, ELVs were always managed by a single PRO. EPR systems for other product categories did not show a clear pattern.

The legal status of PROs also varies widely. PROs can be non-profit organizations (typically), government agencies (rarely), quasi-governmental non-profit organisations (occasionally) and for-profit firms (occasionally). In the study of EPRs in the EU, 13 of the 36 systems involved were for-profit PROs.

An important trend in markets with multiple PROs has been the establishment of clearinghouses. These are neutral bodies that help to co-ordinate the work of PROs by ensuring that collection is provided everywhere that it is needed, that "cherry picking" is avoided, and that there is a level playing field for all competitors. Co-ordination can enhance efficiency by ensuring that competing PROs do not duplicate each other's' activities. Clearinghouses often collect data from producers or service providers and provide a mechanism for managing proprietary data.

The role of municipalities is changing and has generally diminished

In some countries, the increased roles and responsibilities of PROs has led to a redefinition of their relations with municipalities. This is the case for product groups such as packaging and e-waste, where municipalities play an important role.[2] While in many EPRs, municipalities continue to have an active operational role in the collection and treatment of waste, in some systems they do not necessarily have any role. For example, in the packaging EPRs in Austria, Germany and Sweden, full operational and financial responsibility for collection, sorting and recycling has been passed to producers. PROs and municipalities manage separate collection systems – "dual systems". Municipalities may serve as contractors to the PROs, providing local services, but they have no automatic role in the EPR system.

New governance arrangements have emerged since 2001. Since the 2001 Manual, two new governance systems for EPRs have emerged: tradable certificates and government-run EPR systems (see Chapter 3).

In a tradable credit system, a producer's compliance with EPR obligations is accomplished when it collects its allocated number of credits for collected and processed end-of-life products. To date, the United Kingdom packaging EPR is the only tradable credit system in operation. Firms are allocated a share of the relevant obligation according to their position in the supply chain and the recovery target for the specific material used for packaging. The governance arrangement resembles that of a multiple-PRO system, but the role of producers is more modest as they finance only about 10% of the costs and are not involved in operations. Also, unlike in EPR systems involving PROs, the government plays a role in verifying that recycling has occurred. Given the limited experience with this instrument, it is difficult to assess whether it could be more widely deployed. Nevertheless, in view of the theoretical advantages of such market-based approaches, future assessments of the performance of the United Kingdom packaging system will be of interest beyond the United Kingdom.

The concept of EPR implies that producers take responsibility for the end-of-life treatment of their products. Nevertheless, some governments play a direct role in fee collection and disbursement. For example, these types of systems have been established in China for e-waste, Chinese Taipei for all products subject to EPR, and in several US states and Alberta, Canada. The role of the governments in these countries appears to be more interventionist than in countries like Denmark, Hungary and Iceland which levy a tax on products and/or packaging. The revenues from these taxes are only partly allocated to cover the costs of end-of-life treatment of products, with most going to the treasury. EPRs where governments play a leading role may have advantages in economies where industry does not have the capacity to establish and manage its own system. However, they may also delay or prevent producers from taking on the responsibilities implied by the concept of EPR.

1.3.2. Impacts of EPR schemes

With more than 20 years of experience with some of the earliest EPR schemes, there is now some evidence available that allows assessing the environmental and economic performance of these approaches.

Evidence that EPRs have contributed to reducing waste disposal and increasing recycling

Despite data limitations and methodological challenges in attributing trends to specific policy initiatives, there is evidence that levels of waste disposal have decreased and recycling has increased in OECD countries. Figure 1.6 shows that between 1995 and 2011, the amount of municipal waste generated per capita in the OECD area increased from 520kg to 530 kg (OECD 2015). However, the 2011 figure shows a decrease from 560 kg, the level recorded in both 2000 and 2005. Moreover, the amount of material recovery in OECD countries increased from 19% in 1995 to 33% in 2010. Energy recovery increased from 17% to 18% in the same period. Figure 1.7 indicates that the levels of material recovery varied widely among OECD countries. This suggests that there is further scope in many OECD countries to increase recycling levels. Well-designed EPR systems could contribute in this regard.

Figure 1.6. **Trends in municipal solid waste management in the OECD**

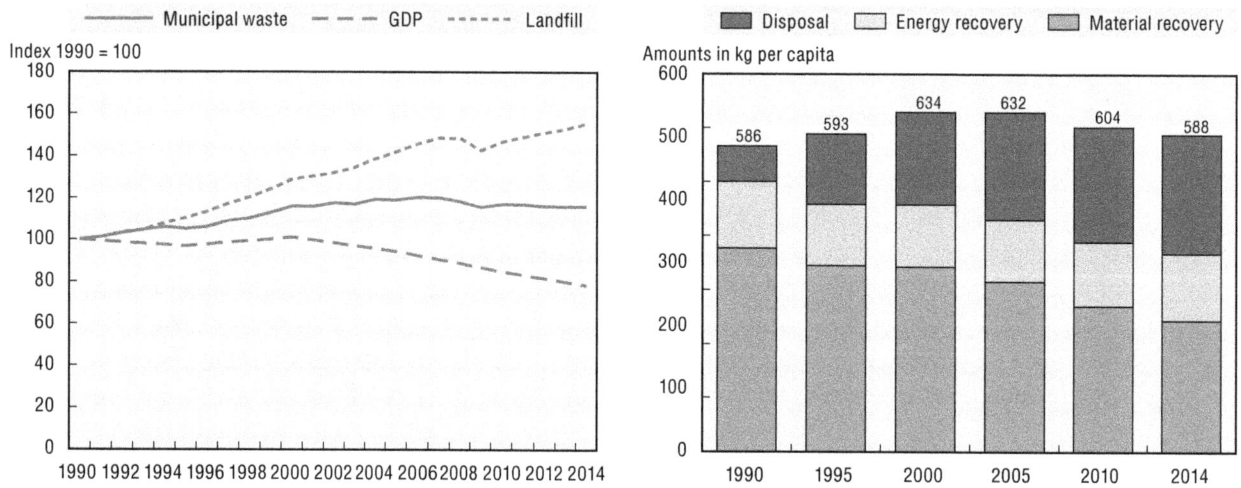

Note: Municipal solid waste only covers packaging and e-waste, but other types of waste covered by EPR, such as ELVs and tyres are not reflected in these numbers.

Source: OECD (2016), "Municipal waste generation and treatment", *OECD Environment Statistics* (database).

Waste collection rates vary between countries and across product types. In Western Europe the annual collection rates for e-waste are below 10 kg/capita; the Australian scheme for television and computer recycling collects about 2 kg/capita/year; while in the US between 0.3 kg and 4 kg e-waste per capita per year is collected (see Chapter 4). This compares with average annual purchases of electric and electronic equipment per inhabitant of 25 kg in Western Europe, close to 30 kg in Australia and above 30 kg in the US.

Another study focused on EU Member States also concluded that collection rates for oils, batteries and WEEE varied significantly among countries (see Table 1.1 below and European Commission, 2014). The rates generally did not reach more than 80%, apart from oils where 100% collection was not unusual. The same study concluded that EPRs had helped to achieve variable but reasonably high recovery targets. Evidence from Japan also

Figure 1.7. **Trends in municipal solid waste management by country**

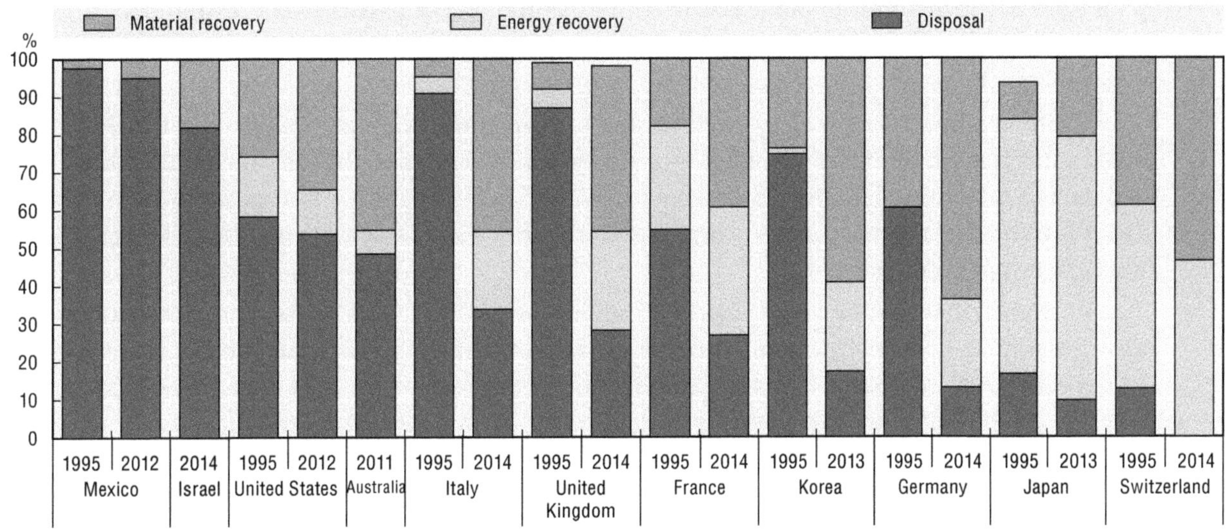

Note: The sum of the categories presented here might not add up to 100% because "Other recovery" and "Other disposal treatments" are not presented. Germany and Italy, break in time series.

Source: OECD (2016), "Municipal waste generation and treatment", *OECD Environment Statistics* (database).

Table 1.1. **Performance features of selected EPRs in the EU**[3]

	Collection (C) or recycling and recovery rates (R)	Average producer fees
Batteries	5-72% (C)	240-5400 EUR/t
ELV	64-96% (R)	0-66 EUR/Vehicle
Oil	3-61% (C)	42-231 EUR/t
Packaging	29-84% (R)	20-200 EUR/t (average 92)
WEEE	1.2-17.2 kg/cap (C) (average 6.6)	68-132 EUR/t

Source: European Commission (2014), *Development of Guidance on Extended Producer Responsibility (EPR), final report.*

suggests that EPRs contributed to increased rates of recycling of containers and packaging waste; a 27% increase between 1997 and 2000 from1.25 to 1.59 million tonnes (OECD, 2014).

Although inter-country comparisons are fraught with methodological problems, these data suggest that there is probably scope in many countries to boost collection and recycling rates, for example through more ambitious targets and better product monitoring.

Reduced waste disposal and increased recycling will have helped to mitigate a range of environmental problems traditionally associated with waste management such as air pollution from waste dumps or incinerators, and contamination of land and water. More recently, evidence has emerged that improved recycling can also contribute to mitigating other environmental problems such as the generation of greenhouse gases (Menikpura et al., 2014). A broader analysis of the environmental benefits of EPR policies would be a useful contribution to the assessment of these policy instruments.

The impact of EPRs on eco-design has been less than originally hoped for

One of the objectives motivating the establishment of EPR systems was to encourage producers to design their products in a way that reduces their environmental impact. This can be achieved in several ways, including by: reducing the environmental impact of materials used (e.g. avoiding hazardous substances or using recycled resources), reducing the amount of material used (e.g. development of lightweight products or reducing packaging), optimising the potential lifetime of products (e.g. increasing durability or re-using components), and reducing environmental impacts at the end of the product cycle (e.g. design for disassembly or producing mono-material goods).

Many academic and policy reviews have examined this issue and, despite the methodological challenges involved, generally concluded that the impact of EPRs on eco-design has often been limited (see Chapters 2 and 4). The consensus appears to be that EPR systems contribute to eco-design but that other factors sometimes have a more important triggering role.

In theory, EPR systems organised by individual firms provide better incentives for eco-design than collective systems: since the waste management costs accrue exclusively to the individual company, that company has a stronger incentive to reduce these costs through eco-design than firms participating in collective systems. However, there does not appear to be any empirical information shedding light on the relative effectiveness of individual and collective EPR systems in promoting eco-design. In Japan, some positive experience has been gained with combining collective systems with elements of individual producer responsibility as well as developing improved communication between upstream and downstream actors in the value chain (e.g. Japan's EPR for household appliances, see Box 1.3 and Annex H).

Box 1.3. **Japan EPR for home appliances**

The Act for Recycling of Specified Kinds of Home Appliances was enacted in June 1998 and enforced in April 2001. It aims to achieve a reduction in the volume of waste and to improve material recovery from this source. The Act covers four categories of home appliances: air conditioners; TV sets; electric refrigerators and freezers; and electric washing machines and clothes dryers.

Under the act, each home appliance Manufacturer has the obligation to set up designated collection sites for taking back and recycling their products when they become waste. To implement their obligations, Manufacturers have set-up two competing groups, each of which brings together three or four of the most important producers and represents a similar share of the market. One of the groups has set-up its own recycling facilities, while the other contracts with existing operators.

Consumers and businesses that dispose end-of-life home appliances have to pay both, the collection/ transportation and the recycling fee. Traceability is ensured by the use of home appliance recycling tickets (manifests) that are issued to consumers who paid the recycling fee. This manifest system ensures that waste home appliances are delivered to the original Manufacturers of the products.

While there is no differentiation between the costs to manage individual brands within each of the producer groups, potential cost savings through efficient processing or product design remain with producers. Hence, there is competition between the two manufacturer groups over the minimisation of recycling costs.

The advantage of the vertical integration that results from the implementation of Japanese EPR for Household Appliances is that it creates a strong link between downstream management of end-of-life products and the producer. There is some empirical evidence that the system does provide tangible incentives for eco-design.

Source: See Annex H, Case study: Japan EPR for home appliance; Dempsey et al. (2010) and Tojo (2004).

The way in which the PROs of collective systems establish fees can have an important bearing on incentives for eco-design. PROs are financed either on a variable- or fixed-fee basis. Fixed-fees are typically used in PROs for complex goods such as electronic equipment, cars or furniture where it would be difficult to link the fee to the product's environmental impact. In such cases, it is easier to apply a common fee. However, this results in the lack of a link between the fee and waste management costs associated with specific products and it provides only weak and indirect incentives for eco-design.

Variable-fee PROs are mainly used for mono-material products with limited durability, such as packaging or graphic paper. Typically they calculate the fee by weight which provides a financial incentive to make products lighter. Some systems also aim to provide incentives to simplify recycling, for example by charging higher fees for multi- as opposed to mono-material products, or by targeting other design parameters. In Eco-Emballages, France's packaging PRO, incentives for eco-design were provided by increasing the fee for glass packaging with ceramic caps and other material mixes that were difficult to separate by 50%.

Table 1.2 below shows that the weight of food packaging in Europe decreased, 2000-10. The variable fee structure may have influenced this trend, but it is difficult to disentangle the contribution of EPRs from other factors (e.g. the financial gains from less packaging material).

Table 1.2. **Reduction in food packaging weight in Europe, 2000-10**

Packaging and product	Weight 2000 (kg)	Weight 2010 (kg)	% change
PET bottle of 1.5 L still water	0.0318	0.0280	-12
Aluminium can of 330 ml for soft drinks	0.0158	0.0131	-17
Glass bottle of 250 ml for olive oil	0.2236	0.2002	-10
Tin can of 125 gr for fish	0.0343	0.0319	-7
Plastic bag for 1 kg of pasta	0.00903	0.00785	-13
Cardboard box for dry food	0.01388	0.01132	-18

Source: http://proeurope4prevention.org/packagings-trends.

Some evidence that EPR systems have helped reduce financial burdens on public budgets and taxpayers

One of the main objectives of EPR is to shift the financial responsibility for treating end-of-life products from local public authorities to producers (and ultimately consumers), thereby reducing the burden on public budgets and taxpayers. There is some evidence that this has occurred. For example, in France in 2012, total spending on municipal waste management was estimated at EUR 9.7 billion. In the same year, EUR 630 million of collected "eco-contributions" were allocated to local authorities to cover the costs of collecting and treating waste products. In addition, producer organisations spent EUR 230 million on the treatment of used products.

While EPRs appear to have reduced the burden on public budgets, assessing their cost effectiveness is more challenging. This task is hampered by a considerable lack of data. PROs generally do not publish financial data, sometimes on the grounds that it is commercially sensitive (for the PRO or its members). The most extensive analysis on this issue was carried out in the EU (European Commission, 2014). It found that, to the extent that information was available, the fees paid by producers varied greatly for all product categories (see Table 1.1 above). This reflected differences in scope, cost coverage, and the actual costs of collection and treatment. The study concluded that the best performing schemes, in most cases, were not necessarily the most expensive. No single EPR model emerged as the best performing or the most cost-effective.

The cost-effectiveness of EPRs can also be viewed from the point of view of instrument selection. OECD (2013) reviewed the economic literature related to EPRs and concluded that "for a given target of waste reduction, the deposit/refund is the least cost policy followed by the advanced disposal fee, both of which were calculated to perform better than the non-EPR recycling subsidies. Comparing the marginal costs of waste reduction against the social benefits, ... modest increases in waste reduction would be efficient." These two policy instruments together accounted for just over one-quarter of EPR systems surveyed; deposit refund, 11%, and advanced disposal fees, 16%. However, the paper indicates that there is a gap in the academic literature regarding the cost-effectiveness of other instruments such as take-back which accounted for 72% of the instruments reviewed. Empirical analysis of the cost-effectiveness of the various EPR instruments in practice could provide useful guidance to policy makers.

Some evidence that EPRs generate economic opportunities as well as environmental benefits

In addition to environmental benefits, EPR programmes generate a range of economic opportunities. Again, information is sparse, but they include increased technological and organisational innovation, a diversification of sources of material supply and, hence, resource security, and a better organisation of supply chains. Given the lack of economic analysis of EPR systems, including their cost-effectiveness, it is not surprising that no attempt appears to have been made to assess their costs and benefits. Given the importance of EPRs as instruments for achieving waste management and resource productivity policy objectives, such analysis would clearly be of value for policy makers (OECD, 2005).

The economic opportunities associated with environmentally sound waste management and recycling can support green growth. One study (Chalmin and Gaillochet, 2009) estimated that at the global level, the waste sector, from collection to recycling, represented a market in excess of EUR 300 billion. Of this, municipal waste accounted for about EUR 150 billion, with non-hazardous industrial waste accounting for the remainder. A study by the European Environment Agency (2011) suggested that in the period 2004-08, turnover in the European recycling sector increased by 100% and reached at least EUR 60 billion. These sectors also create and sustain thousands of jobs, though, again, the data is patchy. Further analysis of EPR systems' contribution to the recycling sector would be a valuable contribution to discussions on green growth.

Notes

1. About 420 responses were received, mostly from people with a long experience of working with EPRs in both the public and private sectors. About 30% were from Japan, 28% from Europe, 11% North America (Canada and US), and 9% from Asia (excluding Japan).
2. Municipalities usually do not play a role in the collection of other waste streams, such as used oil, end-of-life vehicles and lead-acid batteries.
3. Data presented in the table is mostly from 2013-14.

References

Chalmin, P. and C. Gaillochet (2009), "From waste to resource: an abstract of world waste survey 2009", *www.veolia-environmentalservices.com/veolia/ressources/files/1/927,753,Abstract_2009_GB-1.pdf.*

Dempsey M. et al. (2010), "Individual producer responsibility: a review of practical approaches for implementing individual producer responsibility for the WEEE Directive", *INSEAD Faculty & Research Working Paper, http://sites.insead.edu/facultyresearch/research/doc.cfm?did=45054.*

European Commission (2014), *Development of Guidance on Extended Producer Responsibility (EPR), final report, http://ec.europa.eu/environment/waste/pdf/target_review/Guidance%20on%20EPR%20-%20Final %20Report.pdf.*

European Environment Agency (2011), *Earnings, jobs and innovation: the role of recycling in a green economy,* EEA Report No. 8/2011, *www.eea.europa.eu/publications/earnings-jobs-and-innovation-the.*

Menikpura, S.N.M., A. Santo and Y. Hotta (2014), "Assessing the climate co-benefits from Waste Electrical and Electronic Equipment (WEEE) recycling in Japan", *Journal of Cleaner Production*, Vol. 74, No. 2014, pp. 183-190, *www.sciencedirect.com/science/article/pii/S0959652614002613.*

OECD (2016), "Material resources" and "Municipal waste generation and treatment", *OECD Environment Statistics (database), http://dx.doi.org/10.1787/env-data-en.*

OECD (2013), *What Have We Learned About Extended Producer Responsibility in the Last Decade? A Survey of the Recent Economic EPR Literature*, ENV/EPOC/WPRPW(2013)7/FINAL, *http://spot.colorado.edu/ ~daka9342/OECD_EPR_KO.pdf.*

OECD (2005), *Analytical Framework for Evaluating the Costs and Benefits of Extended Producer Responsibility Programmes*, OECD, *http://dx.doi.org/10.1787/oecd_papers-v5-art13-en.*

OECD (2003), *Voluntary approaches for environmental policy – Effectiveness, efficiency and usage in policy mixes*, OECD Publishing, Paris, *http://dx.doi.org/10.1787/9789264101784-en.*

OECD (2001), *Extended Producer Responsibility: A Guidance Manual for Governments*, OECD Publishing, Paris, *http://dx.doi.org/10.1787/9789264189867-en.*

Tasaki, T., N. Tojo and T. Lindhqvist (2015), *International Survey on Stakeholders' Perception of the Concept of Extended Producer Responsibility and Product Stewardship*, IIIEE and NIES Joint Research Report, *www-cycle.nies.go.jp/eng/report/epr_eng.html.*

Tojo, N. (2004), *Extended Producer Responsibility as a Driver for Design Change – Utopia or Reality?*, IIIEE dissertations 2004:2, Lund, IIIEE, Lund University.

Extended Producer Responsibility
Updated Guidance for Efficient Waste Management

PART I

Chapter 2

Towards more effective producer responsibility

This chapter integrates the main elements of OECD's 2001 Guidance Document with the findings and recommendations emerging from the most recent analysis of EPRs. It finds that most of the original guidance remains valid and adds guidance in the areas where recent analysis has focused, particularly on the governance of EPR systems, the competition concerns that have arisen, opportunities to strengthen design-for-environment incentives and the role of the informal sector in EPR.

This section aims to integrate some of the key elements of guidance from 2001 together with the more recent experience gained. Specifically, it covers:

- Key factors in designing EPRs
- Governance of EPRs
- Financing, free-riding and orphan products
- Trade, competition and EPR
- Eco-design
- EPRs in emerging economies: the role of the informal sector.

Box 2.1. **2001 OECD Guiding Principles for EPRs**

The 2001 OECD Guidance Manual is more than 150 pages, divided into 8 chapters and with 15 annexes. It includes 6 checklists for policy makers as well as the following set of guiding principles for the design and development of EPR policies and programmes:

- EPR policies and programmes should be designed to ***provide producers with incentives*** to incorporate changes upstream at the design phase in order to be more environmentally sound.
- Policies should stimulate ***innovation*** by focusing more on results than on the means of achieving them, thus allowing producers flexibility with regard to implementation.
- Policies should take into consideration a ***life cycle approach*** so that environmental impacts are not increased or transferred somewhere else in the product chain.
- ***Responsibilities*** should be well defined and not be diluted by the existence of multiple actors across the product chain.
- The ***unique characteristics and properties*** of a product, product category or waste stream should be factored into policy design. Given the diversity of products and their different characteristics, one type of programme or measure is not applicable to all products, product categories or waste streams.
- The ***policy instrument(s)*** selected should be flexible and chosen on a case-by-case basis, rather than setting one policy for all products and waste streams.
- Extension of producer responsibilities for the product's life cycle should be done in a way to ***increase communication*** between actors across the product chain.
- A ***communication strategy*** should be devised to inform all the actors in the product chain, including consumers, about the programme and to enlist their support and co-operation.
- To enhance a programme's acceptability and effectiveness, a ***consultation of stakeholders*** should be conducted to discuss goals, objectives, costs and benefits.
- ***Local governments*** should be consulted in order to clarify their role and to obtain their advice concerning the programme's operation.
- Both ***voluntary and mandatory approaches*** should be considered with a view on how to best meet national environmental priorities, goals and objectives.

Box 2.1. **2001 OECD Guiding Principles for EPRs** *(cont.)*

- A ***comprehensive analysis*** of the EPR programme should be made (e.g. which products, product categories and waste streams are appropriate for EPR, whether historical products should be included, and the roles of the actors in the product chain).
- EPR programmes should undergo periodic ***evaluations*** to ensure that they are functioning appropriately and are flexible enough to respond to these evaluations.
- Programmes should be designed and implemented in a way that environmental benefits are obtained while domestic economic ***dislocations*** are avoided.
- The process of developing and implementing EPR policy and programmes should be based on transparency.

2.1. Key factors in designing EPRs

A key message in the 2001 Manual is that there is no single "right approach" when designing EPR systems. Solutions need to be found depending on the specific objectives to be achieved and taking account of the economic, political and cultural context. Accordingly, it was recommended that EPRs be established in accordance with general good governance principles, including:

- Clearly *define objectives*, based on analysis and consultation with all relevant stakeholders. EPRs usually aim to achieve one or more of four main goals: reducing the use of (virgin) resources and materials; waste prevention; reducing the environmental impacts of products; and closing material use loops ("circular economy"). Each EPR should clearly specify which of these goals it aims to achieve.
- Ensure *consistency and coherence with related policies*, in particular waste management and product policies. A life-cycle approach was recommended to ensure that environmental impacts are not increased or transferred somewhere else in the product chain.
- The *scope* of the EPR should be clearly defined and the unique characteristics of the product, product category or waste stream factored into policy design. Products with the greatest potential environmental impacts should be the main target. However, a range of other factors influence which product/waste to focus on and how the EPR should be designed, including: the durability and composition of the product, the primary and secondary markets in which they are traded, and their distribution networks and supply chain.
- The *producers* of the products subject to an EPR should be clearly defined. The "producer" was defined as the entity with the greatest control over the selection of materials and the design of the product. This could be the brand owner or the importer, or the filler of the packaging rather than the firm that produces the container.
- A *consultation process* should be organised when establishing an EPR system with a view to enhancing its acceptability, transparency and effectiveness. Subsequently a communication strategy should be developed to keep all stakeholders informed of the EPR's operations.
- Specific challenges may arise in the *start-up phase*, such as uncertainty about waste volumes and the need for large capital investments in collection and treatment facilities. Consideration should be given to specific measures that may be required to help phase-in the introduction of the EPR.

- Mechanisms for *reporting and monitoring* should be established to enable the results of the EPR to be evaluated and the adjustments made as required.
- When EPRs are based on law, which is not always the case, there will be a need for appropriate *enforcement mechanisms and sanctions*.

The selection of policy instruments should be based on clear criteria such as environmental effectiveness, economic efficiency, political acceptability, ease of administration, and the incentive they provide for innovation. Effective implementation can be enhanced by a clear allocation of responsibilities, and by using EPR instruments in conjunction with other waste-related instruments such as pay-as-you-throw systems, landfill bans and taxes, bans or restrictions on products/materials and green public procurement. Indeed, the 2001 Manual emphasises that EPR systems should be designed using a mix of instruments that target different points in the product chain: EPR as a policy concept is intended to address the lack of co-ordination that often exists among policies at different points in the product chain.

2.2. Governance of EPRs

Governance refers to the organisation of the EPR system and the roles and responsibilities of the stakeholders involved. Again, the 2001 Manual emphasises that the governance arrangements should be determined as a function of objectives, coverage of the EPR, the instruments used, and context.

There are several key actors in EPR governance systems. While each should have a defined role, the ways in which the various actors co-ordinate or share responsibilities may vary considerably and may also require specification. Decisions on the allocation of responsibility should be made in view of the policy goals, product characteristics, market dynamics, actors in the product chain, and the resources needed to implement the policy. The governance arrangements in developing and emerging economies will generally be quite different from those in developed countries.

- Given their technical and managerial know-how, the leadership role of *producers* is fundamental to the success of any EPR. Producers are usually ultimately responsible for achieving EPR policy objectives, whether individually or collectively, and whether through a single or competing PROs.
- *National governments* are generally, though not always, responsible for providing the legal framework, as well as for monitoring and enforcement. They can also contribute to the effectiveness of EPR by eliminating conflicting policies and implementing supportive policies.
- Defining the role of *municipalities*, particularly vis-à-vis PROs, is often particularly challenging: the establishment of an EPR involves a change in responsibilities between local authorities and producers as well as the definition of new revenue streams. In some cases, municipalities continue to play a role in the collection and treatment of end-of-life products, and in others they may simply oversee the actions of PROs. Municipalities may be required to play several additional roles such as stimulating the recycling market, assisting firms to build appropriate recycling capacity, and facilitating information flows and dialogue among stakeholders.
- *Consumers* play an important role supporting the collection of various products and waste streams. This needs to be clearly explained and the return of products made as convenient as possible for the consumer.

- *Retailers* can be an important conduit of information to the consumer and bridge the information gap between producers and consumers. They may also be involved in the collection of end-of-life products.

Two studies have recently reviewed experience with the design and governance of EPRs, both with a view to drawing lessons learned and identifying good practices (see Chapter 2 and European Commission, 2014). However, both concluded that it was not possible to identify good practices because of the diversity of goals and situations in which EPR systems operate, as well as the lack of data and comparability of EPR systems. Nevertheless, these reviews, as well as the assessment of trends above, allow several general conclusions to be drawn about how the design and governance of EPR systems could be improved.

First, many of the recommendations regarding good governance identified in the 2001 Manual are still relevant and should be applied more systematically. In particular, the main objectives of EPR systems have not always been as clearly defined as they could or should have been. Similarly, the roles and responsibilities of key stakeholders have not always been delineated sufficiently. European Commission (2014) found that in most of the EPR systems examined, no specific dialogue mechanism had been established which sometimes resulted in contentious relationships among stakeholders.

Second, there appear to be opportunities to improve the environmental effectiveness of EPR systems in various ways. Wide variations in collection and recycling rates suggest that the ambition of some EPR systems could be strengthened. Target-setting is an approach to enhance the effectiveness of EPRs. The establishment of binding targets should be informed by an assessment of costs and benefits as well as consultation with stakeholders. There is also a wide recognition that the environmental effectiveness of EPR systems could also be enhanced by better enforcement (see Box 2.2). The European Commission (2014) suggests

Box 2.2. **Enforcement of EPR obligations**

Failure to provide consistent enforcement creates unfair advantages for those who do not meet their obligations and increases the burden on those facing collective targets. Producers, for example, that evade payment of fees to PROs both reduce their own expenses and increase the costs that are borne by other producers. If evasion of fees is extensive, the financial viability of an EPR system can be at risk. Similarly, if collection or treatment service providers operating outside an EPR system do not face the same degree of enforcement as those working within, EPR systems are disadvantaged and the level of services may be reduced. Inadequate enforcement also makes it easier for illegal trade to take place.

Since 2001, registers of producers and accreditation of PROs have become important means of promoting compliance with EPR obligations. Registers provide PROs with the means to compile information needed to set fees and to identify free riders. Accreditation provides governments with a means to ensure that PROs meet specified performance criteria and to monitor their activities.

An enforcement issue specific to EPR arises when a jurisdiction has a single PRO for a given product category. In such cases, the threat to revoke a PRO's accreditation is less credible than it might otherwise be. In Ireland, this issue has been addressed by requiring PROs to establish a contingency fund equal to approximately one year's operating cost. In Austria, when a PRO does not comply with the obligation of free take-back, the Ministry of Environment has the ability to organise the collection and treatment of ELVs and charge the costs to the PRO.

Box 2.2. **Enforcement of EPR obligations** *(cont.)*

The legal authority to impose penalties is typically the task of governmental agencies. In the European Union, there is widespread agreement among stakeholders that member states and PROs should both be responsible for the monitoring of EPR schemes, and should ensure that there are adequate means for enforcement. However, there is less consensus on the division of responsibilities and costs. European Commission (2014) suggests that in some cases, the creation of an ad-hoc, independent authority for surveillance and regulation may be appropriate, and that it could be financed by a tax on PROs.

Sanctions can range from criminal and civil penalties, fees, revocation of the right of a PRO to operate, public disclosure of non-compliant entities (e.g. free-riders) and prosecution. As explained further below, the activities of PROs may also be reviewed by competition authorities, which may require PROs to modify activities that they consider to be anti-competitive.

Source: See Chapter 2 and European Commission (2014), *Development of Guidance on Extended Producer Responsibility (EPR)*, final report.

that enforcement capacity is lacking in some EU Member States and that unauthorised facilities and collection points are in operation. Inadequate enforcement can undermine not only the effectiveness but also the financial viability of EPR systems. It also fosters the export of hazardous waste.

Finally, the governance of EPR systems should be made much more transparent: this would provide a more effective means for assessing their performance and holding them accountable for their activities. There is a serious lack of both technical and financial data. The reasons for this vary among EPRs but may be due to unclear reporting requirements, the commercial sensitivity of some information, and/or anti-competitive behaviour on the part of the producers concerned. European Commission (2014) suggests that, at a minimum, EPR systems/PRO's should be obliged to provide information on:

- their fees
- the amount of products put on the market by their members
- the amount of waste collected and treated (reused, recycled, recovered (including energy recovery) and disposed of), so that the final destination of all collected waste is identified.

Where municipalities are involved, they should publish information on their activities and the related costs.

In the case of EPR systems working within a national or supra-national (such as the EU) framework, performance monitoring would be helped if definitions and reporting modalities were harmonised, and some mechanism established to check the quality and comparability of the data.

Updated recommendations on the design and governance of EPRs

- Fully implement the recommendations on the good governance of EPR systems in the 2001 OECD Guidance Document, particularly concerning the need to establish clear objectives, to specify the roles and responsibilities of stakeholders, and to establish platforms for dialogue among stakeholders.

Updated recommendations on the design and governance of EPRs *(cont.)*

- Periodically review the targets of EPR policies and adjust their ambition in line with waste management and resource productivity policy objectives; take account of the costs and benefits of proposed targets and establish them in consultation with stakeholders;
- Consider extending the scope of EPRs, particularly to cover more environmentally sensitive end-of-life products which are inappropriate for landfill disposal or incineration.
- In mandatory systems, governments should establish consistent and credible means for enforcing EPR obligations, including registers of producers, accreditation of PROs and appropriate sanctions.
- Governments and industry should co-operate to establish effective, adequately-resourced monitoring systems; in some circumstances, they may consider establishing an independent monitoring body financed by a tax on PROs.
- Mandatory EPR systems should be required to report regularly on the technical and financial aspects of their operations; their performance should be regularly audited, preferably independently; to the extent possible, definitions and reporting modalities for EPR systems operating in the same jurisdiction should be harmonised, and a means for checking the quality and comparability of data established; voluntary EPR systems should be encouraged to be as transparent as possible and periodically to undergo independent evaluations of their operations.
- The sharing of experience among EPRs, nationally and internationally, should be encouraged with a view to improving collection and recycling rates, disseminating information on eco-design, and enhancing the cost-effectiveness of EPR systems.

2.3. Financing, free-riding and orphan products

The 2001 Manual stressed that since producers are best placed to reduce the environmental impacts of their products, EPR policy should provide them with incentives to minimise the related environmental costs. Any remaining environmental costs, including those at the post-consumer stage, should be incorporated into the price of the product, and ultimately paid for by consumers. Responsibility for financing the post-consumer treatment of products should be clarified when the EPR is established.

While the principle that producers should be responsible for the post-consumer treatment of their products is well-established, its application in practice is more problematic. The study of EPR systems in the EU (European Commission, 2014) found that most EPRs covered, partly or fully, the net operational costs. These include the costs for collection and treatment as well as the administrative, reporting and communication costs required for the operation of the systems, minus revenues from the sales of recovered materials. However, the full costs of treating end-of-life products include a broader range of activities including the costs of public information and awareness campaigns, waste prevention actions, and monitoring and surveillance. Many EPRs did not cover this fuller range of costs.

To the extent that the full costs of EPRs are not covered by producers, they fall on municipalities and taxpayers. In France, the aim is for producers to bear 80% of the (optimised) costs of household packaging waste (if the collection rate is 75% or above), with 20% covered by municipalities. On the other hand, in Korea, the fees paid by producers fully cover the collection, treatment and administrative costs. 70%-90% of the fees are used to remunerate recyclers, and 1-5% allocated for information campaigns. In the United

Kingdom system for packaging, producers only cover about 10% of the costs, i.e. the difference between the costs of the initial recycling performance (before introduction of EPR) and those required to achieve the EU recycling targets.

In keeping with the Polluter-Pays Principle, producers should pay the full net costs related to the separate collection and treatment of end-of-life products (see Box 2.3). Where municipalities carry out some of the EPRs operations, this will involve calculating the appropriate level of compensation that PROs should provide to municipalities. That said, it should be recognised that it is not always easy to calculate net costs and that they may fluctuate significantly with changes in the price of secondary materials.

Box 2.3. **Full cost recovery from producer fees**

There is a broad consensus among stakeholder, supported by economic theory, that the full end-of-life costs should be internalised in EPR producer fees in order to apply the polluter-pays-principle. However, there is a debate about how this principle is implemented in practice and in particular about the scope of the costs that should be considered. There are diverging views as to whether littering costs and the costs of waste treatment of EPR products not collected by the EPR systems, but treated as part of mixed municipal waste should be part of the costs to be covered from producer fees. Ultimately the most adequate approach depends on who is best influence these aspects of end-of-life management and, hence, should receive an economic incentive to do so.

For littering some cost sharing between producers and municipalities would seem appropriate as both are able to influence littering. At the municipal level a range of measures can be taken to improve waste collection infrastructure in public spaces, as well as strengthening enforcement measures that discourage littering. In some cases, producers are also able to improve product design so that littering is minimised, e.g. when the element to open a metal can stays with the can rather than being pulled and discarded.

For EPR products treated as part of municipal waste, this would depend on who is responsible for separate collection. If producers are in charge of separate collection these costs should be covered from producer fees so that there is no incentive to diminish collection rates and to ensure that end-of-life costs are reflected in product prices. If municipalities are in charge of separate collection, these costs should be borne by them as they are best placed to improve collection rates.

"Free-riding" is a key challenge for most EPR schemes. Free-riders are those producers who benefit from EPR systems without contributing their share of the costs. The challenge of reducing free-riders generally increases with the number of producers and the length of the product chain. The problem can be acute when the valuable fractions of the waste stream are diverted to other destinations and threaten the financial viability of the EPR system. On the other hand, the cost of the measures required to eliminate free-riders for some product groups may make it not worth trying to entirely eliminate them. The 2001 Manual suggests several ways to address free-riding: keeping fees (and hence the incentive to free-ride) low, peer pressure and strict enforcement with suitable sanctions. However, keeping fees low may not be consistent with the objective of setting producers' fees at a level that covers the full costs of post-consumer treatment of products. A new issue is related to the large share of online sales and the possibilities for free-riding that arise therefrom. Further work is required to better understand these issues.

Box 2.4. **Optimising cost effectiveness – The example of the Japanese Packaging Recycling Act**

This Act foresees the payment of a commission by PROs to municipalities corresponding to actual recycling costs. If the municipality provides high-quality well-sorted waste to recyclers, the costs are decreased. Producers reimburse 50% of the difference between the actual and initially estimated costs. This approach based on variable contributions is intended to incentivise municipalities to provide high-quality and properly sorted packaging waste to recyclers, thereby reducing overall recycling costs.

Source: Based on Annex I.

Available information suggests that the scale of "leakage" – EPR systems not capturing all the products they were established to manage – has now reached a significant level for some waste streams (European Commission, 2014). Data in this area are based on estimates and should be treated with caution. Nevertheless, it appears that nearly half of the large home appliances subject to EPR in Japan are estimated to move through non-EPR channels, including reuse through the second-hand product market or collection as municipal waste. More than half of the WEEE collected in Europe probably leaks to improper treatment and illegal exports, or is treated properly but that the amounts treated are not reported. In France, it is estimated that only about one third of WEEE generated on French territory is handled by EPR systems, and between 45% and 75% moves through alternative channels and/or is exported. The growing volume of internet sales creates additional opportunities for leakage and free-riding.

Various approaches have been suggested to deal with leakage. To combat leakage in the form of illegal exports of waste, many environmental authorities are now working more closely with customs authorities. In Finland, producers have priority access by law to designated wastes. In the revised EU WEEE Directive, WEEE formally collected outside the EPR system is included in the calculations of regulatory targets, and non-PRO facilities are required to meet the same environmental standards as the PRO operating in the EPR system. The law on energy transition, recently adopted in France, aims to avoid leakage by forcing professional holders of waste of electric and electronic equipment to sign a contract with an approved compliance scheme, which could enable to better monitor and control the collection and treatment of waste. In Japan, estimates have been made of leakage and an order issued to local governments to regulate informal collectors.

"Orphans" are products subject to EPR requirements but whose producers are no longer in operation (e.g. due to bankruptcy). The challenge is how to equitably cover the post-consumer costs. The costs could be significant in the case of widely-used, environmentally hazardous products (e.g. some oils or chemicals) or virtually zero (e.g. if the main objective of the EPR is to influence how products are designed). Thus the choice of mechanism depends on the nature of the challenge. The 2001 Manual identifies various possible approaches that should be adapted to the specific nature of the challenge involved, including: current producers covering their own costs as well as those of former producers; ADF; fees paid at purchase; last owner pays; and insurance.

Updated recommendations on financing, free-riding and orphan products

- In mandatory systems, governments should establish consistent and credible means for enforcing EPR obligations, including registers of producers, accreditation of PROs and appropriate sanctions.
- Governments and industry should co-operate to establish effective, adequately-resourced monitoring systems; in some circumstances, they may consider establishing an independent monitoring body financed by a tax on PROs.
- The cost of end-of-life treatment ideally should be internalised into the price of the product and paid for by consumers; Producers should be responsible for financing the end-of-life costs of their products
- Free-riding should be addressed through peer pressure and strict enforcement with suitable sanctions.
- Orphan products should be addressed by opting for an approach that is adapted to the specific nature of the challenge involved, including: current producers covering their own costs as well as those of former producers; ADF; fees paid at purchase; last owner pays; and insurance;
- Governments should exchange experience on, and identify ways in which EPR systems can be financed in a sustainable manner; this should include analysis of how risks such as price volatility, leakage, etc. could be managed.

2.4. Trade, competition and EPRs

Since EPR policies impose requirements on products they necessarily have implications for related product and secondary markets. In particular, they may affect competition and trade, either intentionally or unintentionally. EPR policies may affect trade simply as a result of the higher administrative and transport costs faced by importers. The 2001 Manual argued that EPR systems that are compatible with competition and trade policies can be environmentally effective, more economically efficient, and subject to less opposition from the business community.

In order to avoid conflicts with trade policies, the 2001 Manual recommended that EPR systems should involve importers in the design of the system and ensure that proposed provisions are compliant with World Trade Organization (WTO) requirements. Specifically, when EPRs that may be trade-sensitive are being established, importers should be invited to participate and given an adequate comment period, foreign trade associations and the WTO should be informed, and consideration given to providing developing countries with technical assistance to comply with requirements. To ensure compatibility with WTO agreements, the requirements of a proposed EPR should be checked to ensure that they do not discriminate against importers, restrict trade more than necessary to achieve its purpose, prescribe process or production methods, or risk generating surplus recycled materials that may be dumped on international markets.

In order to promote competition in the markets served by EPRs, the 2001 Manual recommends that, where possible, competition authorities should be invited to provide advice on the likely impact on competition of alternative EPR approaches – as well as alternatives to EPR. The potentially anti-competitive behaviour of PROs was highlighted in this regard. By establishing PROs, producers can achieve significant economies of scale and even help smaller companies to stay in business. However, PROs may also provide opportunities for producers to collude illegally in order to eliminate competition among themselves, or to disadvantage their competitors. Thus competition authorities could help:

- to ensure that PROs do not abuse market power through excessive or opaque pricing or other anti-competitive practices;
- to avoid regulatory barriers to entry in post-consumer materials markets, including barriers to other PROs entering an EPR market;
- to require the PRO to contract out collection and recycling services on a competitive basis; to establish contracts that are not unduly long; and to use bidding procedures that are open, competitive and fair.

Since 2001 the size and complexity of EPRs and the related markets have increased. As a result, the potential financial gains for producers, as well as the welfare losses for society, that result from anti-competitive behaviour have become even more significant.

In addition, there have also been a number of decisions taken by courts and competition authorities that have addressed competition policy and EPRs. The number of cases is limited and they do not cover all competition-related issues. In some cases, the criteria used to determine if behaviour is anti-competitive vary among countries. As a result, decisions about competition and EPRs are not fully comparable or consistent across jurisdictions. Nevertheless, the experience gained enables some conclusions to be drawn about better managing trade-offs between competition and environmental policies when designing and implementing EPRs.

This section highlights several types of restrictions to competition in the context of EPR-related markets, based on a more detailed discussion in Chapter 3:

- Horizontal agreements among competing producers to establish one or more PROs; this includes the potential anti-competitive behaviour of PROs; barriers to entering markets served by PROs; the impact of higher costs when switching providers; and agreements among PROs operating in the same market that may limit competition.
- Vertical agreements between PROs and suppliers of collection, sorting or treatment services that limit competition.
- Anti-competitive behaviour in markets for waste treatment and secondary materials, and for products.

2.4.1. *Horizontal agreements among competing producers*

Horizontal agreements among competing producers to establish one or more PROs is another issue. This section focuses particularly on the advantages of individual versus multiple PROs and the potential barriers to entering markets that are served by PROs.

Individual and multiple PROs

Many EPR systems were initially established with a single PRO operating as a monopoly. Over time, some of these systems were broken up, in some instances following scrutiny of their anti-competitive conduct; in others, in response to competition authorities' critique of the regulations establishing the monopoly. As a result, an increasing number of EPRs involve multiple rather than single PROs.

The classic arguments against monopolies are that they can extract higher prices and are subject to lower incentives to seek and pass-on cost-savings. Both effects raise the cost of products. There is some empirical evidence that this has been the case with some monopoly PROs. The cost of handling waste packaging in Germany fell significantly with the introduction of competitive tendering of services at the level of PROs and the introduction of competition between PROs. The Norwegian competition authority identified several examples where producer-owned collecting/recycling monopolies incurred excess costs because they could be passed on to consumers, and that this formed part of a pattern of inefficiency.

On the other hand, it has been argued that monopoly PROs may have some advantages in terms of exploiting economies of scale, addressing free-riders and reducing the costs of regulatory oversight. While these arguments have been challenged, establishing a PRO as a monopoly in a start-up phase is a special case. It might be more efficient initially to set up a PRO as a monopoly if its establishment involves high sunk costs, and where there is also uncertainty about future costs and revenues. The European Court of Justice considered that the grant of exclusive rights to a PRO in Copenhagen to process non-hazardous building waste was justified because the establishment of capacity to manage this waste would not otherwise be built. This was considered to outweigh the cost in terms of restricted competition.

There is insufficient empirical evidence to determine the conditions in which a monopoly PRO is more efficient than competing PROs. The arguments for and against monopoly have been considered in different national jurisdictions on the basis of their cost and benefits. However, jurisdictions differ in terms of which costs and benefits should be considered and, hence, on when, or if, a monopoly PRO would be considered acceptable.

Barriers to entry

Barriers to entering a market will tend to reduce competition in that market. Such barriers may be structural or strategic. Structural entry barriers often take the form of high and irreversible costs of entering a market. The high costs may result from the nature of the market itself or a legal requirement. In contrast, strategic barriers are *deliberately* created or enhanced by the market incumbent. In some cases, it may be difficult to distinguish the two.

An obligation to enter a market nationwide is an example of a structural entry barrier; it increases sunk costs if the best entry strategy, absent the obligation, would be to enter at a small scale, in a limited area. Some jurisdictions have imposed a universal service obligation on PROs, often to prevent new entrants from "cherry picking" the most profitable areas. As the Swedish case illustrates (see Box 2.5), one way around this is to require incumbents to share infrastructure with the new entrants. However, this may not always be possible. In such cases, the benefits of a universal service obligation should be reviewed.

Denying entrants access to "essential facilities" is a strategic entry barrier that has featured in competition cases in PRO markets. Although the definition of "essential facilities" differs somewhat between jurisdictions, the basic idea is that there is something to which access is necessary in order to compete in a market, it cannot be feasibly duplicated, it can be feasibly shared, and it is controlled by a monopolist or a dominant firm.

A long-term exclusive contract is another example of a strategic entry barrier. This involves one or both parties agreeing to deal in a certain product only with the other party for an extended period of time. The European Commission has ruled against such contracts and required that the duration of contracts be restricted.

Box 2.5. Shared collection infrastructure in Sweden

In Sweden, waste packaging PROs were required to serve the entire country. A company trying to enter the market faced significant barriers to establishing collection infrastructure. In rural areas, it was very costly, and in urban areas the incumbent used municipal sites that could not be duplicated. The entrant complained to the competition authority, accusing the incumbent of denying access to infrastructure and abuse of dominance. After consultations with the competition authority, the parties entered into commercial negotiations that led to a solution whereby the two firms shared the collection infrastructure at issue, and shared the costs. This enabled both PROs to offer nationwide service.

Source: See Chapter 4.

2.4.2. *Vertical agreements between PROs and suppliers: markets for collecting and sorting waste*

Markets for the collection of waste may differ for households and businesses. Economies of population density and economies of scale may result in collection markets for households being local natural monopolies, most efficiently served by a single entity. This is often the case for kerbside collection of waste from households. In contrast, markets for the collection of waste from businesses are generally oligopolies, markets subject to competition from a small group of rivals.

The geographic extent of collection and sorting markets depends *inter alia* on transport costs and on legal restrictions on trade in waste. Sorting is generally more efficient when it is organised on a larger scale than collection. As a result, bundling these services together may result in less efficiency than managing them separately.

Many PROs procure waste collection and sorting services. Even where it is a monopoly, there can be benefits when waste collection is subject to competition "for" the market. There is evidence that the use of competitive tenders significantly reduces collection costs (see Box 2.6). However, the tender rules and procedures used by PROs can have an important impact on the cost of services they procure. Efficiency is enhanced when procurement is fair and competitive. This means *inter alia* that incumbents are not given advantages; when potential bidders receive the same information, at the same time, and with sufficient time to prepare their bids; and when a sufficient number of potential bidders submit bids for competition to occur. Contracts should not be too short (costs may

Box 2.6. Promoting more competitive tendering in Germany

In 2003, a first call for tender by the German packaging PRO, DSD, did not result in effective competition in many contracting areas. In about half the contracting areas, only one bid was received. In these contracting areas, prices were on average 70% higher than the lowest price offered in contract areas where two or more bids were submitted. DSD modified the tender conditions to attract bids from more small- and medium-sized companies. By 2005, the costs of collecting and sorting had been reduced by 20-30% compared with 2003. More recently, the German competition authority has expressed the view that "ensuring the separate tendering of collection [and sorting] services is particularly important. This safeguards that competition in sorting is not distorted".

Source: See Chapter 4.

not be recovered) or too long (some of the benefits of competition, such as the adoption of more efficient technology, may be lost). In two decisions, the European Commission decided that a duration of three years for contracts between waste packaging collectors and PROs was indispensable.

2.4.3. *Markets for products*

Promoting competition in product markets has been an important issue at the interface of competition policy and EPR schemes. Markets for products such as cars or tyres are large, so uncompetitive practices in these markets can have a particularly big impact on economic welfare.

One issue that has arisen in several jurisdictions is the agreement to pass onto consumers the fee charged by a PRO. Even if the fee is "visible", and legislation specifically requires the fee to be passed on, it may still be viewed as illegal price-fixing. In some instances, small "visible" fees were found to be necessary and not to cause a discernible harm to competition. In others, they were viewed as not necessary to the broader co-operation agreement and, hence, as anti-competitive.

There are several other concerns regarding the effect of PROs on competition in product markets. First, the PRO may create opportunities for the exchange of information and thereby facilitate price-fixing or other distortions to the product market. Second, if a PRO is a monopoly or dominant, and if it significantly increases producers' costs, and if the fees constitute a substantial part of producers' variable costs, then the intensity of competition in the product market could be reduced. Third, recycling fees may be increased to raise consumer prices and the profits of suppliers, or discriminate against some groups, for example foreign suppliers or small suppliers. Finally, concerns about competition may arise when the state provides subsidies or grants exclusive rights.

2.4.4. *Towards a better integration of competition policy and EPRs*

The OECD Council adopted a Recommendation on Competition Assessment in 2009. Amongst other things it states, in the section on revision of public policies that unduly restrict competition, that "Governments should adopt the more pro-competitive alternative consistent with the public interest objectives pursued and taking into account the benefits and costs of implementation". This recommendation is relevant when considering possible trade-offs between competition policy and EPRs.

Several issues concerning potential anti-competitive behaviour by EPRs have been discussed by competition authorities but no consensus has been achieved. They merit further consideration:

- The benefits of requiring the vertical separation of producers, waste collectors and waste treatment firms.
- Whether requiring collection services to be provided nationwide represents a barrier to entry for new PROs, particularly when sharing of infrastructure is not feasible.
- Whether the non-portability of financial reserves affects competition.

Updated recommendations to further promote the integration of competition policy and EPRs

- Competition impact assessments should be integrated into the design of EPR policies, taking account of the 2009 OECD Council Recommendation on Competition Assessment (2009), and the 2005 Council Recommendation on Regulatory Policy and Governance.
- Competition authorities periodically should issue easily-accessible guidance or information regarding their consideration of EPRs.
- Agreements to establish a PRO should be assessed by competition authorities within the jurisdiction's general framework for assessing horizontal agreements. Contracts between service providers and PROs should be assessed on a case-by-case basis within the jurisdictions general framework for assessing vertical agreements.
- Competition authorities should not distinguish between voluntary and government-sponsored agreements.
- EPR schemes should allow single PROs only when it can be demonstrated that the benefits (for example the capacity to manage the waste would not otherwise be built) outweigh the costs of less competition; the operations of monopoly PROs should be kept under review and competition encouraged when the benefits of single PROs no longer outweigh their costs.
- Any restrictions on competition intended to support the introduction of the EPR (such as allowing a PRO exclusive rights to a market) should be phased out as soon as possible.
- Services such as waste collection, sorting, and treatment, should be procured by transparent, non-discriminatory and competitive tenders. Factors that should be taken into account in this regard include providing for sufficient but not excessive contract duration, sufficient scale to provide incentives for investment, and sufficient scale and level of aggregation to facilitate bidding by all qualified firms.
- Tenders should not oblige collectors and recyclers to contract exclusively with one PRO. Other possible distortions, such as those that may result from bundling collection together with sorting and processing, should also be assessed.
- Post-consumer materials should not be allocated in a way that raises barriers to entry or expansion in the product market; for example when material is allocated at below market prices according to historical product market share.
- PROs, national registers or other clearinghouses should be designed so as to prevent the sharing of confidential market information that could result in anti-competitive behaviour.

2.5. Eco-design

In Chapter 1 there is a section that examines the impact of EPR systems on eco-design. It concludes that while EPR systems have helped to stimulate eco-design in some countries and sectors, their overall impact has been less than originally hoped for. Various proposals have been made and initiatives taken to address this issue.

First, higher fees, as well as ambitious objectives and robust enforcement, will generally strengthen incentives for eco-design. Setting fees at a level where they recover the full cost of the end-of-life management of the products covered by the EPR would better internalise the environmental costs associated with those products and thereby stimulate eco-design. Fees generally leave producers with more flexibility to find cost-effective solutions than regulations. The incentive effect increases as the PRO fee becomes a significant fraction of the production cost, which is the case in some sectors. For instance, in France, the EPR packaging fee is equivalent to about 4% of the sales revenue of packaging producers.

Second, the way in which the PROs of collective systems establish fees can have an important bearing on incentives for eco-design. PROs are financed either on a variable- or fixed-fee basis. Fixed-fees are typically used in PROs for complex goods such as electronic equipment, cars or furniture where it would be difficult to link the fee to the product's environmental impact. In such cases, it is easier to apply a unit-based fee, but the lack of a link between the fee and waste management costs associated with specific products provides only weak and indirect incentives for eco-design. To address this, some EPRs have introduced modulated fees linked to specific product design features other than weight. For example, the French Eco-Systèmes applies higher fees for mobile phones that do not have a standardised charger in order to discourage excessive production of peripheral devices.[1] Emerging technologies such as Radio Frequency Identification (RFID) may provide new opportunities to link fees with environmental impacts and improve cost allocation among producers. However, establishing modulated fees involves additional administrative costs so they should be proportional to the environmental and/or financial benefits. Moreover, modulated fees create winners and losers so it may be difficult for producers to agree on a modulated fee structure. There is also a risk of collusion among producers and the abuse of market power. Thus, it would be prudent for competition authorities to assess the establishment of modulated fees.

Third, the scope of EPRs could be enhanced. Some systems are weakened by exemptions. For example, in France, the printed press is exempt from the EPR for graphic paper even though it accounts for about one-third of this market. Another challenge is the potentially perverse incentives for products that are difficult to recycle: as long as products are "non-recyclable" they are not subject to an EPR and producers may have a perverse incentive not to develop their products so that they become potential subjects of EPR with take-back requirements. Korea has addressed this issue by applying an ADF to products such as chewing gum, disposable diapers, cigarettes, non-packaging plastics and kitchenware.

Fourth, there is scope to promote greater international harmonization of eco-design incentives for some global consumer products such as mobile phones. The challenge here is that whereas EPR systems are organised within (supra-) national or sub-national contexts, some consumer products (and packaging) are designed for the global market. Thus, in the absence of harmonization, individual schemes will provide weak and diverging signals about product preferences to producers. Harmonization can provide more consistent and effective incentives for eco-design. One example is the EC Directive on Restrictions on Hazardous Substances which has induced design changes worldwide thanks to its transparent focus on six hazardous materials. However, the obstacles and costs of achieving harmonization should not be underestimated.

In addition to the stimulus to redesign products that EPR systems are intended to provide, some producers have used EPRs to enhance co-operation to promote innovation (OECD, 2014). In Japan, for example, automobile manufactures have reportedly developed their own "3R evaluation" technological systems as an answer to the End-of-Life Vehicles (ELV) Recycling Law. The aim is to simulate recycling rates and costs, and to assess design improvements that can facilitate recovery. These systems are based on life-cycle-assessments (LCA) that help vehicle designers to improve the recyclability and environmental performance of their products.

The **Nordic Waste Group**[2] is developing new business models for plastic and textiles waste collection and waste treatment with a view to promoting more sustainable supply chains. The Resource Efficient Recycling of Plastic and Textile Waste initiative encompasses six projects including one aimed at developing a region-wide EPR model. The project will suggest innovative business models for the Nordic textile industry that enable more sustainable life-cycles of textiles. The projects were open for procurement last year, and will contribute to further efforts in the Nordic region to develop the recycling of plastic waste as a profitable Nordic industry along the entire value chain.

There is evidence that PROs are increasingly present in research consortiums, for example for improved eco-design of paper. Extending their role beyond information sharing to providing research funds could help to induce more effective eco-design (see Chapter 5).

Updated recommendations on incentives for design for environment

- Ensure the full costs of end-of-life management are covered by producer fees in order to maximise design-for-environment incentives.
- Variable rather than fixed producer fees should be applied in collective schemes where this is feasible;
- Consider the use of innovative approaches such as modulated fees (e.g. according to content of hazardous substances) or the use of new technology that may allow to link fees with end-of-life costs for specific products and improve cost allocation among producers;
- Enhance information flows from downstream sectors and users to manufacturers with a view to enhancing design for environment.
- PROs should support R&D efforts intended to improve the eco-design of their products by sharing their experience and, when cost-effective, by providing financial support.
- International harmonization of the design of globally-traded products should be encouraged with a view to improving their eco-design.

2.6. EPR systems in emerging and developing economies: The role of the informal sector

EPR systems in emerging economies is a relatively new phenomenon that was not covered in the 2001 Guidance. A common challenge faced by emerging and developing countries in applying EPR policies is the absence of well-established waste management systems. These systems as well as important stakeholders such as manufacturers, PROs, municipalities and recyclers that are usually available in developed countries and have potentially significant roles in EPRs are not always available in developing countries. As a result, EPR approaches in these countries generally place a heavier reliance on financial incentives than those in OECD countries. Accordingly, EPR systems in emerging economies make more extensive use of economic instruments such as price support and diversion credits than OECD countries. A further consequence is that regulatory requirements are generally less important. In marked contrast to OECD countries, the role of take-back requirements appears to be so far negligible in many emerging economies, except as a means to reinforce buy-back channels for low-value or non-recyclable materials. Waste in emerging economies has also attracted a large number of actors to recycle end-of-life products with a positive economic value. These informal waste workers are now estimated to be about 20 million (see Chapter 5). While the informal sector is usually relatively small in OECD countries and is frequently associated with interfering negatively with well-functioning formal waste management systems (e.g. by illegally removing high-value products and materials from the waste stream), it often performs useful functions that are not provided by the formal sector in middle-income countries with limited waste management systems (such as collecting valuables and recovering the material from them). When an EPR system is introduced in these countries, it interferes with the livelihoods of informal waste pickers who will compete for valuable materials.

The informal recycling sector is diverse, heterogeneous and context-specific which makes it difficult to characterise. In some cases, it may operate in the absence of a formal waste management sector; in others, informal workers may compete with it. Figure 2.1 illustrates the complexity of activities that can be involved. About one quarter of informal workers are estimated to be women, and there is a significant amount of child labour. While waste picking can be financially rewarding, it is hard and dirty work, and operators are often willing to accept risks to their health and safety for the rewards. It can also be socially precarious, with workers not covered by social or health protection. Informal sector operations typically involve high levels of working capital and high transaction costs. Unaccountable inventories leave operators vulnerable to theft.

Figure 2.1. **Schematic representation of the informal waste sector in emerging economies**

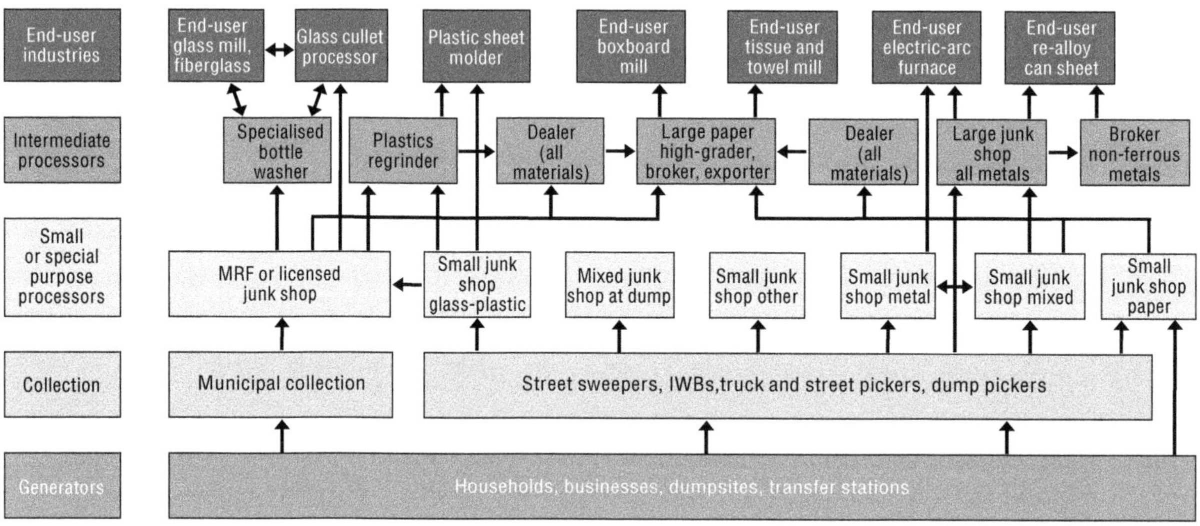

Note: MRF is a material recovery facility.

Source: Based on WASTE (2010), Training materials in integrated sustainable waste management, WASTE, Gouda, The Netherlands.

While informal waste pickers can perform useful functions by collecting and sorting waste with positive economic value, there are serious concerns about downstream informal dismantling and recycling which can generate negative economic and environmental impacts. Poor recycling processes usually fail to prevent emissions of hazardous substances resulting in occupational and environmental risks, and are relatively inefficient in recovering valuable material (Akenji et al., 2011; Williams et al., 2013; Romero, 2014). Furthermore, residuals that have no economic value are improperly managed in general and can be discarded or dumped leading to negative environmental impacts (Akenji et al., 2011). The challenge for middle-income countries is to find ways to integrate informal waste pickers and secure the positive contribution they can make, while mitigating environmental impacts from downstream informal waste processing.

Until about 10 years ago, the typical attitude of policy makers to informal waste pickers was that they were social victims who needed "rescuing" from dangerous and socially precarious activities. Accordingly, they were invited – in many cases pushed – to exit waste picking. However, this approach failed to recognise that waste picking provided an important means of livelihood. The employment opportunities available when exiting the sector generally did not come close to replacing the income of a waste picker. As a result, projects to "rescue" waste pickers generally failed and most returned to picking waste when the projects ended.

There is also now recognition that informal waste pickers can have positive economic and environmental impacts including by: reducing the amount of waste going to landfill, providing an alternative, "free" collection service, and helping to reduce collection and disposal costs for municipalities. When taken together with other actors in the informal value chain, there is some evidence that informal systems in emerging economies recover more materials than formal recycling systems when they work in parallel. There are also cases that show how failure to effectively include the informal sector in EPR systems can undermine the efficiency and effectiveness of EPR systems (see Box 2.7).

Box 2.7. **Exclusive and inclusive approaches to informal workers: the cases of Bulgaria and Colombia**

Bulgaria established an EPR system in 2004. It involved 100% producer responsibility for end-of-life management of packaging waste. The estimated 10 000 pre-existing Informal workers were not consulted in the establishment of the system which effectively aimed to take over their activities, thereby threatening their livelihoods. This resulted in overt and covert conflict between the formal and informal systems. The informal workers undermined the operations of formal system, including by removing materials from drop-off containers that had been set up to collect packaging waste. The operators of the formal system called for the arrest of informal workers and the closure of the centres to which they sold recyclable material (Ministry of Environment and Water, Republic of Bulgaria, 2003; Doychinov and Whiteman, 2013).

Failure to engage the informal sector resulted in the establishment of a highly dysfunctional EPR system. The recycling targets that were set were lower than the amounts that were actually being collected by informal recyclers. Households and businesses preferred to continue working with the established recycling channels. As a result, about 90% of materials continued to flow through informal channels. Since the formal system captured such a small share of recyclable material, unit cost were high and revenues were insufficient to cover even operational costs (ISWA/EXPRA/RDN, 2014).

Colombia provides an example of more inclusive approach. About 20 private sector organisations established CEMPRE as an NGO to promote inclusive recycling. Using member fees, CEMPRE conducts a wide range of activities including: promoting policy and legal reform to protect the rights of informal workers; supporting the establishment and operations of waste picker associations; capacity building and training; facilitating dialogue and co-operation between waste pickers and public authorities; and helping to mobilise finance. Within the framework of a product stewardship system for packaging materials, the city of Bogota pays a diversion credit of USD 50 per tonne to waste pickers selling to authorised junk shops. Authorised junk shops register the quantities of all materials purchased from individual recyclers, and forward this information to the city administration. Junk shops only trade with waste pickers that are members of recognised

Box 2.7. **Exclusive and inclusive approaches to informal workers: the cases of Bulgaria and Colombia** *(cont.)*

associations, providing waste pickers with an incentive to join accredited associations. The city administration deposits the diversion credit directly in the waste picker's bank account. Thus waste pickers must be part of the formal economy in order to receive their credit.

The Colombian system is inclusive but incomplete. Not all producers are members, not all packaging materials are covered, not all waste pickers are members of co-operatives, and many co-operatives and junk shops remain independent and unauthorised. Further analysis is needed to assess how well the system works in terms of diverting packaging waste from disposal to recycling, and on the impacts of inclusive recycling. A national observatory has been established to examine some of these questions.

Source: See Chapter 5 for more detail.

The failure of traditional policies for informal waste workers, and the recognition of the positive role they can play, has helped foster a variety of new approaches to integrate informal workers into formal waste management systems. They include:

- *Rights-based interventions*, typically to support groups of waste pickers and their families to claim labour or citizenship rights and to build associations that strengthen this claim.
- *Informal sector integration*, sometimes also referred to as inclusive recycling. This involves recognising the role of informal waste pickers, for example through formal contracts or agreements, identity cards, or licensing. This approach may also stimulate the formation of co-operatives and business associations in which informal recyclers are considered as entrepreneurs or enterprises.
- *Formalisation* involves providing support to waste pickers to comply with tax laws and business norms, to register as enterprises, and to agree to follow laws and rules. It does not imply any reciprocal measures by the public authorities.
- *Professionalisation* and access to financing. This approach aims to support informal recyclers as autonomous micro- or family enterprises which, because of their ethnic, social, or immigration status, have limited or no access to financial and business services. It is primarily demand-driven and focuses on increasing knowledge, capacity, business skills, and access to materials and financing.
- *B2B (business to business) value chain activities*. This approach involves value chain support in the form of pre-financing of inventory and/or providing infrastructure and equipment.

Ideally the goal should be to establish arrangements whereby the informal sector is not working against the formal sector, and where safe and viable employment opportunities exist. However, the obstacles should not be underestimated; for example, "cherry-picking" and material theft by waste pickers can undermine the financial viability of the formal waste management sector, and informal recyclers may circumvent regulations regarding the dismantling of products containing toxic substances. Learning lessons from current efforts to integrate waste pickers into formal waste management systems would provide useful guidance for the further development of policy in this area.

Recommendations for integrating informal workers in EPRs in emerging and developing economies

- The role that informal recyclers play should be recognised: in many emerging economies, they are responsible for most of the materials that are captured, processed and sold in the recycling value chain.
- Cities in emerging economies should consider how they could best draw on the knowledge of waste pickers and junk shops; they are often the only stakeholders with practical experience, knowledge to maximise recycling under local market conditions, and incentive to adapt quickly to new value chains and market opportunities.
- Informal recyclers should be invited to contribute their experience and expertise in all relevant public decision making processes. They should be engaged in the design, monitoring and evaluation of recycling and valorisation systems, as well as the definition of quality standards.
- Producers, city authorities and informal recyclers should work together (experiment) to strengthen, or introduce, upstream separation of recyclables, organics and residuals at the level of businesses and households. Upstream separation provides important support for EPR systems. Downstream activities such as dismantling and recycling are potentially more problematic and authorities need to enforce environmental standards in such operations.
- Public authorities should work with informal recyclers to collect data on waste generation and recycling rates. It should not be assumed that no recycling is taking place.
- The insights and ambitions of informal recyclers should be combined with international good practice approaches for integrating informal workers into formal waste management systems, and take full account of relevant health and safety, social protection and financial considerations.
- EPR systems in emerging economies should avoid becoming involved in the recycling of materials where private value chains are likely to work well. EPR systems provide more opportunities for stakeholders, including informal recyclers, when they address market failures, including: environmentally sensitive waste streams, low-value materials, recyclables difficult to dismantle, or recycling in areas where there are few value chain buyers within reasonable transport distance.
- Priority should be given to developing business partnerships with informal, and micro and small, recycling enterprises over Public Private Partnership approaches government more than the host community.
- In developing EPRs, engage local authorities, municipal associations, national governments, regional economic communities, and bilateral and multi-lateral institutions; evaluate, disseminate, and use good practices of partnerships involving informal recyclers to inform public policy and legislation; and use these partnerships and activities to promote recognition of the informal recycling sector.

2.7. Concluding remarks

EPR has become a well-established policy instrument since 2001. If anything, its potential importance has increased with the greater emphasis now placed on policies for resource productivity and the circular economy in recent years.

There is some evidence that EPR systems have achieved their main goal of shifting the burden of treating end-of-life products from municipalities and taxpayers to producers. It is also likely that they have contributed to the decreased share of waste that is destined for final disposal and to increased rates of recycling. On the other hand, they appear to have been less effective in promoting more environmentally-friendly design of products.

However, lack of data, methodological challenges and lack of comparability make it difficult to assess the impact of EPRs with any precision. Equally, it is not possible to identify good practices or the most cost-effective models for EPRs. Nor is enough known about the contribution that EPR has made, and could make, to promoting green growth. Arguably, the single most important recommendation arising from this review is that EPR systems should be more transparent and make available the information that is needed to assess their performance and, thereby, help to identify ways in which they can be more efficient and effective.

Many of the recommendations from the 2001 Manual are still relevant and should be applied more systematically. There are many opportunities to make EPRs more effective including by: increasing their level of ambition; broadening the scope of products covered; better internalising environmental costs; and strengthening enforcement, particularly to reduce free-riding and leakage. Further efforts are particularly needed to strengthen the incentives for eco-design of products. Better eco-design may also be achieved by linking EPRs with broader innovation initiatives and, in the case of globally-traded products, by harmonising environmentally-sensitive design features.

Continued vigilance is needed to ensure that the product markets that EPR systems serve remain competitive. The potential financial gains for producers, and the welfare losses for society that result from anti-competitive behaviour, have increased along with the growth and concentration of the recycling and waste management industries. Lack of transparency increases concerns about anti-competitive behaviour.

The EPR systems being established in emerging economies do not necessarily follow the same models as those in OECD countries. Many focus on products with an economic value. These EPR systems are more dependent on financial transactions, and have fostered the emergence of an informal waste work force that now numbers about 20 million. While new approaches are being applied to integrate these workers into formal waste management systems, the livelihoods of many remain hazardous and precarious.

The global context has evolved significantly since the development of the first EPR policies. New economic powers have emerged in the global economy, product value chains have become more complex and extended across national boundaries, technological changes are altering patterns of communication and consumption, not least due to the internet, and markets for some materials and waste streams have been highly volatile. In such a context, EPR systems will have to continue to evolve if they are to become more effective waste management policy tools and to support the transition to more resource-efficient economies.

The following four chapters present the analysis that was developed to support the development of more up-to-date and policy-relevant guidance. Four issues were examined in more depth. These are: design and governance of EPR systems (Chapter 3); the anti-competitive behaviour that has been observed in EPR systems, a concern that has increased with the growth and increased concentration of the waste and recycling sectors (Chapter 4); the role of EPRs in promoting more environmentally friendly design of products (Chapter 5); and the operation of EPR systems in emerging economies, particularly the important role played by the informal waste sector (Chapter 6).

Notes

1. However, since the increase was only from EUR 0.01 to 0.02 per unit, the effect of this differentiation is probably negligible.
2. The **Nordic Waste Group** works under the Nordic Council of Ministers (including Prime Ministers from Norway, Sweden, Denmark, Finland and Iceland) to support sustainable processing of waste in the Nordic countries and Europe.

References

Akenji et al. (2011), EPR Policies for electronics in Asia – A phase in approach, Slide presentation IGES, *www.iges.or.jp/en/archive/wmr/pdf/activity20111111/IGESEPR.pdf.*

Chalmin, P. and C. Gaillochet (2009), "From waste to resource: An abstract of world waste survey 2009", *www.veolia-environmentalservices.com/veolia/ressources/files/1/927,753,Abstract_2009_GB-1.pdf.*

Dempsey M. et al. (2010), "Individual producer responsibility: a review of practical approaches for implementing individual producer responsibility for the WEEE Directive", *INSEAD Faculty & Research Working Paper, http://sites.insead.edu/facultyresearch/research/doc.cfm?did=45054.*

European Commission (2014), *Development of Guidance on Extended Producer Responsibility (EPR), final report, http://ec.europa.eu/environment/waste/pdf/target_review/Guidance%20on%20EPR%20-%20Final %20Report.pdf.*

European Environment Agency (2011), *Earnings, Jobs and Innovation: The Role of Recycling in a Green Economy,* EEA Report No. 8/2011, *www.eea.europa.eu/publications/earnings-jobs-and-innovation-the.*

Menikpura, S.N.M., A. Santo and Y. Hotta (2014), "Assessing the climate co-benefits from Waste Electrical and Electronic Equipment (WEEE) recycling in Japan", *Journal of Cleaner Production,* Vol. 74, No. 2014, pp. 183-190, *www.sciencedirect.com/science/article/pii/S0959652614002613.*

OECD (2014), "The state of play on extended producer responsibility: Opportunities and challenges", Issues Paper, Global Forum on Environment, Tokyo, 17-19 June 2014, *www.oecd.org/environment/waste/Global%20Forum%20Tokyo%20Issues%20Paper%2030-5-2014.pdf.*

Plastics Europe (2015), *Plastics – The facts 2014/15, An Analysis of European Plastics Production, Demand and Waste Data, www.plasticseurope.org/Document/plastics-the-facts-20142015.aspx.*

Romero Victoria (2014), *Steps Towards EPR Implementation in Developing Countries,* MSc thesis, Imperial College London.

Tasaki, T (2016), "Negative effects caused by shared responsibility and the significance of goal setting in recycling systems", *NIES-CMW Policy Brief* N°1, Published 6 April 2016, *www-cycle.nies.go.jp/eng/report/policy_brief_001.html.*

Tasaki, T., N. Tojo and T. Lindhqvist (2015), *International Survey on Stakeholders' Perception of the Concept of Extended Producer Responsibility and Product Stewardship,* IIIEE and NIES Joint Research Report, *www-cycle.nies.go.jp/eng/report/epr_eng.html.*

Tojo, N. (2004), *Extended Producer Responsibility as a Driver for Design Change – Utopia or Reality?,* IIIEE dissertations 2004:2, Lund, IIIEE, Lund University, *https://lup.lub.lu.se/search/record/1277476.*

Waste (2010), *Training materials in integrated sustainable waste management,* WASTE, Gouda, The Netherlands.

Williams et al. (2013), "Linking Informal and Formal Electronics Recycling via an Interface Organization", *Challenges,* Vol. 4, No. 2, pp. 136-153.

Websites:

http://proeurope4prevention.org/packagings-trends.

PART II

Analysis and key issues

Extended Producer Responsibility
Updated Guidance for Efficient Waste Management

PART II

Chapter 3

Governance issues and extended producer responsibility

This chapter begins with a discussion of why government is involved in EPR systems[1] *that, on first glance, are intended to lessen the role of the public sector in end-of-life management. Four typical governance structures are then presented and their advantages and disadvantages discussed. The role and status of producer responsibility organizations (PROs) are characterized and the debate over non-profit versus for-profit status for PROs is discussed. From there, the chapter looks at governance functions in EPR and their allocation among different stakeholders, as well as the resources needed for government participation in EPR systems.*

3.1. Introduction

The concept of "extended producer responsibility" (EPR) has become an established principle of environmental policy in many countries since the term was first introduced in the early 1990s (Lindhqvist 2000). It is an environmental policy strategy in which the responsibility of producers for their products is extended to include the costs and sometimes the management of end-of-life products, especially the recycling of their products.[2] The concept of EPR includes a wide range of instruments, from eco-design measures to information programs to financial contributions towards separate collection and recycling. Some of these systems are put in place by industry on a voluntary basis, while others are based on legislative obligations requiring companies to pay fees to support the costs of organising the (separate) collection, sorting, and recycling of specific waste streams to meet sub-national (state or provincial),[3] national, or supra-national (European) targets.

Many OECD countries have EPR policies in place for packaging, electronics, batteries, tires, or end-of-life vehicles, and have accumulated experience on the design and implementation of the resulting systems. At the same time, an increasing number of developing countries are either considering or have already undertaken steps to introduce EPR into their materials management policy frameworks and these are often an integral part of policies to achieve a circular economy, improve materials security, and increase competitiveness.

The OECD carried out a significant body of research on extended producer responsibility systems in the 1990s and early 2000, including publication of *Guidance Manual for Governments* in 2001 (OECD, 2001). This manual defines EPR as having two separate objectives: i) to shift responsibility upstream to the producer and away from municipalities and ii) to incentivise producers to incorporate environmental considerations in the design of their products. Since then, a significant body of experience of the real-world use of these policies has been built up. In general, while EPR has been successful at shifting responsibility and cost burdens to producers, effects on product design have been harder to identify.

Recent discussions have highlighted diverse experience and opinions on the levels of government involvement in the administration of EPR. Based on these experiences and on feedback provided by governments at the *Global Forum on Environment: Promoting Sustainable Materials Management through Extended Producer Responsibility (EPR)* (Tokyo, June 2014), selected governance-related topics are addressed in this report. Unfortunately, systematic or critical analysis and even information on many governance approaches and functions is lacking in both academic and policy research. While there are many attempts to assess the policy design of EPR systems (policy goals, performance targets, fees, operational roles) and policy outcomes (recycling rates, cost-efficiency), little has been written about how the nitty-gritty of governance and administration is conducted or how well various approaches work.[4] This report is thus an initial foray to clarify issues, identify lessons learned, and determine where further research is needed.

This chapter begins with a discussion of why government is involved in EPR systems[5] that, on first glance, are intended to lessen the role of the public sector in end-of-life management, examining the role of market failure as it relates to voluntary and mandatory approaches to EPR, to markets where some end-of-life products have value, and to EPR in developing countries. Four typical governance structures are then presented. The role and status of producer responsibility organizations (PROs) are characterized and the debate over non-profit versus for-profit status for PROs is discussed. From there, the chapter looks at governance functions in EPR including policy formulation and evaluation; operations; stakeholder consultation; registration and accreditation; collection and disbursement of fees; co-ordination through clearinghouses; monitoring; and enforcement. The allocation of governance functions and advantages and disadvantages of alternative governance structures are discussed, with particular attention to the role of municipalities. Resources needed for government participation in EPR systems are described including the role of outsourcing and delegation of governance functions to non-governmental entities. The chapter concludes with observations about patterns and best practices in EPR governance.

3.1.1. *Why is government involved in EPR systems?*

A key premise in EPR is a shift of responsibilities for waste management from the public sector to the private sector, raising the question of why – or how much – government is involved in such systems. In most countries, government is involved in waste management as a regulator and as a service provider. Typically, national or regional sub-national (state or provincial) governments act as regulators, whereas it is local governments that are assigned operational responsibility for waste management. In some cases, those local governments provide services themselves using government employees. In others, the local government contracts out those services to commercial providers. In still others, individual waste generators contract directly with collection services (with varying degrees of oversight by local governments). Thus, EPR represents not so much an entirely new intervention of government in the market for waste management as a different form of intervention.

This new role of government involves a different characterisation of the problems posed by waste and a reassignment of responsibilities entailing involvement of additional stakeholders. The result is that new forms of governance of waste management have emerged. Governance in this context[6] refers to the organisation and administration of matters related to waste generation, management and policy by decision-makers at all levels including governments, businesses, advocacy organisations and other elements of civil society, members of the informal economy, and consumers/waste generators. Government is one element in the system of governance, albeit a pivotal one.

EPR represents, at least conceptually, a potential reduction of government involvement in waste management and yet even a cursory examination of EPR as it is practiced around the world indicates that governments are involved. The nature and extent of government involvement in EPR systems reflects the need to respond to market failures and, as EPR systems evolve, to problems that emerge with policies and program that are in place. Government involvement also exists because of the familiar requirement for monitoring and enforcement of public policy. Put more simply, government is involved because EPR systems require not only sound policy design, but also effective governance to operate effectively.

3.1.2. *Voluntary approaches to EPR*

Voluntary approaches to producer responsibility for end-of-life management of products exist in a variety of markets, adding a further dimension to the discussion of market failure and EPR governance. Voluntary product take-back – the most salient form of voluntary

EPR – is most commonly found in markets for durable commercial products and/or where end-of-life products have value (positive prices), often under the rubric of "asset recovery". Photocopiers are a long-cited example (Murray and Vietor, 1993; Reinhardt, 2000). Voluntary take-back may also be pursued by a producer seeking to prevent acquisition, refurbishment, and resale of its own products by third-parties (Ferguson and Toktay, 2006). Concern over loss of market share to refurbishers is thought to spur take-back by producers of toner cartridges and lenses from single use cameras. Reputational advantages and pre-emption of mandatory obligations can also motivate firms to engage in product take-back (Toffel, 2004; Nash and Bosso, 2013; Agrawal et al., 2015).

Voluntary approaches have been advocated by regulated and potentially regulated industries as a superior approach to the mandatory approaches discussed in this chapter, in part, because of their potential for flexibility and lower cost (Galeano, 1996; Renckens, 2008). While little systematic quantitative information is available on the proportion of end-of-life products managed through voluntary EPR, it appears confined to a few, specific products or product categories. That is, market forces will lead firms to take back products where it is profitable to do so. The question then becomes one of whether the amount of product take-back that occurs as a result of market forces is sufficient for policy purposes. Do products with lower end-of-life value, hazardous substances, or which are costly to collect and process also need to be the responsibility of producers? In the latter case, the issue is whether producers will adopt ongoing "beyond compliance" practices through a voluntary commitment to the extent deemed necessary by policy makers. Stated another way, will corporate social responsibility provide the level of continuing end-of-life management that is sought or is mandatory EPR required?[7]

Voluntary approaches to EPR can be hindered by the failure to progress beyond business as usual, regulatory capture, poor monitoring, free riding, and transaction costs (Barde, 2004). Analysis of voluntary and mandatory approaches suggests that voluntary approaches do not have the scope and transparency that can achieve ambitious policy targets (Quinn and Sinclair, 2006). That is, voluntary approaches not only face challenges in terms of performance, but also with respect to governance – data availability, measurement, and transparency in particular. The potential effectiveness of voluntary EPR is likely to be a function of the characteristics of the relevant product, the market conditions in the associated industry sectors, and the extent of transparency and oversight in the voluntary programs.

Governance of voluntary environmental initiatives has been studied (King and Lenox, 2000; Potoski and Prakash, 2013), but no research has been found on governance of voluntary EPR specifically. More broadly, voluntary approaches to environmental policy in general have been widely criticised in the research literature as ineffective (Harrison, 1998; OECD, 2003; Morgenstern and Pizer, 2007; Darnall and Sides, 2008). The effectiveness of voluntary approaches appears to be contingent on credible threats of regulation (Khanna, 2001). Arguments against mandatory EPR, however, often are not focused on the possibility of wide-scale voluntary EPR, but rather on the use of other policy approaches (often municipal responsibility for recycling (e.g. SAIC Energy, 2012a)) or challenge the need for enhanced end-of-life management (e.g. Scarlett et al., 1997). This shifts the policy debate away from questions of the adequacy of voluntary approaches to one of policy justification and/or instrument choice, matters outside the scope of this report.

3.1.3. EPR and market failure

From an economic perspective, inadequate end-of-life management arises because of the failure to price externalities. "Getting the prices right," that is, internalising externalities so that market prices reflect environmental impacts, is thus the remedy that economic theory suggests as the approach of first resort. However, a variety of factors can make getting the prices right impractical or ineffective. Most notably, quantity-based user charges ("pay-as-you-throw"), a crucial element in a get-the-prices-right strategy, can provoke illegal dumping (Ino, 2007); without variable fees, the price signal to generators for appropriate waste management is muted.[8]

In concrete terms, there are two consequences of the market failure in end-of-life management that motivate EPR. First, producers face insufficient incentives to (re)design products and packaging to facilitate end-of-life management, i.e. to pursue design for environment (DfE). As noted above, however, because of the structure of many EPR systems, incentives for DfE have been limited. The evidence for the impact of EPR on product design is sparse (Tojo, 2004; Gottberg et al., 2006; Røine and Lee, 2006),[9] although there is some indication that EPR has an impact on material choice for packaging (see Chapter 4, Annex I and Bell 2013). Second, when levels of recycling or other appropriate forms of end-of-life management are seen as inadequate, EPR is one among a portfolio of policy strategies that can be used to raise the levels (Hickle, 2014a).[10]

It has been argued that governments turned to EPR in the early 1990s in part because of the failure of voluntary industry initiatives to remedy end-of-life management problems (Eichstädt and Kalhlenborn, 2002). This comports with a view that industry has the technical and managerial capacity, and notably, the resources, to change products and end-of-life management that are out of reach of government (Hickle, 2014a). That is, the argument for EPR in this view rests on matters of funding, institutional competence, and policy deficiencies rather than market failure.

The extension of the responsibility of producers to include end-of-life management is premised on a degree of privatisation of solid waste services. EPR was envisioned in part as a performance- and market-based environmental policy instrument (Lifset, 1993). In simple terms, producers would be assigned the responsibility for (some degree of) collection and processing for recycling, often with a quantitative target. Businesses would be left to organise the specific elements of those activities, using, it was hoped, the superior managerial acumen of the private sector. Municipalities and taxpayers would be relieved of the financial burden of recycling and related activities because the burden would be transferred to producers and consumers.

As EPR systems have been established around the world for various types of products, the notion of assigning a policy target and stepping aside to let producers self-organise to meet those goals confronted a variety of challenges and existing institutions. The result is that government involvement in the administration of EPR has taken a diversity of forms. In some instances the issues have involved misaligned incentives, in others the need for conventional government oversight. That governments play an overt role in EPR systems is not surprising. As noted in a recent review of market-based instruments in sustainability transitions (Parson and Kravitz, 2013), market instruments do not avoid authority based state regulation. Any policy is designed, enacted, implemented, and enforced by state authority and requires multiple design decisions by government... Governments must also decide how to monitor and respond to targeted actors' behaviour, including what mix of rewards, punishments, and other modes of influence to use.

3.1.4. EPR *when some products have value*

The question of market failure and the role of government intervention when some, but not all, end-of-life products have value raises complicated challenges for EPR. Over the past decades some types of waste electrical and electronic equipment (WEEE) have become valuable, that is, the discarded products have positive, rather than negative, prices in market transactions. This has been the case for (many) mobile phones and commercial information technology products for many years. More generally, it is business-to-business goods, frequently disposed in the institutional, commercial, and industrial waste streams, which are likely to have positive value. More recently, however, consumer WEEE such as desktop PCs and printers have become valuable primarily because of the potential for sale into second hand markets and the metals contained in their circuit boards as well as improvements in recovery technologies. Some types of packaging and printed paper also fetch positive prices.

At a foundational level, the presence of end-of-life products and materials with positive prices in waste streams regulated through EPR raises the question of whether there is in fact market failure and whether and how such wastes might be excluded from EPR obligations (Kunz et al., 2014). There are, however, practical challenges in differentiating EPR obligations according to whether market failure has occurred[11] or even on the basis of the current price of end-of-life products. These challenges include the tracking of wastes managed outside EPR systems and the definition of thresholds that might trigger the release from EPR obligations. A key challenge for governance and EPR policy is the potential for increased cost as the valuable fractions of the waste stream are shifted away from EPR systems leaving only the costly to manage. The same applies to collection of waste from remote areas. A second challenge is whether EPR systems could be designed to accommodate changes in the end-of-life product streams that they manage as the prices and profitability of various wastes shift over time. An approach that differentiated among waste according to value would depend as well on the effectiveness of the regulation of entities providing collection and processing services outside of EPR systems in order to ensure that positive prices of end-of-life products do not arise from inadequate environmental management.

Defining a positive price for end-of-life products

Most secondary materials (i.e. materials from end-of-life products that have been processed and made ready for use by industry) have a positive price. What are at issue in EPR are end-of-life products that have a positive value *prior* to collection and/or processing. A discussion of the challenges of defining what constitutes a "positive price" for the purpose of EPR is presented in Appendix 3.A.1.

Leakage in EPR systems

The value of the end-of-life products has led to legal, semi-legal, and illegal diversion from EPR-managed systems. Many EPR systems report large flows of end-of-life products outside of EPR systems. In France it is estimated that only about one third of WEEE generated on French territory is handled by EPR systems, with between 45% and 75% moving through alternative channels and/or exported (Monier et al., 2013). Nearly half of the home appliances subject to management via EPR in Japan are estimated to move through non-EPR channels (see Annex H). According to Countering WEEE Illegal Trade (CWIT), a recent study commissioned by the European Union, of the 9 450 kilotons of WEEE generated in 2012 in Europe, approximately 3.3 million metric tonnes are officially reported as collected and recycled, 3.15 million tonnes are collected and processed without reporting and/or under other non-compliant conditions, and 1.3 million is exported without reporting[12] (Huisman et al., 2015).

The leakage in the system occurs at many stages in the product life cycle. In some cases, municipalities formally opt out of EPR systems, choosing instead to contract with waste firms and recyclers to manage their recyclables. In Germany, for example, as much as 65% of the WEEE collected was done by municipalities that elected not to participate in EAR, the EPR system (see Figure 3.1).

Figure 3.1. **Share of German municipalities choosing not to manage WEEE through EPR (%), 2009-13**

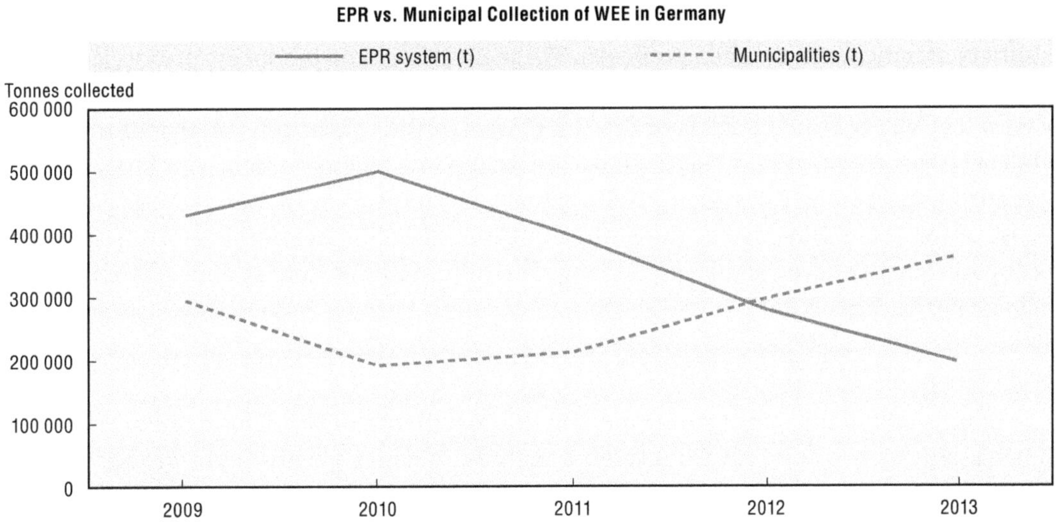

Source: Stiftung Elektro-Altgeräte Register (2014), *www.stiftung-ear.de/en/*.

Leakage also occurs through "cherry-picking", that is, collecting or managing only the valuable components of the waste stream, leaving the costly portion to others. This occurs when municipalities, retailers, or other entities providing collection or drop-off sites bypass producer responsibility organisations by selling, formally or informally, the more valuable fractions of the collected recyclables to recyclers not participating with the PRO program. Still more informal cherry-picking occurs when brokers intercept consumers en route to drop-off or civic amenity sites, purchasing items that are valuable. It can also occur when generator-segregated waste is scavenged prior to collection. Diversion of end-of-life products from EPR systems also occurs on a more systematic basis in developing countries where peddlers purchase discards from households and other waste generators at a price higher than that offered by the formal EPR system. This, for example, has been a long-standing challenge in the establishment of the Chinese EPR system for WEEE (Yu et al., 2010a).

In the US, bottle deposit laws, considered by some but not all stakeholders as a form of EPR, have experienced leakage. Containers discarded in a state without deposit legislation are transported to nearby states that have such legislation (see, CalRecycle, 2015). Deposits are falsely redeemed causing considerable financial impact on the deposit redemption system (Paben, 2015).

Finally, leakage can occur through illicit export of end-of-life products, including WEEE sent to developing countries and end-of-life vehicles (ELVs) transferred to Eastern Europe from other countries in Europe. The impetus for illegal exports ranges from lower labour costs available for dismantling of end-of-life products to lower, and thus less costly, environmental standards for processing to unadorned dumping.

The CWIT study found that offences related to illegal trade in WEEE involved 13 different types of actors across the EPR chain. These offenses include inappropriate treatment and violations of WEEE trade regulations, theft, lack of required licenses/ permits, smuggling, and false declaration of the load. The study also identified financial crimes including fraud, money laundering, and tax evasion connected to illegal WEEE trade (Huisman et al., 2015).

Leakage of various sorts is often accompanied by inadequate handling and treatment of wastes. It also poses a direct problem for EPR systems insofar as the systems find it difficult or costly to collect sufficient quantities of end-of-life products to meet regulatory obligations. A variety of remedies have been proposed. These include giving producers priority access by law to designated wastes as in Finland (Kalimo et al., 2015). Another remedy is to include wastes formally collected outside of the EPR system in the calculations of regulatory targets while also ensuring that non-PRO facilities meet the same environmental standards as those operating as part of the EPR system. This approach has recently been mandated as part of the recast of the WEEE Directive in the European Union (EU DG Environment, 2014). In Japan, local governments have been pressed to regulate informal collectors as a means of reducing leakage (Hotta et al., 2015).

3.2. Governance structures in EPR

To understand the role of government in the governance of EPR, it is useful to look at the EPR systems as a whole. All EPR systems include roles for government, producers, and providers of collection and processing services, but differ in the number of producer responsibility organisations (PROs) involved; which entities collect and distribute funds, and monitor and enforce compliance; and the responsibility and autonomy granted the participating entities. The organisation of EPR systems across the world, however, has tended to follow a handful of approaches: single PRO, competing PROs, tradable credits, and government-run.

Figures 3.2, 3.3, 3.4, and 3.5 depict, in a simplified manner, the legal and reporting relationships in the four representative approaches. These figures represent EPR systems for an individual product category in a jurisdiction[13] that has enacted EPR policies.

A country may have not only different EPR systems for WEEE and ELVs, for example, but those systems may have different governance structures. Not surprisingly, the amount of responsibility or activity assigned to each type of entity varies across EPR systems, e.g. monitoring may be an extensive and rigorous process in one country and much less elaborate in another. The diagrams show main types of entities involved in EPR and the legal and regulatory, data and reporting, and financial processes that connect them.

These diagrams are meant to be illustrative of generalised, representative approaches and do not reflect the variation and complexity present in systems across the world. Similarly, the boxes in these diagrams do not delineate details such as when government functions are separated among agencies (e.g. collection of fees in one, disbursement of funds in another as in China) and the many ancillary entities that can be involved, as discussed later, are not represented. To keep the diagrams comprehendible, governance relationships related to individual producer responsibility (IPR) and to end user industries are not included. With as many as 400 EPR systems in existence worldwide (OECD, 2013b; Lifset, 2014), there are many variations of each of these approaches and many more legal relationships and data flows than shown.

3.2.1. *Single-PRO systems*

Particularly in the European Union and Canada, many EPR systems have assumed a similar structure: establishment of a PRO to organise and execute producer obligations; registries to identify obligated producers and to gather data for allocation of responsibility; government agencies to certify PROs; payment of fees by producers[14] to the PROs often on the basis of market shares of the obligated products; contracting for collection and processing of designated products by waste and recycling companies or municipalities; and oversight by governments. Examples of the single PRO approach include FOST Plus in Belgium for household packaging and Multi-Material Stewardship Manitoba for household packaging and paper. This is portrayed in Figure 3.2.

Figure 3.2. **One PRO with commercial and/or municipal collection and processing services**

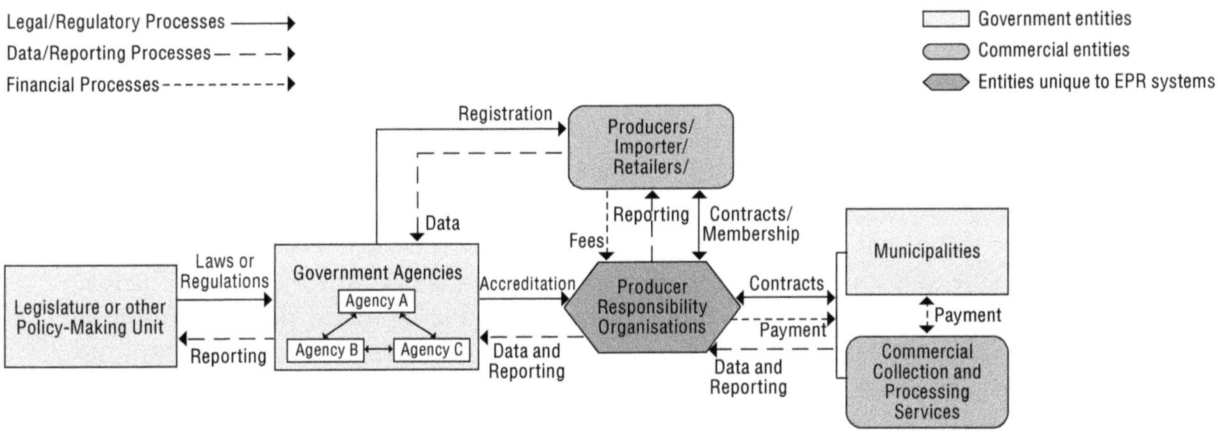

Note: Material flows are not included for simplicity. The relationships related to individual producer responsibility (IPR) or end user industries are not shown. Municipalities may pay commercial providers for services and/or be paid for recyclables depending on the arrangements in the EPR system. Registration can be viewed as something the governments perform (i.e. they "register the producers") or something the done by producers (i.e. the producers register with the government). The depiction of the directionality is thus somewhat arbitrary.

This structure is typical of many of the EPR systems established in the first generation of EPR systems (Monier et al. 2014d). Collective management of EPR compliance and services – as opposed to individual compliance via individual producer responsibility (IPR; Kalimo et al., 2012; Kalimo et al., 2014)[15] – minimises administrative and logistical costs (Atasu and Subramanian, 2012). A single, monopoly PRO for an industry sector provides administrative simplicity for the producers and for the government regulatory entities and in smaller jurisdictions may reflect an effort to capture economies of scale. Contracting out for waste collection and processing by PROs reflects the business focus and/or absence of expertise in waste management services by both individual producers and PROs. Registries (discussed in § 3.3.6 on the role of clearinghouses) emerged because of the need to identify obligated producers and to obtain data on their sales in order to assess fees. In some cases, these functions are performed by a stand-alone organisation as shown in Figure 3.3. In others, the registry function is assumed by government agencies, as in Sweden where producers register with the Swedish Environmental Protection Agency for WEEE, or by PROs, as in the Canadian province of Quebec where Éco Enterprise Québec registers producers for packaging.

Figure 3.3. **Multiple PROs with clearinghouse and commercial and/or municipal collection and processing services**

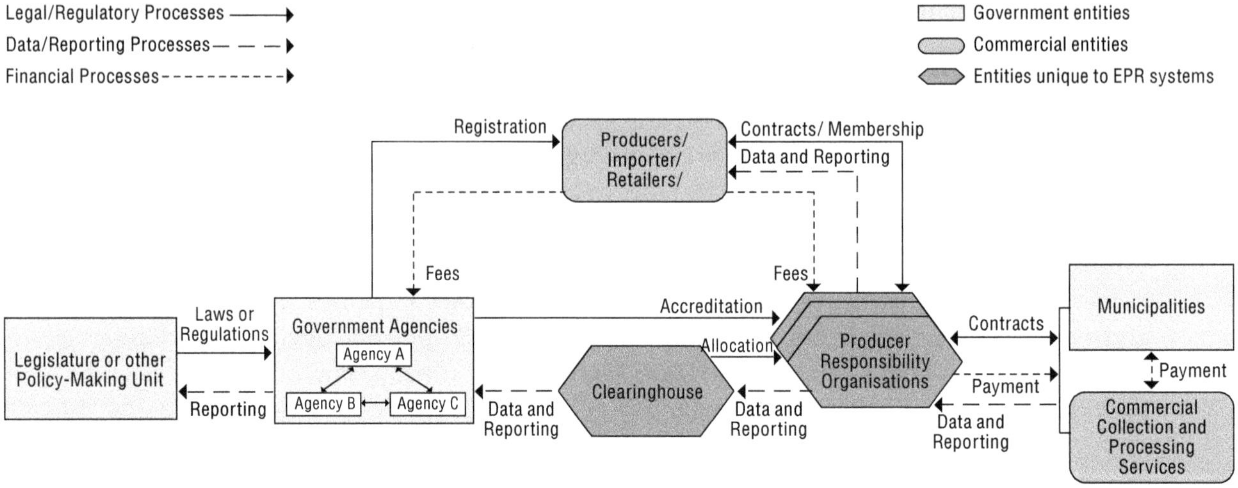

Note: Material flows are not included for simplicity. The relationships related to individual producer responsibility (IPR) or end user industries are not shown. Municipalities may pay commercial providers for services and/or be paid for recyclables depending on the arrangements in the EPR system. Registration can be viewed as something the governments perform (i.e. they "register the producers") or something the done by producers (i.e. the producers register with the government). The depiction of the directionality is thus somewhat arbitrary. PROs may be non-profit or for-profit.

The relationship between the PROs and municipalities can take several forms (Quoden, 2014). The "dual" model, most well known in the German EPR system for packaging, places full operational and financial responsibility for collection, sorting, and recycling in the hands of producers. A collection system separate from and parallel to the existing municipal system – thus the term "dual" – is a conspicuous part of this approach. Municipalities may serve as contractors to the PROs, providing local services, but have no automatic role in the system. This approach is used for packaging in Austria, Germany, and Sweden.

In what is sometime called the "shared responsibility"[16] model, municipalities are responsible for collection and sometimes for sorting. The division of financial responsibilities varies across EPR systems. In some jurisdictions, operational responsibility may reflect pre-existing laws defining the role of local government in solid waste management, rather than a choice the design of EPR systems.

These approaches are discussed further below in the section on the role of municipalities.

3.2.2. *Competing PRO systems*

The number of arrangements for governance of EPR is vast and thus many do not follow the pattern shown in Figure 3.2. One important variant occurs when there are multiple PROs competing to service the same territory or population of producers (Figure 3.3) as in Germany for both packaging and waste electrical and electronic equipment (WEEE). (Note that this does not refer to jurisdictions with multiple PROs that service different, non-overlapping product categories within a larger industry sector or waste stream and thus do not compete. Such instances of multiple, noncompeting PROs occur, for example, when household and commercial packaging wastes are managed by different EPR schemes.)[17] In systems with multiple PROs, waste management companies sometimes own the PROs (Spasova, 2014). In most cases a clearinghouse is established in order to organise data collection and ensure appropriate burden sharing of collection (see discussion on clearinghouses in § 3.3.6).

In Germany, packaging for EPR was managed through a single centralised PRO, the Dual System Deutschland (DSD), from the initiation of EPR in 1991 until the beginning of the 2000s when the system shifted to multiple packaging PROs (10 as of 2015). The introduction of multiple PROs occurred in response to decisions by the German Federal Cartel Office and European Commission regarding anti-competitive practices (OECD, 2011). Changes in the EPR system for packaging were realised in 2003/4 through the dismantling of governance and contractual arrangements that had been established in 1993 in response to a financial crisis at the DSD (OECD, 2006, 2013b).

In contrast, in the United Kingdom competing PROs were an aspect of the initial design of the EPR systems for packaging and WEEE. Governance structures in the United Kingdom reflect shared responsibility among entities in the product chain and a market-oriented framework for targets specified in EU directives (Cahill et al., 2011). The United Kingdom EPR system for packaging subsequently evolved into a tradable credit system, discussed below.

3.2.3. *Tradable credit systems*

A third, distinctive variant in the organisation of EPR systems is based on the use of tradable credits. In this approach a producer's compliance with EPR obligations is accomplished when a producer is in possession of the number of credits for collected and processed end-of-life products equal to a set target. Accredited actors in the system collect a specific amount of the relevant products, and process or export them and thereby generate a credit.

Tradable credits or permits as a market-based instrument have been widely discussed and used for a variety of environmental policy goals including air pollution regulation, management of water supply and quality, and grazing rights allocation (Tietenberg, 2007). They have not, however, been widely used for solid waste management (Salmons, 2002). In the early 1990s in the US they were proposed as a means to meet recycled content obligations for newsprint (Dinan, 1992). Tradable credit systems can minimise compliance cost and thus the burden to society by allowing those that can more cheaply perform a given task – e.g. reduce emissions or collect recyclables – to exceed regulatory targets and sell the credits to those for whom the task is expensive.

Among EPR systems, the United Kingdom packaging regime is notable in that it employs tradable credits.[18] In that system, national level obligations for collection and recycling of packaging are set to meet the requirements of the EU Packaging Directive[19] though some targets exceed the requirements of the Directive (OECD, 2014a). The obligations are divided among members of the packaging supply chain: 6 % to the packaging material manufacturers, 9% to converters (those that fabricate packages from the packaging materials), 37% to the packers or fillers (those put the product into the package), and 48% to sellers of the packaged product. Firms are allocated a share of the relevant obligation according to their position in the supply chain and the recovery target for the specific material used for packaging (ranging from 22% of wood up to 81% of glass).

The credits, called packaging waste recovery notes (PRNs) or packaging waste export recovery notes (PERNs), are generated when an accredited collection or processor (called re-processors in the United Kingdom scheme) processes or exports a tonne of packaging waste for recycling. Producers join PROs which purchase credits on their behalf from re-processors who in turn have acquired waste packaging from collection services or municipalities or private entities (generators).

Some aspects of the resulting governance system are similar to the multi-PRO systems shown in Figure 3.4. There are approximately 30 accredited PROs as of 2014 (Monier et al., 2014h). A clearinghouse, the National Waste Packaging Database, is employed, but rather than allocating collection routes, it facilitates information flow. It registers producers, accredits re-processors, issues the PRNs and PERNs, manages reporting, and monitors free-riding.

Figure 3.4. **Governance structure for tradable credits system**

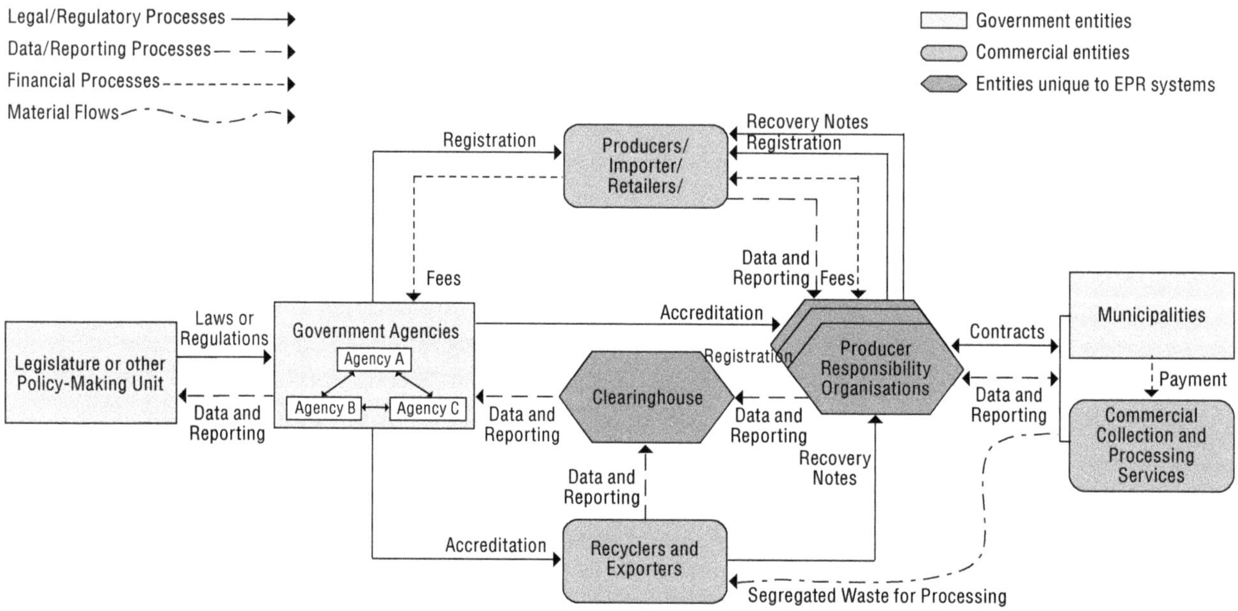

Note: For simplicity, most material flows are not shown nor are export recovery notes. Municipalities play a limited role in sending waste to recyclers within EPR systems. Municipalities may pay commercial service providers for collection. Registration can be viewed as something the PROs perform (i.e. they "register the producers") or something the done by producers (i.e. the producers register with the PROs). The depiction of the directionality is thus somewhat arbitrary. PROs may be non-profit or for-profit.

Municipalities have a diminished and less direct role relative to other types of EPR systems. In the United Kingdom system, targets can be met through the collection of packaging from commercial or household waste generators. It is often cheaper to obtain packaging from commercial generators than from households. In this respect, municipalities have no prerogatives to participate in the system, but if they collect segregated packaging materials they may sell them to re-processors or directly to the market. Figure 3.4 shows a schematisation of the United Kingdom packaging system, which allows multiple PROs (and even individual compliance, not shown on diagram).

The United Kingdom system is notable for the separation between producers and the verification of compliance by re-processors (OECD, 2014). That is, verification that the re-processors have recycled a given quantity of waste and done so appropriately is performed by the government. Neither PROs nor the producers have roles. In this respect, the governance is quite different from the two approaches described above insofar as there is no direct operational or managerial role for producers. The United Kingdom tradable credit system does not encompass producer obligations for public education or design for environment.

3.2.4. *Government-run* EPR *systems*

A fourth variant occurs when the government plays a more direct role in fee collection and disbursement (Figure 3.5) as in China for WEEE (see Annex D), and Chinese Taipei,

Figure 3.5. **Government-run EPR system**

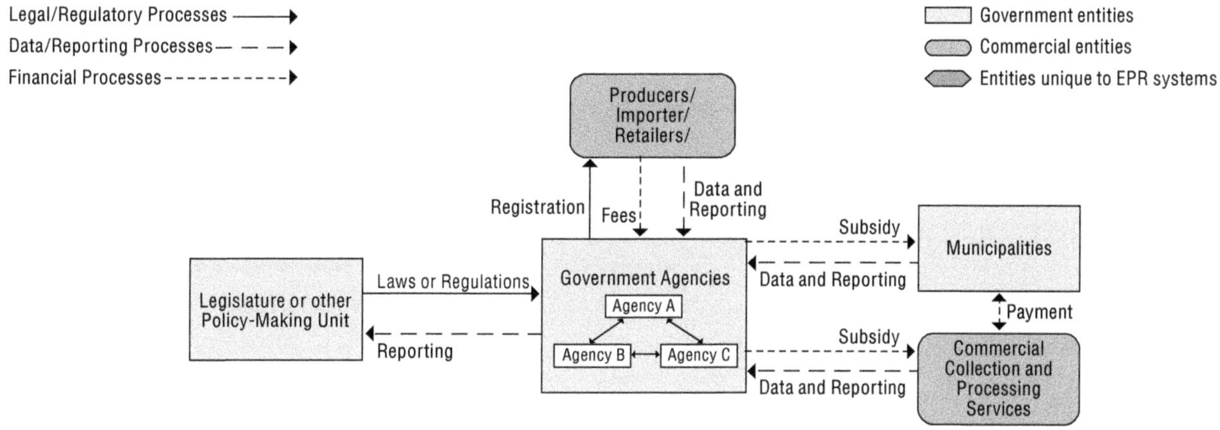

Note: Material flows are not included for simplicity. Municipalities may pay commercial providers for services and/or be paid for recyclables depending on the arrangements in the EPR system. Registration can be viewed as something the governments perform (i.e. they "register the producers") or something the done by producers (i.e. the producers register with the government). The depiction of the directionality is thus somewhat arbitrary.

proposed for Thailand (Manomaivibool and Vassanadumrongdee, 2011), and some US states for all products subject to EPR. The advantages of government collection and disbursement of fees is the control that the government has over the execution of policy, often in the face of an industry that may not be mature enough to organise and manage a privatised system of governance. China considered an approach to EPR for WEEE that relied on PROs to collect fees and contract for waste services. The approach was rejected because of concern that a nongovernmental organisation would not be effective in collecting fees or dealing with substandard recycling operations without strong support from the government. Instead, the government chose to collect fees from producers directly in conjunction with value-added taxes (VAT) using the State Administration of Taxation (SAT) for domestic companies and the General Administration of Customs (GAC) for foreign companies and to disburse the funds via the Ministry of Finance (see Annex D; Tong and Yan, 2013; Tong, 2014). In Chinese Taipei, an EPR system that initially relied on PROs was shifted to a producers-pay-government-disburses model because of difficulties arising from false data reporting, financial scandals, lack of transparency, and poor design of PRO responsibilities (Ching-Wen, 2004; Fan et al., 2005). Thus, in both cases, collection and disbursement of fees reflected difficulties the countries experienced with the PRO-based approach common in Europe and Canada.

Some countries and stakeholders object to the labelling of such government-run systems as EPR. This can reflect a view of EPR as entailing a decision making or managerial role by products and/or a concern that taxes on producers under the auspices of EPR will be diverted by governments to non-EPR purposes. In Canada, government-run systems as shown in Figure 3.5 are not viewed as EPR. Instead, they are viewed as government-run product stewardship programs.[20] Only those systems where producers manage product take-back collectively or individually are considered to be EPR (Environment Canada, 2013; Bury, 2015).

Government-run EPR systems for WEEE also exist in several US states including the state of Washington, where a multi-stakeholder, quasi-governmental non-profit organisation, the Washington Materials Management Financing Authority, performs many of the functions performed elsewhere by PROs including setting and collecting fees from producers (Gui et al., 2013). In Connecticut, the Department of Energy and Environmental Protection selects eligible recyclers and sets acceptable rates. The legislation allows for the participation of PROs, but none have emerged to-date (Metzner, 2015).

3.2.5. *The role and legal status of producer responsibility organisations*

Varied roles of PROs

Producer responsibility organisations, as entities established to execute EPR obligations on behalf of producers, are conspicuous in discussions of the governance of EPR systems. Because of their salience early in the development of EPR in the EU, Japan and Canada, PROs are often seen as one of the signature elements of EPR. As the discussion of EPR governance structures above indicates, however, the role of PROs varies widely and in some systems they do not exist at all.

PROs can be non-profit organisations (typically), government agencies (rarely), quasi-governmental non-profit organisations (QANGOs, occasionally) and for-profit firms (occasionally). In many jurisdictions, as in the Canadian province of British Columbia, PROs are legally obliged to have non-profit status (British Columbia Ministry of the Environment, 2006). For-profit PROs have emerged in some jurisdictions where multiple PROs compete, most conspicuously in Germany and the United Kingdom for packaging and WEEE. Monier et al. (2014d) find that 13 of the 36 EPR systems that they studied in Europe included for-profit PROs. It is not uncommon for such PROs to be owned in whole or part by waste management companies (Spasova, 2014).

In fact, for-profit status is not uniform. In the German system, some for-profit PROs are part of the vertical integration of producer-facing functions – such as collection of fees from producers and reporting to government authorities – with waste management functions. In what is described above as government-run EPR systems as in Connecticut and other US states, the same waste management functions are performed by firms that have no producer-facing functions at all. Further complicating the discussion, conventional for-profit waste firms are widely used by many different entities in EPR systems in traditional roles as contractors providing collection, sorting and processing services. Municipalities with collection and sometimes sorting and processing roles in EPR also frequently contract out those activities to conventional waste firms.

In such cases, the PRO is not paid by producers, but by another entity in the EPR system. A government agency in China, Chinese Taipei, or the state of Washington (US) contracts with such commercial entities directly without a PRO or municipality as an intermediary. In some cases, the relationship with the commercial entities is closer to one of permitting than to contracting-out. In the EPR system for WEEE in Vermont in the US, the state contracts with an entity, which can be for-profit or non-profit, that provides clearinghouse services, registering service providers and providing on-site inspections, training and outreach to the providers. The state also contracts with a WEEE recycler (Vermont Agency of Natural Resources, 2013, 2014). Producers pay fees directly to the state's Agency for Natural Resources, which, in turn, pays the clearinghouse and the recycler.

Thus for-profit PROs or PRO-like entities include:

- vertically-integrated PROs and waste services firms
- PROs without an affiliation with a specific waste services firm
- waste contractors to a government agency or related (non-PRO) intermediary established as part of the EPR system

- waste services firms permitted by a government agency as part of the EPR system
- waste contractors to a PRO
- waste contractors to a municipality.

For-profit vs. non-profit status

There is vigorous debate over the value and propriety of for-profit PROs including those that are vertically-integrated (the first two types of for-profit entity in the list above) and those that only provide producer-facing services. The main argument in favour of for-profit status is that the profit incentive will drive PROs to be efficient and, if the EPR system is properly structured, societal goals will be accomplished in a cost-efficient manner. Thus, the arguments in favour of for-profit PROs are intertwined with arguments supporting competition among PROs. For example, non-profit status could allow very high administrative expenses or inflated salaries that a for-profit endeavour would tend to mitigate because of pressure from shareholders. Some argue, however, that ownership of PROs by producers and/or careful monitoring of PRO finances and operations can drive efficiency (Bury, 2015). Competition among PROs regardless of their legal status could also reduce such unwanted effects.

A particular impetus for the emergence of competing PROs, which are often for-profit PROs, has been concern by producers over what are perceived to be excessively large financial reserves held by PROs. Data from the European Portable Battery Association, for example, showed that in the late 1990s some battery PROs were accumulating financial surpluses at a rate of 15% to 65% of revenue (Mayers, 2007). In another instance, some PROs responsible for WEEE in the European Union received fees based on the quantity of sales of new electrical and electronic equipment (EEE), while not spending commensurate amounts on the management of lower quantities of WEEE that were discarded, leading to large reserves. Some producers saw the size of the reserves as unwarranted and worked to introduce competing PROs as a means of exerting downward pressure on prices (fees).

Some organisations have very specific views in this regard. In January 2013 a group of packaging PROs[21] published an EPR Manifesto which outlined four main arguments against for-profit PROs[22] (EPR Club, 2013; Anonymous, 2013):

1. Distribution of profits would discriminate against non-shareholders within obliged companies.
2. Profits would create an incentive for PROs to offer preferential treatment to larger or more profitable participants.
3. Profits may create an entry barrier into the PRO for small producers or producers that will not increase profits.
4. Non-profit interests in a PRO's operations (such as education or prevention) are better served in the absence of profits as they may be costly.

Later in 2013, these organisations formed the Extended Producer Responsibility Alliance (EXPRA) as an association specifically for not-for-profit PROs. They have since argued, additionally, that for-profit operations may not comply with anti-trust laws, and monopoly charges could be levelled against the PRO. Further, they argue that waste operators should not be a part of a PRO's organisation as they have a profit interest in increasing the waste stream. Opponents of for-profit PROs also argue that the push for profitability generates incentives for, on the one hand, higher targets (Quoden, 2015) and, on the other, lower environmental performance as a means of cost-saving.

The assessment of the impact of for-profit PROs is stymied by the generally poor quality and incommensurable data on PRO performance (Monier et al. 2014d) and more specifically the dearth of detailed data in EPR systems with competing or for-profit PROs arising from concerns about sharing of proprietary data.[23] Monier et al. are the only authors found to explicitly evaluate the question of the effect of the legal status of PROs across multiple jurisdictions. They indicate that the for-profit status of PROs "does not entail any visible incidence on technical performance or on costs." More crucially, they argue that the conditions needed to ensure fair competition are more important than the legal status of the PROs. Ultimately, the status of PROs may not be decisive in influencing PROs pursuing excess profits (i.e. engaging in rent-seeking behaviour). In the most extreme case, the only difference between them may be that excess profits generated in for-profit PROs will end-up in shareholders' pockets, while those generated in non-profit entities will be used to pay for above market salaries, political posts, and "gold-plated" company facilities. Concerns delineated in the EPR Manifesto, in this view, could be addressed through regulation, corporate charters, or similar mechanisms.

This view is further supported by the way competition authorities are looking at this issue. Competition laws in many jurisdictions, including the European Union and United States, apply to undertakings regardless of their ultimate ownership – state, municipal or private – or profit/non-profit status (see Chapter 4). Under EU law, for example, any entity engaged in an "economic activity" is subject to competition law.[24]

Questions of the appropriateness of for-profit status are also tied to the emerging issue, discussed above (§ 3.1.4), of how best to manage waste streams that include both costly and income-producing products and materials. Profitable waste streams will be, unsurprisingly, more attractive to PROs and can provide a basis for commercial operation, sometimes without support from fees paid by producers. As with the question of competition, however, the performance of EPR systems with for-profit PROs where there are valuable waste streams are likely to reflect not only the legal status of the entities, but also the policies and structures of the EPR system. Regulations, for example, specifically targeted to cherry-picking of waste streams would have a strong impact regardless of the legal status of the PRO.

3.3. Key governance functions in EPR Systems

EPR systems include both familiar governance functions such as policy formulation, stakeholder consultation, monitoring, compliance promotion, and enforcement, and some that are more specific to the privatised dimensions of EPR, such as co-ordination through clearinghouses. This section provides summaries of both types of governance functions.

3.3.1. *Policy formulation and evaluation*

It is typically higher levels of government that have formulated and enacted the laws that mandate extended producer responsibility and – to varying degrees – dictate details that specify the structure and function of EPR schemes. In most cases it is the national governments that take on this role (as in Europe and East Asia) while in countries where the national government has not played a central role in EPR (Canada and the US), it is the state/provincial governments that have taken on this function. Conspicuously, the European Union as a supra-national entity has played this role in Europe (mandating EPR for packaging vehicles, EEE and batteries), with some aspects of policy left to member states. In a very small number of places in the US, local governments have enacted EPR laws (New York City; Alameda County, California; and King County, Washington).[25]

The filling-in of details regarding the structure and function of EPR schemes, where those details are not specified in the laws mandating EPR, is often left to administrative agencies (as with Green Manitoba in Canada) or to other levels of government (as with member states of the European Union). The boundary between the making of policy and the implementation of policy, conceptually, and especially in practice, is blurry. In many jurisdictions, governance functions are fully articulated in legislation, whereas in others they are promulgated as administrative regulations, delegated to non-governmental actors, or unspecified. Further, in some countries municipalities have more than an operational role, participating in policy formulation, and may have autonomy with regard to the degree to which they play a service role.

A crucial, though less widely noted, element in EPR governance is the evaluation of policy and program performance. Presumably the data obtained through monitoring (described below) can be used to assess not only the performance of individual actors in the EPR system, but also the performance of the overall system. Recent conspicuous examples of evaluation of EPR are the studies performed in preparation for the Recast of the WEEE Directive in the European Union (Sander et al., 2007; Huisman et al., 2007; Monier et al., 2014d). Policy and programmatic evaluation appear to be occasional rather than periodic in many systems and often hampered by inadequate data.

While policy formulation is primarily the role of governments, in line with the privatised dimensions of EPR, elements of policy may be delegated to non-governmental entities, especially producer responsibility organisations. This often includes the structure of fees paid by producers for the services organised by PROs and the selection of vendors of collection and processing services. While these decisions may not be viewed by some as "policy," the structure of fees in particular can be a central factor in shaping whether EPR generates incentives for DfE.

3.3.2. Operations

The privatising logic of EPR puts many of the functions in EPR systems in the hands of commercial and non-profit organisations. This includes both administrative activities such as tendering for collection and processing services and the operation of those services. Nonetheless, some local governments, primarily municipalities,[26] participate in collection and processing of designated products under the auspices of EPR systems (described above in § 3.2.1 in the discussion of systems of shared responsibilities). The role of local government is also a matter of considerable variation and debate across the world (as discussed further below in the section "*The role of municipalities in EPR systems*").

3.3.3. Stakeholder consultation

Consultation is important in obtaining information and perspectives of the entities in the EPR product chain (government agencies, producers, PROs, specialised organisations established as part of the EPR system, municipalities, waste management companies, recyclers, end user industries, and consumers/waste generators) as well as other interested and affected entities (government agencies indirectly involved, trade associations, civil society). Stakeholder consultation is especially important in EPR, as EPR creates new institutional arrangements among entities that may not have previously interacted. Insofar as EPR constitutes a system of private environmental governance, stakeholder consultation functions tend to shift to PROs and other entities in the EPR chain and are sometimes designed to mirror the administrative rulemaking activities of government bodies (Hickle, 2014a).

The consultation can occur at the formation of an EPR system and on an ongoing basis. Monier et al. (2014d), in their review of EPR in the European Union, found in most cases that there were no specific structures for stakeholder consultation, but that informal dialogue often exists. Consultation practices and organisational structures, where they do have formal status, vary widely. A consultation process may operate within the PRO or it can be external. Within a single PRO, consultation may be directly conducted through the governing board as in the case of COOU and CONOE, non-competing used oil PROs in Italy that have representatives of national government ministries in a minority role on their boards. In France, the EPR for graphic paper has an advisory committee that meets on a quarterly basis to review and approve the activities and strategy of Ecofolio, the PRO that has been accredited by the state. The commission has representatives from three ministries as well as from environmental and consumer associations.

The consultation body may be an organisation external to the PRO as in the cases of packaging in France[27] where the "Commission Consultative d'Agrément", composed of all involved parties (industries, consumers, NGOs, PROs, local authorities, waste managers, etc.) offers non-binding opinions to public authorities with the express intent of securing the proper functioning of the household packaging sector in France. In Belgium, the external organisation overseeing the packaging waste stream (the Interregional Packaging Commission - IPC) is composed of members from the regional governments who, in turn, receive non-binding opinions from the "Packaging Platform" that is composed of industrial associations. The IPC will also seek input from enterprises not represented by the packaging platform and exercises control over PRO activities as well as issuing permits to the PROs.

3.3.4. *Registration and accreditation*

Oversight is typically accomplished through a mixture of ex ante permitting-like processes and through ex post data collection.

While the terminology is not consistent around the world, "registration" often refers to the identification of the participants (producers) in the EPR schemes and acquisition of data on sales of products subject to EPR.[28] It is also a key element in identifying free-riders – producers that fail to participate in PROs or otherwise comply with EPR regulations. Registration is often performed by government agencies as in Finland where paper producers must register with the Centre for Economic Development, Transport and the Environment for Pirkanmaa (ELY Centre Pirkanmaa) (Monier et al., 2014j) or by quasi-governmental agencies as with the Alberta Recycling Management Authority in the Canadian province of Alberta[29] (CM Consulting, 2013). However, the process can also be administered by PROs as in United Kingdom where the registration information is compiled by PROs (called producer compliance schemes in the United Kingdom) and then transmitted to the Environment Agency or the equivalent in England, Scotland, Wales and Northern Ireland (Monier et al., 2014c).

A key means by which governments monitor activities in and the performance of the EPR chain is through the accreditation or certification of producer responsibility organisations. Typically, when a PRO seeks to operate in a given jurisdiction, the government requires that the PRO obtain permission by submitting an application that details the structure of the PRO and how it intends to operate. Many governments require that PROs periodically recertify and that process provides an occasion for the assembly and review of information. A related approach is to require, as a condition of operation, that the PRO submit periodic reports.

Comparison of accreditation criteria is difficult because the criteria are often stated in discretionary terms, that is, the certifying entity has the option to mandate that a PRO meet certain requirements, but is given flexibility as to whether in fact the requirements are imposed. In addition, the language in the relevant statute or regulation is often general, giving the certifying entity flexibility as to the specific form and extent of requirements that PROs face.

In many Canadian provinces, PROs must submit stewardship plans for approval by the relevant provincial government agency. They must also submit an annual report to the provincial government or environment agency on the program's performance. In the province of British Columbia, PROs must submit a product stewardship plan[30] and an annual report (Waste Reduction and Management Division, 2014). The producer[31] must review its approved plan every five years and either propose amendments or indicate that no amendments are necessary. The requirements for stewardship plans are described in the provincial recycling regulations and contain provisions on:

- stakeholder consultation
- collection system and consumer access including recovery rate
- consumer awareness
- management of program costs
- management of environmental impacts
- dispute resolution
- performance measurement.

The provincial recycling regulations provide an extensive list of factors that the director of the Ministry of the Environment of British Columbia may consider when deciding whether to approve the stewardship plan including its timeliness and effectiveness in meeting the goals, the geographical area and market into which the producer sells its products, stewardship plans of other producers for products in the same category, and the structure of financial and operational co-operation with other producers (Province of British Columbia, 2004; British Columbia Ministry of the Environment, 2006).

There is some debate in Canada about whether or not it is appropriate for governments to approve such stewardship plans. Approval implies responsibility which can be problematic for government if the EPR program encounters problems. Some argue that it may be better for governments to concentrate on monitoring the achievement of targets and leave the planning on how best to meet the targets to the PROs (Bury, 2015).

In Ireland, the Minister for the Environment, Community and Local Government is the governing body that approves PROs. Application for PRO status involves submission of documents related to governance and membership, a commitment not to discriminate against any producers with respect to size or location, and an assurance that it will co-operate with other PROs to achieve the environmental targets (Philip Lee, 2014; Gorecki, 2014). The Minister may outline specific conditions that the PRO must meet. These can include meeting specified targets, the composition and representativeness of the management board, expenditures for public education, awareness, and approval of amendments to articles of association, corporate governance and membership rules. The Department of the Environment, Community and Local Government (DECLG) has a management/oversight role ensuring that the PROs meet the targets (Gorecki, 2014).

The Irish approach is notable in that government has the prerogative to review and dictate aspects of the internal governance of PROs. In a recent review of EPR in Ireland, it was proposed the certification process be strengthened through a standardized code of corporate governance and the use of legal contracts ("service level agreements") between the government and the PROs to replace the letters of approval that are currently in place (Philip Lee, 2014).

The regulation of packaging PROs in Lithuania, like many countries grappling with EPR, evolved considerably. Prior to 2013, the statute simply authorized the establishment of PROs. Starting in 2013, requirements were increased to require that PROs conclude agreements with municipalities, packaging waste collectors, and producers. A Board of Control is mandated to provide oversight and enforcement of the agreements. Lithuania requires that a PRO must represent at least 25% of the manufacturers and importers supplying packaging for the domestic market in order to obtain a license to operate (Republic of Lithuania, 2012).

Despite their considerable differences in other respects, the examples above indicate that the requirements that are imposed through accreditation are often quite general, leaving the certifying entity discretion as to the specific conditions PROs must meet.

3.3.5. *Collection and disbursement of fees*

At the core of EPR are the fees paid by producers for collection, reuse processing, recycling of used and end-of-life products. Typically, the fees are collected by PROs, based on fee schedules developed using data obtained from the registries. The PROs contract for collection, reuse, and processing services on behalf of the producers who are their members or clients. The fee structures can be elaborate to reflect varied circumstances for levels of service and co-collection of non-EPR waste.

Where governments collect and/or disburse producer fees as depicted in the third governance structure (Figure 3.4), typically in systems that use forms of advance disposal fees (ADFs), the arrangements resemble conventional systems of government taxation and funding of societal activities. ADFs can be paid by the producer or by consumers at the time of purchase. Collection of fees by governments more directly involves governments in both the managerial and oversight dimensions of EPR systems. Some producers object to the collection of fees by governments worrying that the funds will be diverted to uses other than end-of-life management.

It is important to note that not all ADF-based systems are what might be called producers-pay-government-distributes approaches. While the visible fees paid by consumers in some EPR systems are a form of ADF, many such systems involve neither government collection of funds nor their disbursement. This is the case, for example, with many of the EPR schemes for WEEE in the Canadian provinces (Waste Reduction and Management Division, 2014) and the Specified Home Appliance Recycling Law in Japan[32] (Hotta et al., 2014). In the latter case, consumers pay fees at the time of disposal with the resulting funds managed by PROs (see Box 3.1). Some ADF-based systems, such as California's e-waste law, do not involve producers or producer funding at all (see Annex K) and thus are not a form of EPR.

3.3.6. *Co-ordination through clearinghouses*

In jurisdictions where there are competing PROs seeking to fulfil EPR targets (Figure 2.4), a neutral coordinating entity is needed to make sure that collection is provided

Box 3.1. **The Japanese Specified Home Appliance Recycling Law (SHARL)**

The Act for Recycling of Specified Kinds of Home Appliances was enacted in June 1998 and enforced in April 2001. It aims to achieve a reduction in the volume of waste and to improve material recovery from this source. The Act covers four categories of home appliances: air conditioners; TV sets; electric refrigerators and freezers; and electric washing machines and clothes dryers.

Under the act, each home appliance manufacturer has the obligation to take back and recycle their products when they become waste. To implement their obligations, manufacturers have set-up two competing groups, each of which brings together three or four of the most important producers and represents a similar share of the market. One of the groups has set-up its own recycling facilities, while the other contracts with existing operators.

Consumers and businesses that dispose of end-of-life home appliances have to pay both the collection/transportation and the recycling fee. Traceability is ensured by the use of home appliance recycling tickets (manifests) that are issued to consumers who paid the recycling fee. This manifest system ensures that waste home appliances are delivered to the original manufacturers of the products.

While there is no differentiation between the costs to manage individual brands within each of the producer groups, potential cost savings through efficient processing or product design remain with producers. Hence, there is competition between the two manufacturer groups over the minimisation of recycling costs.

The advantage of the vertical integration that results from the implementation of Japanese EPR for Household Appliances is that it creates a strong link between downstream management of end-of-life products and the producer. There is some empirical evidence that the system provides tangible incentives for eco-design (Tojo, 2004).

Source: Dempsey et al. (2010); Hotta et al. (2014) and Annex G.

everywhere it is needed, that cherry picking is avoided, and that there is a level playing field for all competitors. Without such co-ordination, there is the possibility that collection in some regions – especially rural areas where collection routes may be long and quantities of wastes small – may not be provided service because PROs find it overly expensive. Co-ordination also enhances efficiency by ensuring that competing PROs do not provide duplicative coverage, collecting end-of-life wastes from the same service area or collection points. Clearinghouses, often a separate non-profit organisation, but sometimes a government agency, thus help rectify undesirable incentives generated by competition among PROs. Clearinghouses can also collect data from producers or service providers, and provide a mechanism for managing proprietary data or identification of free riders.

The scope of services and organization provided by clearinghouses varies widely. Denmark, where the same clearinghouse, the DPA-System serves the EPR systems for WEEE, batteries, and ELVs is an example of a clearinghouse with a prominent role and multiple functions (see Box 3.2 for a description). United Kingdom EPR system for packaging, in contrast, clearinghouse activities are more limited, because of the structure of the tradable credit system.

Because the EPR system for packaging in the United Kingdom involves credits for recycling of packaging, two related entities are involved. As noted earlier, the National Waste Packaging Database provides a national on-line registration system for producers;

Box 3.2. Centralised reporting, assignment of obligations, and control in a Danish clearinghouse

In Denmark, the DPA-System serves as a registry and clearinghouse for PROs involved with WEEE, batteries, WEEE, and ELVs. Established in 2006 as WEEE-System and renamed as DPA-System in 2009, it is supervised by a 7 member board appointed by the Minister for the Environment.

The DPA-System administers centralized reporting and oversees the country's four PRO systems for business-to-consumer (B2C) and business-to-business (B2B) WEEE. Its duties include definition of market shares and collection obligations and auditing of PROs and producers if they are not members of PROs (i.e. pursuing individual producer responsibility). The DPA-System assigns WEEE collected by municipalities to producers or PROs. Municipalities report directly to the DPA-System, and producers report to DPA-System either individually or via PRO. PROs also regularly submit data from processors.

EPR for batteries in Denmark functions as a part of the EPR system for WEEE. In the case of batteries, however, SKAT, the Danish Tax and Customs Authority, collects data on the amount of portable batteries put on the market which it then forwards to DPA-System in contrast to the WEEE system where the data are sent directly to the DPA-system by producers.

The DPA-System is overseen by the Danish EPA and funded by producer fees. Because three of the PROs for WEEE also function as PROs for batteries and because the DPA-System oversees both WEEE and batteries, there is thought to be better engagement of the relevant actors (Monier et al., 2014a, 2014b)

accreditation for packaging re-processors and exporters; recording of the credits; and monitoring and reporting. In addition a small number of private internet platforms are maintained, for example by the Environmental Exchange, which provides a marketplace for the trading of credits (Monier et al., 2014h; OECD, 2014a), although the vast majority of credits are traded outside these platforms (Lange, 2015).

Ireland presents an intermediate case where some clearinghouse functions for WEEE are performed by the WEEE Register Society. As the name suggests, the Register Society is primarily a national producer registration body, set up to track which producers are placing EEE on the market. It is an independent, industry-based body, overseen by a management committee that includes including government agencies, and owned by producers (Monier et al., 2014e). The WEEE Register Society is also responsible for notifying the EPA of potential non-compliance with WEEE regulations and for verification of visible environmental management costs (vEMCs) that are applicable to WEEE (i.e. from products put on the market prior to the enactment of the WEEE Directive). As such, it fulfils one of the primary functions of a clearinghouse by supporting the identification of free riders. One distinction from other clearinghouses is that the WEEE Register Society does not determine that market share of individual producers. Rather, determination of market share and financial liability is contracted out to a private entity to ensure confidentiality through data submitted to the "WEEE Blackbox" website (*www.weeeblackbox.ie/website/faq.html*).

In some EPR systems, clearinghouses, rather than PROs, serve as the primary point of contact for municipalities. According to a recent study for the European Union, national clearinghouses, either as independent entities or government operations exist for all EPR systems for WEEE and for packaging in Austria and the United Kingdom (Monier et al., 2014d).

3.3.7. Monitoring

Monitoring, both public and private, is needed to ensure effective policy implementation, as a prerequisite for enforcement, and to inform subsequent policy assessment and debate about the effectiveness and future development of EPR. Monitoring is also a component of co-ordination. Key functions in EPR systems such as assignment of service territory and pick-up routes require information for operational purposes. Monitoring focuses on three key entities in the EPR chain: producers, PROs, and collection and processors including municipalities. Information about producers mostly relates to participation (to address free-riders) and sales or market activity (to allocate responsibility and fees). The information needed about PROs is more extensive and focuses on the extent of activities and performance. Information about service providers focuses on operational results, environmental compliance and financial activities. Monitoring of an EPR system thus requires both technical and financial data. Monitoring of non-EPR actors is also important in addressing leakage of end-of-life products and materials outside the EPR system (see §1.4.1). The effectiveness of the system will depend on transparency, auditing, harmonisation, and enforcement (described below).

Technical performance data

Information on technical performance is mainly related to collection and processing quantities but can also include environmental performance such as emissions and practices at processing facilities. Such data are likely to include (at least):

- *Quantities of products placed on the market by producers or discarded by waste generators* – Such data are usually the basis for allocation of physical or financial responsibility and thus calculation of producer fees. Data on the quantity of products placed on the market can be obtained from individual producers and/or from statistical agencies and industry associations.
- *Collected quantities* – A central task of PROs is to collect and report information about the quantity of recyclables managed under its auspices. In EPR systems where producers are permitted to exercise responsibility without joining a PRO (i.e. a form of individual producer responsibility, IPR), provisions need to be made as well for the reporting of data directly from such producers. Reporting of quantities collected through a PRO or by an individual producer does not provide data on collection and processing of recyclables handled outside the EPR scheme. As noted above, many jurisdictions estimate that a large proportion of targeted recyclables that are collected do not go through the PRO chain; thus such data are important for policy assessment and increasingly a concern in management and EPR systems.
- *Collection activities and accessibility* – The number of collection sites, frequency of collection, and related factors indicate accessibility and convenience of EPR services for waste generators.
- *Sorted and processed quantities* – As with the collected quantities, data on the sorting and/or processing of recyclables are often assembled by PROs (or individual producers under IPR) from operators for transmittal to government agencies. Particularly important here is clarity about whether data refer to quantities arriving at a given facility (inputs) or quantity departing the facility (outputs) with the difference being the residues from the processing activities (Waste Diversion Ontario, 2010).

- *Destinations of various waste flows (collection, reuse, recycling, end use, incineration, landfill, etc.)* – Information on all flows in an EPR chain improves the opportunities for government oversight. Traceability is a key concern as it can be difficult for government agencies to identify all the entities in the chain, especially processing facilities and end users (Monier et al., 2014d). In the Canadian province of British Columbia, mass balances are included in PRO reporting for WEEE to check for discrepancies in reporting.[33] Traceability can be hindered by governments' lack of legal authority to require tracking outside of jurisdictional borders.
- *Residue quantities* – Residues can include both materials that are inappropriately captured in collection (non-recyclables) and appropriate materials in a form and shape that cannot be further handled (small fragments). Residues have an impact on costs insofar as they increase processing activities and require disposal. They also serve as an indicator of the quality of collection practices and processing technology.
- *Consumer awareness* – Knowledge of the availability and functioning of the EPR system is an important driver of collection and contamination rates and of public participation in policy formulation. In Canada, surveys of consumer awareness of EPR for WEEE are conducted regularly (Kerr, 2015).

These data can then be used to calculate recycling and related rates and are an obvious and critical component of evaluations of cost-effectiveness.

Financial and programmatic data

Financial information for monitoring is more varied because of the differing configurations of EPR systems and pressure for protection of proprietary commercial information from business participants. Such data are likely to include:

- costs incurred by entities in the EPR chain
- payments made between entities in the chain and revenues from sales of used goods and secondary materials.

Cost information is important because it serves as the basis for subsidies and payments between members of the EPR chain, provides information on assets and financial reserves that may accrue in the systems, and supplies a key ingredient in the oversight of the system. Data on payments provide information complementary to technical data on how the EPR system is functioning and where policy intervention might be needed.

Non-quantitative information is also important regarding the structure and rules of organisations, corporate governance, membership criteria, management of financial resources, reporting procedures, R&D and design for environment initiatives, and outreach activities.

Transparency and auditing

To be effective, monitoring needs to provide transparency and be reliable. Transparency involves documenting and disclosing the manner in which entities involved in EPR operate and how effectively and efficiently they do so. This in turn means that the data must be generated, collected, collated, stored, verified, and made accessible or distributed to the appropriate stakeholders. The reliability of the information is often ensured through auditing performed by governments or independent third parties. In the Austrian EPR system for packaging, for example, the government audits not only PROs but also individual producers (Monier et al., 2014d). In Chinese Taipei, rigorous auditing of producers and of recyclers is a key component of the EPR system because of difficulties experienced in the 1990s with false data reporting and financial irregularities. The subsequent use of individual item counting, mass balances, 24-hr closed-circuit television monitoring, and other methods have helped to maintain low rates of fraud among recyclers (Fan et al., 2005).

Like other administrative processes, a trade-off exists between detail which enables better monitoring and understanding of the performance of EPR schemes and the cost that the detail engenders. In recent guidance proposed for the EU, a set of actions was recommended regarding transparency and monitoring of PROs (Monier et al. 2014d):

- provide a formal authorisation (or recognition) procedure for PROs
- provide monitoring procedures and audits over PROs, including self-control procedures
- set up a system of compliance promotion and enforcement that effectively discourages free riders
- define ambitious targets and develop the indicators and reporting obligations to allow their monitoring
- ensure the quality of statistics reported
- define and enforce monitoring procedures on quality of recycling for exported materials.

As noted earlier, a central tension regarding transparency relates to the disclosure of financial and technical information in systems with for-profit PROs and with respect to commercial collection and processing firms. For-profit PROs argue that disclosure of data about their fees and their business relationships in the EPR chain jeopardises their competitive positions and that the goal of appropriate performance is met through monitoring of technical outcomes (e.g. quantities collected or processed) and the competition in the market. Sirena (2013) argues that this tension could be addressed by requiring disclosure of costs when a PRO has a market share exceeding a specific threshold (e.g. 50% of the products placed on the market). Collection and aggregation of data by a neutral third-party organisation and/or use of non-disclosure agreements by the organisation collecting data could also be used to address this challenge, as was done with a recent study prepared for the Irish government (RPS et al., 2014).

Harmonisation

Harmonisation of reporting requirements and especially data format and definitions is needed if benchmarking of performance is to be feasible and to minimise the cost of compliance of the entities in the EPR chain (Monier et al., 2014d). This is a form of co-ordination between, rather than within, EPR systems and addressing either the same or different product categories. At the government level, the Canadian province of British Columbia and the US state of Maine have enacted a product stewardship framework law that defines key parameters of EPR systems as new product categories are considered. Eight other US states have considered such legislation (Hickle, 2014b). Inter-state harmonisation has been pursued through a model law for paint EPR, adopted by 4 states in the US. Canada and Australia both have federal guidance that, while largely voluntary, provide impetus for consistency across provinces/states and product schemes (CCME, 2009). In Canada non-governmental entities have also played a role in harmonisation: the electronics and retail industries have established a new governance structure, Electronics Product Recycling Association (EPRA), which has taken over the management of the PROs responsible for WEEE in eight of ten provinces, bringing common standards, management structures and reporting to most of the country (Waste Reduction and Management Division, 2014).

Organisations such as the WEEE Forum and EXPRA, which represent non-profit PROs in EPR systems for WEEE and packaging respectively, function like trade associations providing platforms for sector-wide communication and initiatives as well as political representation.

Efforts at harmonisation are also occurring with respect to treatment standards in recycling. WEEELABEX, an initiative of the WEEE Forum, is creating a common set of European standards for the treatment and recycling of electronic and electrical waste and for the monitoring of the processing companies (WEEELABEX, 2013). CENELEC, the European standards body for the information technology industry, is also working on collection, logistics, and treatment requirements for WEEE (CENELEC, 2014). Each European country will separately determine whether to require this standard or others, in order to participate in the EPR scheme. These initiatives are an attempt to both raise the floor with respect to recycling practices for WEEE as described in the section on enforcement below, and to level the playing field for PROs. Facilities operating outside the EPR chain may perform at a lower standard, putting the EPR schemes at a competitive disadvantage. Harmonised standards also help reduce potential spill over effects and eliminate the incentive to buy and discard in different jurisdictions. Mandatory compliance with standards arising as a result of legislation, contracts, or market practice, however, may be constrained by competition law and international trade rules (see Chapter 4). An example of efforts towards consultation, co-ordination and harmonisation in the Belgian packaging EPR system is provided in Box 3.3.

Box 3.3. **Consultation, co-ordination and harmonisation in the Belgium packaging EPR system**

In Belgium, the Interregional Packaging Commission (IPC) is responsible for harmonised management of packaging waste among the 3 regional governments (Flanders, Wallonia and Brussels). The IPC is in charge of permitting of the 2 packaging PROs, inspections of service providers, and aggregation and reporting of data (Monier et al., 2014g). The establishment of the IPC reflects the need for co-ordination among semi-autonomous regions of Belgium (Marques et al., 2012), but also creates a platform which provides a ready basis for oversight, harmonisation, and management of proprietary information. Other entities in the Belgian system operate between the municipalities and the PRO: inter-municipal organisations ("intermunicipalities") performing inter-municipal co-operation (IMC). These organisations seek to increase the efficiency of waste services by sharing know-how and operations. The legal form of the IMC is governed by regional law. Studies of the cost efficiency of the IMC have been conducted, but correlation between governance and efficiency was not assessed (De Jaeger and Rogge, 2014).

3.3.8. *Enforcement*

Enforcement involves the identification of violations and imposition of sanctions. In the context of EPR, enforcement is principally concerned with failure to achieve collection and related targets, free-riding by producers, compliance by PROs with respect to managerial and fiduciary responsibilities, diversion of recyclables to inappropriate channels, and conformity with environmental regulations by collectors and processors of the relevant wastes. Consumers, importers (acting as producers), retailers (if obligated to accept end-of-life products), municipalities and exporters can also be non-compliant and the subject of enforcement actions. A recent study for the Irish government argues that the extent of non-compliance problems depends on the design of the system and the type of product involved. "Systems with a large number of producers have a higher potential for noncompliance than more concentrated markets. The scope for non-compliance is greater and more complicated to deal with when a large number of producers are part of a long production chain" (RPS et al., 2014). Packaging systems are especially vulnerable in this regard.

Enforcement is important to increase compliance with EPR-related obligations, but also to ensure a level-playing field among competing producers, PROs, and service providers. Failure to provide consistent enforcement creates undue advantages to the entities failing to meet obligations and also can increase burden on those facing collective targets. Producers, for example, that evade payment of fees to PROs both reduce their own expenses and increase the costs that are borne by other producers. If evasion of fees is extensive, the financial viability of an EPR system can be at risk (Kalimo et al., 2012).[34] Similarly, if collection or processing services operating outside of the EPR system do not face the same degree of enforcement, EPR schemes are disadvantaged and, as mentioned above, the level of services may be reduced. (See section "*The role of municipalities in EPR systems*" for additional discussion in the context of leakage). Enforcement in such cases can include monitoring of collection points maintained by retailers and municipalities (RPS et al., 2014, 343).

The legal authority to impose penalties is typically reserved to governmental agencies. In the US, however, a EPR model law for batteries under development by the Product Stewardship Institute and other stakeholders includes a right of private action, that is, producers who would not otherwise have legal standing to sue entities involved in EPR for failure to perform, are granted the right to do so (Nash and Bosso, 2013). In a related vein, entities in the EPR system can play a role in identifying violators even without having formal legal authority. This is especially the case with respect to free-riding by producers (i.e. avoiding payment of fees and participation in EPR schemes).

In the European Union, there is widespread agreement among stakeholders that member states and PROs should both be responsible ("co-responsible") for the monitoring of EPR schemes, and should ensure that there are adequate means for enforcement , but there is less consensus as to the specifics of the division of responsibilities and costs ("co-funding" of enforcement) (Monier et al., 2014d).

Sanctions can range from criminal and civil penalties, fees, revocation of the right of a PRO to operate to public disclosure of non-compliant entities (e.g. free-riding producers) and successful prosecutions. There is little systematic and especially cross-national research on enforcement mechanisms and especially on the use of sanctions and penalties in EPR systems.[35] Analyses of illegal export of WEEE, ELVs and other wastes, however, have been conducted (e.g. Miller et al., 2012; Bisschop, 2012), but are typically not tied specifically to such exports arising from EPR systems. An issue with respect to sanctions that is specific to EPR arises, however, when a jurisdiction has a single PRO for a given product category. In such circumstances, the threat of revocation of a PRO's license to operate (i.e. its accreditation) is less credible than it might otherwise be. A contingency fund held in reserve by PROs equal to approximately one year's operating costs is used in Ireland to mitigate the difficulties in replacing a PRO (RPS et al., 2014).

In some countries, municipalities have enforcement roles. This can include monitoring the performance of EPR schemes, especially shared sites and activities (e.g. collection points and civic amenity sites), and permitting and sanctioning collection and processing activities within their borders. For example, in Ireland the collection points used by distributors must be registered with local governments (RPS et al., 2014). At the same time, municipalities that play an active role in the EPR chain are sometimes monitored by PROs primarily to control costs. National antitrust and competition authorities may also monitor the activities of PROs and service providers in an effort to ensure competitive markets in PRO services, collection, sorting and processing of wastes. Competition authorities may take participants in an EPR system to court or otherwise raise objections to the existing structure and practice of an EPR system (see Chapter 4).

3.4. Governance structures and allocation of tasks in EPR systems

3.4.1. Patterns in the allocation of functions

Typical allocation of governance functions

The four typical governance structures described earlier in this report – single PRO, competing PROs, tradable credits, and government-run – are only partially correlated with the allocation of governance functions in EPR systems. More broadly, irrespective of the governance structure used, the assignment of governance tasks related to accreditation, monitoring, reporting and financing of administration described in this report shows wide variation across EPR systems and jurisdictions. Notably, several functions are executed by government agencies, clearinghouses and PROs without a conspicuous pattern across jurisdictions:

- registration of producers
- collection of sales and import data
- certification of PROs
- allocation of collection territory; and
- monitoring of recyclers and auditing.

In contrast, the assignment of some tasks has an obvious logic: PROs cannot certify other PROs,[36] so this task falls to government agencies or, where they exist, sometimes clearinghouses. There is obvious need for clearinghouses in a multi-PRO system, and in many cases in government-run systems there are no PROs. Enforcement, because of the legal powers required, is almost always the prerogative of governments. Table 3.1 summarises the variety of assignments of governance tasks.

Table 3.1 indicates variability of the assignment of key tasks in governance among entities in EPR systems. For example, multiple types of entities may play a role in co-ordination and monitoring. That variability and the absence of commentary in cross-jurisdictional analyses of EPR systems about optimal arrangements suggest that those arrangements need to reflect local conditions and that best practice may lie elsewhere. Put another way, for several key governance tasks, with modest exceptions there is not clear evidence as to which sort of entity should take on a given function in an EPR system.

At the same time, analysts of EPR have argued that clear delineation of roles among stakeholders is important to good governance (e.g. Kalimo et al., 2012; 2014). Monier et al. (2014d) suggest that while there is no "one size fits all" solution in the allocation of responsibilities, "achievement of good...EPR performance is a result of each stakeholder's contributions toward a common goal and precise roles should be defined at the national scale, in accordance with the respective financial and/or operational obligations." Quoden (2015) points to the "Basic Document on the Monitoring of Packaging",[37] a document with legal status in the Netherlands that defines roles in detail, as a model for clear allocation of functions.

Table 3.1. **Typical allocation of governance tasks in EPR Systems**

Governance Function	Government*	PROs	Clearinghouses	Commercial Waste Companies	Municipalities	Specialized External Entities
Policy formulation and evaluation		**		**	**	
Operations						
Stakeholder consultation						
Registration (of producers)						
Accreditation (of PROs)						
Collection & disbursement of producer fees						
Co-ordination						
Monitoring						
Enforcement						

Key:
- Primary entity performing functions
- Entity sometimes performs function
- Entity rarely performs function
- No examples of entity performing function found

* Government refers to public sector entities involved in policy formulation and oversight. Does not include local governments solely involved in service delivery.

** Governments take the lead in policy formulation, but key stakeholders are often involved.

***Municipalities have ongoing interaction with citizens even when they don't have a central role in stakeholder consultation processes.

The role of municipalities in EPR systems

The role of municipalities in EPR systems is a contentious issue in many jurisdictions. One of the motivations for EPR is to relieve local governments (and tax payers and rate payers) of the burden of managing many waste streams. However, the form of this shift raises important questions for all parties in the EPR chain as it also implies a shift in control and in revenue and cost streams.

Municipal control. Municipalities argue that they bear legal or other obligations for waste management and public health in their communities and thus should have a major say in how waste is managed under EPR (Council of European Municipalities and Regions, 2013; Municipal Waste Europe, 2013). Municipal waste collection is also often an important source of local employment, one which local governments are reluctant see shift to the control of outside entities (Tojo and Hansson, 2004). In addition, local governments may have stranded assets if control shifts to PROs. That is, existing capacity in government-run or -financed facilities for material sorting or disposal may be displaced by facilities run by other entities; the existing facilities may be idled with a loss of revenue and on-going financing costs. The issues of employment and stranded assets are likely to pose a greater challenge for local governments that provide waste services directly than for those that contract with commercial providers for those services.

The central issue in most cases revolves around whether local governments provide EPR-related collection services and the extent to which their costs are covered by PROs. Where the cities see their role as crucial in ensuring public health and democratic control, others argue that efficiency and innovation are more likely to occur where industry has control (Valiante, 2015).

Other aspects of the municipalities' role, such as funding for public outreach and the extent of consultation regarding policy and operations in the EPR systems, are also at issue. Participation of municipalities in the different stages of EPR varies in several ways; it may:

- be mandated or optional
- be an exclusive prerogative of the municipality or may be conducted in competition with other collection and processing service providers
- include an oversight function as well as an operational role
- be financed by the industry or by the municipality itself; financing may be contingent on goals or performance.

With respect to service provision, municipalities typically function in one of two ways. In some EPR systems, described as "shared systems" in § 2.2.1, they have the obligation or prerogative to provide collection and sometimes sorting of designated wastes, that is, they have the option to provide (or contract for) the services. This is often the case in the packaging arena where municipal collection and recycling programs sometimes pre-date EPR regulations and producer obligations. Frequently the municipalities are reimbursed in whole or in part for the cost of their services and with rules governing amount of reimbursement. In other EPR systems described earlier as "dual systems", they may compete to provide collection and sorting, but the PROs are not obliged to use their services. These two approaches are illustrated in Figure 3.6.

Costs and Cost Control. A key point of contention is cost. In systems where municipalities have the prerogative to provide services, producers (and PROs as their agents) may not be able to control costs to their satisfaction. As a result, producers have contested the very assignment of costs as with the local collection of WEEE.[38] In some instances producers may seek to exercise extensive control of local collection and processing, as in the EPR system for packaging and printed paper in the Canadian province of British Columbia where producers have been given full financial and operational responsibilities for collection and processing (Elliott, 2014; Sinoski, 2014). In the Canadian province of Ontario, a formal arbitration process was conducted between the municipalities and the packaging PRO because of conflicts over what municipal costs were eligible for producer funding (Armstrong, 2014). The arbitrator found in favour of the municipalities, but the fundamental issues were not resolved (Bury, 2015).

Tensions over the cost of services provided by municipalities reflect both questions of whether a given type of service is delivered in a cost-effective manner and differences over what level of service should be provided. In the latter case, a municipality may want to provide frequent collection services or provide a dense network of drop-off (bring) sites, greater than a PRO judges to be necessary to meet EPR targets.

Some EPR systems have established reference costs for municipal services which provide a benchmark or limit regarding what constitutes reasonable costs. Reference costs can be configured based on a variety of strategies:

- reimbursement of the municipality tied to quantity or percentage of designated materials collected
- adjustments to reimbursements tied to quality of material collected (i.e. level of residues)
- definition of a standard cost for individual elements of municipal activities
- definition of a standard level of service beyond which reimbursement is not provided or limited, or
- bonuses for additional services (e.g. public outreach).

Figure 3.6. **Municipal roles in collection and sorting**

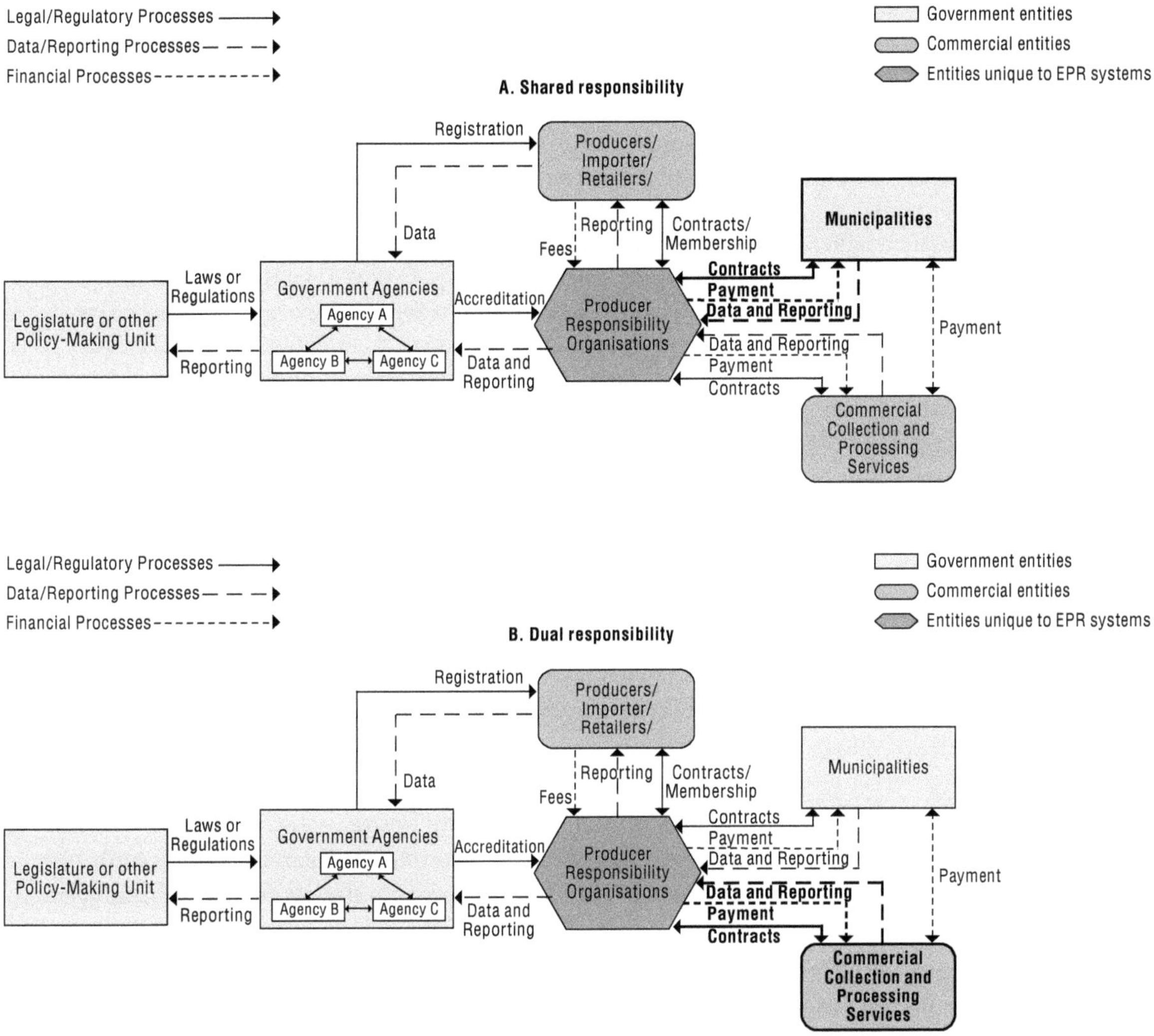

Note: Commercial waste services may be used by PRO when municipality chooses not to participate or by municipality when it contracts out services. Registration can be viewed as something the governments perform (i.e. they "register the producers") or something done by producers (i.e. the producers register with the government). The depiction of the directionality is thus somewhat arbitrary.
Note: PRO may contract with municipality but is not obligated to do so. Bold figures indicate primacy of municipality (2.6a) or commercial services (2.6b).

3.4.2. *Advantages and disadvantages of governance structures*

The PRO operating in a single-PRO governance structure (Figure 3.2) is almost always a non-profit PRO (one of the few exceptions is the Italian PRO for used oil). The single-PRO structure has the advantage that monitoring, reporting, and consultation are simplified and transparency tends to be greater than with multiple PROs (EXPRA, 2013). At the same time, the absence of competition among PROs in such a structure is argued to lead to higher prices (i.e. higher fees charged by the PROs to the participating producers). There are also concerns that a single PRO may exert market (monopsony) power over waste service providers (see Chapter 4) or may favour large producers at the expense of smaller ones. The strengths and weaknesses of a single PRO model are difficult to disentangle from those that arise from the choice as to whether PROs should be non-profit or for-profit (see section "*For-profit vs. non-profit status*" above).

Box 3.4. **Assignment of costs and level of service: Packaging EPR in France**

In the EPR system for packaging in France, municipalities are assigned the tasks of separate collection and sorting of waste packaging from households. As of 2012, a target of 75% recycling of packaging waste had been adopted and 80% of the net reference costs for an optimized service of collection and sorting incurred by the municipalities are reimbursed by Eco-Emballages, the PRO and French Green Dot system responsible for household waste packaging provided that the 75% target is achieved. Most of the remuneration paid to municipalities consists of i) a payment varying according to the level of performance achieved (kilograms of recycled packaging per inhabitant per year) (Cabral et al., 2013), ii) a payment for public outreach activities and iii) a payment for the reporting of their costs regarding environmental, economic and social targets so as to incentivize municipalities to monitor better their performances. Contracts between municipalities and service providers have a term of six years (French Ministry of the Environment 2014). Municipalities can issue tenders for take-back of materials (i.e. manage the contracting for services themselves), but Eco-Emballages will also provide a take-back guarantee with a take-back operator designated by the PRO. In that case, material take-back prices and contractual arrangements are applied in the same manner to all municipalities and prices and terms are not negotiable (Monier et al., 2014e). A recent review suggests that there may be too many sorting centres, that there is a lack of transparency regarding municipal costs, and that the reference cost calculations do not reflect variation in costs across municipalities. The result is that the reference costs appear to be higher than actual costs (Monier et al., 2014e). The long service contracts (six years) make changes in the organisation of EPR chain slow.

According to the recent study by Monier et al. (2014d), multiple PROs manage WEEE in all seven of the EPR systems examined. Batteries are also frequently managed by multiple PROs. In contrast, end-of-life vehicles (ELVs) in the systems studied were always managed by a single PRO. EPR for other product categories did not show a clear pattern.

Competition authorities have sometimes been the impetus, in whole or part, for the shift from single-PRO to multiple-PRO governance structures (see Chapter 5). As noted earlier, the German Federal Cartel Office played a key role in shifting the German EPR system for packaging from a single, non-profit PRO to multiple, competing for-profit PROs. More recently, the ARA, the monopoly, non-profit PRO responsible for household packaging waste in Austria, was warned by European Commission that it was abusing its dominant position in the market. The Austrian law governing waste management, already in revision, came into force and a new packaging ordinance with clearer definition of conditions was implemented in 2015, resulting in multiple, competing PROs for household packaging waste (Monier et al., 2014f; Wollman, 2015).

It is difficult to ascertain whether single or multiple PRO systems are preferable. Data are sparse because governance structures for multi-PRO systems as shown in Figure 3.3 often have for-profit PROs which inhibit disclosure of financial and other market information. In Germany, the introduction of competition among packaging PROs led to significant reductions in cost. Some stakeholders argue that the competition led to efficiency and flexibility and thus to the decline in costs. Others argue that the decline reflects primarily competition at the level of collection and processing rather than PRO services (Monier et al., 2014d, 103). Further, as Monier et al. (2014d) point out, there are few cases of real competition, because, where there are multiple PROs, a large PRO usually dominates. They looked at 8 EPR systems with multiple PROs and found that the market share of the dominant PRO in all cases was at least 48% and in most cases was above 70% (Monier et al., 2014d). In an asnalysis of competition law and EPR under Chapter 3, it is argued that competition should be the norm once EPR systems reach an appropriate level of maturity.

Systems with tradable compliance certifications, such as the United Kingdom packaging system, will tend to be efficient and flexible. Economic efficiency is promoted because those with lowest costs of compliance have an incentive to "over-comply" and sell their surplus certification notes to those with highest costs, reducing average costs. Flexibility arises because actors alter their participation strategy based on the cost of acquiring versus producing the notes.

As noted earlier, the United Kingdom system separates producers from the process of compliance verification. Some criticise this separation because producers are quite removed from the achievement of environmental goals and the meeting of environmental requirements. OECD (2014a) argues that firms are not prevented from putting their own standards in place in order, for example, to maintain reputation. Further, the separation allows regulatory compliance to be focused on a smaller number of (re-processor) firms. In this respect, the virtue – or drawback – of the system is that it does not exploit the business acumen of the producers that some argue is an important element of EPR and that others dismiss as irrelevant.

The United Kingdom system is criticised for its impact on municipalities (Cahill et al., 2011). The system is not designed to cover the full cost of the collection and processing of packaging waste, with coverage typically ranging from 5-10% (Monier et al., 2014c). Because compliance in the system can be achieved by collecting waste packaging from sources other than municipalities, and because municipalities may nonetheless operate recycling programs, they benefit little from the system because few of the funds generated through producer financing are paid to them.

EPR systems where governments play a direct role in collecting and disbursing fees as shown in Figure 3.5 are less common. They can be found in China for WEEE, Chinese Taipei for all products subject to EPR, Denmark and Hungary for packaging and in several US states and in Alberta, Canada. Government control can facilitate the integration of strategies with other waste management objectives. They are also often considered by developing countries contemplating EPR (Manomaivibool, 2009). The absence of industry involvement in the fee collection and disbursement functions however, means that i) societal obligations for waste management have not been shifted to producers ii) any business acumen that producers might bring to those tasks is foregone, and iii) market forces are not exploited as an impetus for efficiency in the EPR system.[39] In addition, the public sector may be in no better position than industry to execute these responsibilities. Nonetheless, generalisations about the effectiveness of governance structures where governments collect and disburse fees as displayed in Figure 3.6 are difficult because many systems are only well documented in the native language of the relevant country, because many are new and lack a track record, and because of especially wide variation in circumstances in the countries adopting this model.

Hickle argues that EPR represents not only an incentive-based policy strategy but also a systematic reframing of the roles of producers and local governments and a transformation of the governance of waste management that can reflect the dynamism of the global market and deploy technology for enhanced collection, sorting and processing. In that respect, governance structures that maintain a central operation role for government may reflect a transitional stage to a newer, more ambitious strategy.

3.5. Government resources

Considerable attention and analysis in the literature on EPR is paid to costs and fees assigned to producers and incurred by others in the EPR chain – with good reason, as these details go the core of the policy objective and functioning of EPR. However, governance of EPR systems requires action by government agencies – policy formulation and evaluation, data collection and management, accreditation, enforcement, etc. – which, in turn, means that resources for those activities are needed. These activities and the resources they require are separate from traditional waste management services provided by local government. These resources needed for EPR governance include authority, staff, and funds.

EPR systems that are cost-effective lower resource demands insofar as they deliver more social and environmental net benefits than if other, less cost-efficient waste management strategies are pursued. It is worth noting that, when comparing cost-effectiveness across EPR systems, Monier et al. (2014d) found that the most expensive programs were not necessarily the most effective. While measuring cost-effectiveness is crucial for optimal use of resources, its measurement will not necessarily be straight forward and it may or may not make government administration of EPR systems less expensive: it is the cost to society as a whole that is relevant in assessment of cost effectiveness. Nonetheless, the administrative cost per societal benefit delivered (e.g. tonnes recycled) is a key consideration in design of governance systems.

3.5.1. *Authority*

Government agencies charged with the oversight of EPR schemes need sufficient statutory or other legal authority to collect data, establish requirements for, and engage in enforcement regarding entities in the EPR chain. Many EPR systems, as part of the application process for certifying PROs, will require formal agreements between the government agency and the PRO which spell out obligations regarding governance, operations, targets, financial management, and data collection and reporting. The government agency must have the authority to mandate the agreements and to review, revoke, vary, and replace them (Philip Lee, 2014) if effective oversight is to be provided.

3.5.2. *Staff*

No studies documenting staffing levels in government agencies charged with oversight of EPR schemes have been found. It appears that many governments use or expand existing staff in environmental or related agencies to perform these duties.

3.5.3. *Funds*

Funding for government oversight activities related to EPR is also not well documented. It is not uncommon for governments to require payment of a fee by producers at the time of registration and/or payment of fees at the time of certification of PROs (separate from fees paid by producers to PROs). In some cases, the fees are mandated by statute. In California, for example, the Paint Stewardship Law includes provisions for an administrative fee imposed on the producer or PRO to fully cover the cost of administering and enforcing the statute (California Paint Stewardship Law, 2010). The fee is not charged separately, but is part of the fee paid by producers. The agency charged with overseeing the EPR system, the Department of Resources Recycling and Recovery (CalRecycle), prepares an administrative budget and invoices the PRO for its expenses (Zarrehparvar, 2012).

According to a report prepared by SAIC (SAIC Energy, 2012b), costs[40] to state governments in the US for EPR systems for WEEE as indicated by registration fees ranged from USD .05 to USD 0.28 per pound (lb) collected. For states with higher collection rates, the report indicates that the costs are closer to USD 0.01-0.02/lb collected. It is not clear whether registration fees pay all or only a portion of the governmental costs.

In some EPR systems, funding for what might be labelled ancillary activities – public outreach, waste prevention initiatives, research and development related to DfE, anti-litter campaigns, etc. – is assigned to government agencies (or required of PROs, clearinghouses, or producers).

Fees from enforcement actions imposed upon noncompliant entities in the EPR chain can be additional sources of funds. Reliance on such fees can be problematic to the extent that it creates awkward incentives around enforcement and can be an unpredictable revenue stream.

In some EPR systems, governments use a portion of producer fees as a source of funds to support administrative activities. Chinese Taipei imposes a tax on producers which is then used by the government to develop or subsidise waste services and infrastructure. A portion of the tax revenue is used for the government activities (Ching-Wen, 2004). China currently also imposes a tax on EEE producers but the tax is deliberately set such that the amount collected is less than the amount needed to support collection and processing so that the authorities distribute the funds without retaining any surplus (see Annex D).

3.5.4. *Delegation and outsourcing*[41]

In many cases, governments reduce their administrative burden – and thus indirectly the cost – by delegating monitoring, co-ordination, and other administrative tasks to quasi- and non-governmental entities. This aspect of delegation is largely distinct from the shift of waste management activities from municipalities to producers.

PROs are typically responsible for arranging for collection and processing services by waste companies and municipalities. As part of that activity, monitoring and reporting are often performed with the statistics then being provided to the government. The extent of the services that are in effect outsourced to PROs by the government agency ranges widely. For example, the PRO dealing with battery recycling in Japan, the Japan Portable Rechargeable Battery Recycling Center (JBRC), monitors the delivery of waste batteries from participating retailers, businesses, and municipalities to those who receive and recycle the batteries and also audits recyclers (see Annex G). In Denmark, by contrast, PROs have a more limited role. The battery (and WEEE) producer responsibility system is controlled by the Ministry of the Environment through a clearinghouse, the Danish Producer Responsibility System (DPA, Dansk Producentansvarssystem) system. The amount of batteries collected is reported to DPA by a combination of producers, PROs, and municipalities. The amount of batteries treated is reported by the waste treatment company directly to DPA (Monier et al., 2014h).

As described earlier, enforcement is generally not outsourced to non-governmental entities because the legal authority typically cannot be delegated. However, other entities in the EPR system can be enlisted formally or informally in the identification of non-compliant participants. In the Czech Republic, the EPR law holds retailers and distributers responsible for take-back obligations in cases where their supplier does not comply with the packaging law. This has proven to be a very efficient anti-engee riding measure because retailers have strong incentives to ensure that suppliers comply with the law (Monier et al., 2014g). See Box 3.5 for more detail.

Box 3.5. **Delegation of EPR Governance to external organisations**

In Korea, the Korea Environment Corporation (KECO), a public entity established by statute provides oversight for multiple activities in the EPR chain for all products subject to EPR (see Annex J). These include assembling data on sales of products, monitoring and approval of recycling results, and administration of low-interest loans to small and medium-sized recycling businesses. KECO evolved from earlier organisations established by the Korean government. The Korea Resources Recycling Corporation (KORECO) focused on organisational and technological development in 1980s. It was followed by the Korea Environment and Resources Corporation (ENVICO) in 2002 which focused on regulation of the EPR system. In 2008 ENVICO became KECO which added authority for enforcement of design for environment and toxics reduction requirements. This evolution reflected the development of the Korean EPR system. It started as a producer-based deposit refund system in the 1990s arising from a concern about difficulties that might arise from newly granted autonomy for municipalities. It changed to a mandatory EPR system with financial and physical responsibility similar to those in Europe from 2002 onward. In 2008 design for environment incentives were added to EPR systems previously focused on collection and recycling (Chung and Murakami-Suzuki, 2008; Manomaivibool and Hong, 2014).

In some EPR systems, there are additional quasi-governmental and non-governmental entities that play a role in the administration of the system, effectively shifting tasks away from traditional environmental agencies. In Germany, the Chamber of Industry and Trade acts as a data repository for packaging put on the market as required for the Packaging Ordinance. The German Society for Packaging Market Research (Gesellschaft für Verpackungsmarktforschung, GVM), a B2B market research institute specialising on packaging, collects and collates data from across German government agencies as part of the process of verifying the calculation of packaging quantities.

3.6. Observations on patterns and best practice in EPR governance

3.6.1. *The difficulties in deducing best practice*

As noted in the discussion of allocation of functions in EPR governance systems (§ 3.4.1), there are only limited commonalities across EPR systems regarding the particulars of the involvement of government. Perhaps more important, it is difficult to identify best practices among the patterns that do exist. This reflects the diversity of goals and situations in which EPR systems operate across the globe. It also reflects the conspicuous lack of data and comparative analysis regarding EPR governance. This is compounded by the multiple factors influencing outcomes: with many factors at play, it is difficult to ascribe the performance of an EPR system to a particular governance practice. As Monier et al. (2014d) indicate, there is both a lack of data and lack of harmonisation of data (and practice) across EPR systems. Teasing out causes and effects in governance outcomes requires especially good data. With the vast number of EPR systems in operation around the world, generalisations are precarious.

The boundary between policy and governance is hazy as well. Issues such as which end-of-life products should be included in an EPR scheme – e.g. those from residential waste streams versus those from commercial sources or a few types of WEEE versus an expansive list – are presumably matters of policy design. However, in some EPR systems they are a matter of choice, left to the discretion of producers and PROs. Thus, to whom decision making authority is delegated can be a matter of both governance and policy. It also adds to the list of factors shaping governance choices and outcomes in a particular EPR system along with population density, the historical configuration and maturity of waste management systems, legal structures, and the like. The good news is that the choices are many; the bad news is that analysis is difficult.

Even governance practices that appear to be especially attractive bear careful scrutiny. There is a temptation to look for self-implementing mechanisms that will generate desirable results without extensive government involvement. The most obvious example is the incentives facing producers participating in EPR schemes who are prompted to identify other producers that are free-riding, shirking their obligations. Other mechanisms exist such as making one interest responsible if another fails as when retailers bear the burden of EPR if producers do not meet obligations. These strategies, however, require political will at the time of policy formulation because the interest that is targeted for the assignment of such an additional responsibility will object.

3.6.2. *Opportunities for better governance and increased understanding*

All is not opaque nor out of reach, however. Much of what appears to be important in this admittedly brief scan of EPR governance practice are basic elements of good governance – clear assignment of roles, systematic monitoring and data collection, transparency, enforcement, stakeholder consultation, adequate resources for oversight. Some of the common features have EPR-specific coloration. Transparency is a particular challenge where for-profit PROs operate, hindering well-informed evaluation of EPR. Enforcement, especially as it relates to entities that collect and process end-of-life products outside of EPR systems, is critical not only for environmental protection, but also for the financial viability of EPR.

Some of the generic gaps in knowledge regarding EPR governance systems can be addressed with moderate investment of time and effort. Many systems have produced reports evaluating performance for use by policy makers within the relevant jurisdiction. These reports contain some of detailed information that can improve understanding of key governance practices. With additional effort, information about practices across EPR systems can be acquired. For example, compilation of registration fees charged to producers could provide some indication of the funds used for government administration.

Some contested issues could be clarified through detailed comparative case studies. Systems with for-profit PROs could be compared to those with non-profits. Similarly, systems with single PROs could be compared with those that have multiple PROs. If the research is to be useful, it would need to go beyond basic descriptions of the EPR systems and simple calculations of the cost per tonne or kilograms per inhabitant collected to engage the specific claims of competing positions and investigate them empirically in depth. For example, non-profit PROs make several specific claims in their criticism of for-profit PROs (see § 3.3.5). Investigation of those particular claims could advance the discussion of for-profit versus non-profit status. Similarly, competing PROs are argued to lower the cost to producers for EPR compliance (see § 3.5.2). Costs could be examined with careful attention to adjustment of data to reflect the differences in the scope and level of service in EPR systems.

Access to data would be a challenge, but nondisclosure agreements and redaction of sensitive information as used in a recent study of EPR in Ireland (RPS et al., 2014) could mitigate this problem.

Comprehensive comparisons of EPR systems, however, will have to wait for better data and increased harmonisation of systems.

3.7. Conclusion

This chapter focuses on the administration and implementation of EPR from the perspective of public authorities. It examines rationale for EPR and the role of government agencies in EPR systems as well as the functions that are assumed by various stakeholders, the differing ways that the delivery of those functions are organised, and how those activities are funded. The primary focus is on activities related to government oversight – monitoring, co-ordination, enforcement, and government resources. The role of municipalities and the legal status of producer responsibility organisations (i.e. non-profit vs. for-profit) are examined as well.

As governments have embraced and implemented EPR, they have found that simple delegation of end-of-life tasks to producers has not been sufficient to achieve policy goals. Along with policy revision in some cases, governance structures have emerged to address the need for monitoring, enforcement, co-ordination, and consultation. Governments in different countries have used a variety of forms of governance for EPR schemes across varying product types. Some aspects of government involvement in EPR systems are a matter of the typical need for government oversight while others are responses to misaligned incentives that have become evident over time.

Four governance structures for EPR are common: single PRO, competing PROs, tradable credits, and government-run. A few key types of organisations are common in EPR systems: government agencies to set, enforce, and evaluate policy and to monitor and certify participants; producer responsibility organisations to organise and manage collection and processing of end-of-life products; clearinghouses to gather data and to allocate service territory; waste management companies and municipalities to provide collection and processing services; and in developing or emerging economies, informal recyclers. Variations on these approaches are many with respect to the type and number of organisations involved and especially with respect to the allocation of specific governance functions to those organisations.

The impact of the legal status of PROs – for-profit vs. non-profit – is widely disputed. For-profit PROs have emerged in EPR systems with competing PROs, particularly for WEEE and packaging in Europe. In other jurisdictions, PROs are, in some cases, required by law to be non-profit. The advantages and disadvantages of for-profit PROs are difficult to disentangle from the effects of the choice of single versus competing PROs.

Governance functions in EPR include policy formulation and evaluation; operations (collection and processing); stakeholder consultation; registration of producers and accreditation of PROs; collection and disbursement of fees; co-ordination of service delivery; monitoring of technical and financial performance of producers, PROs, waste companies, recyclers, and municipalities; and enforcement. Obtaining, verifying and using data on performance and compliance is key element of EPR governance that in many cases needs vast improvement. Increased transparency and harmonisation of data definitions and acquisition practices are central to improvements in cost-effectiveness and to good governance.

Government resources required for oversight of EPR systems include legal authority, staff, and funds. Fees charged for registration of producers appear to be a common source of funding, but little systematic information exists about the resources employed by governments for administrative purposes. Governance functions in some EPR systems are outsourced to non-governmental and non-environmental agencies in diverse ways, reducing resource demands on the central agencies tasked with oversight of EPR.

The role of municipalities in EPR systems is also a matter of contention. The central debates relate to whether and in what manner they should be involved in collection of end-of-life products in EPR systems and especially how the costs are best managed when they do play such a role. "Reference costs" to benchmark or limit costs incurred by municipalities and reimbursed by PROs are increasingly used by EPR systems.

This review of EPR governance belies the notion that EPR is basically a form of privatisation and shedding of responsibility by government. Regulatory authorities have an important role to play and municipalities, in diverse ways, can expand under EPR their waste-related activities as readily as they can shed them. Under EPR some municipalities may collect wastes that were previously not managed. They also may assume a regulatory role in overseeing producer and PRO activities within their jurisdictional boundaries.

Familiar elements of good governance – clear delineation of roles, systematic monitoring and data collection, transparency, enforcement, stakeholder consultation, and adequate resources for oversight – are both crucial and lacking in many EPR systems. Best practices specific to governance of EPR systems are less clear. Commonalities and insights regarding the role of government are both limited and blurred by the multiplicity of factors at play. This review suggests that there is much to be learned about strategies and structures for governance of EPR.

APPENDIX 3.A.1

Defining a positive price for end-of-life products and materials

The presence of products and materials with commercial value in waste streams managed through EPR raises the question of what constitutes a waste with a positive price.

For recycling of a waste to be economically viable without government regulation, subsidy or other form market intervention, the value of the waste must cover the collection, sorting, processing and residue disposal costs. Collection is a particularly important hurdle as it can account for a disproportionate share of the overall cost of recycling.

In discussions of valuable end-of-life products, there is often ambiguity as to whether the price in question includes the cost of collection or whether it simply refers to the price available in the recycling market once the end-of-life products are in the possession of sorters (material recovery facilities), brokers, or processors. Coverage of collection costs is complicated to make into a universal criterion because some end-of-life products are brought to a collection point by households and other waste generators (e.g. WEEE) whereas others are typically collected at the site of discard, as with curb-side collection of packaging waste.

As shown in Table 3.A.1, products and materials vary as to the point in the end-of-life chain when they become valuable.

Table 3.A.1. **Value of end-of-life products and materials by stage in end-of-life chain**

Value of Products/Materials by Stage in End-of-Life Chain

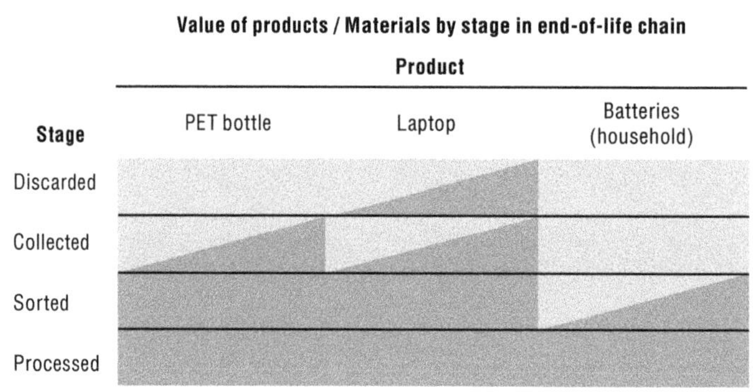

Note: Grey cells indicate negative market values; blue cells indicate positive market values; divided cells indicate mix of negative and positive values depending on condition, components/materials, and markets.

Table 3.A.1 presents positive or negative market values typical of three products often managed through EPR. The table indicates whether used PET bottles, laptops, and batteries (dry cells) from households can be sold for a positive price absent government intervention in the market (e.g. deposit-redemption systems) in OECD countries. In developing countries, where peddlers and scavengers collect end-of-life products, the situation is quite different and not meant to be represented here. In the case of a PET bottle, when a household wishes to discard a used (i.e. empty) container, it is rarely the case that the container can be sold. If collected in a separate stream, the resulting aggregation of bottles in some markets may fetch a positive price (at the front door of the sorting facilities as it were). As part of a mixed stream of recyclables, whether the PET bottles will have a positive market as part of the aggregate is very sensitive to market conditions. Once sorted, washed and then flaked or pelleted, secondary PET can demand anywhere from USD 770 to USD 1 650 per metric tonne.[42]

Unlike the case with PET bottles, a household may have the opportunity to sell a used individual laptop depending on the model, age, and condition. When brought to a collection point and kept separate from other types of WEEE, an unsorted batch of used laptops may either demand a positive price or require payment to be taken by the next actor in the end-of-life system. Once sorted by model, etc., laptops are more likely to have positive value depending on market conditions and model in question. Once dismantled and/or shredded, some components or materials will typically have positive values.

Like PET bottles, used household batteries have no market value for consumers seeking to discard them. Nor do many such batteries have positive value once sorted by type. It is only after processing to separate constituent materials that some of those materials have market value.

Notes

1. The term "EPR systems" is used in this document to refer to the laws, policies, governance structures, and the ensemble of organisations and operations involved in EPR in a given jurisdiction. It includes government agencies, and the EPR chain, i.e. collectors, processors, end use industries, producer responsibility organisations, and clearinghouses. "EPR schemes" is intended to be a narrow term, referring to entities in the EPR chain, especially producer responsibility organisations (PROs).
2. Unless otherwise noted, the term "products" is used in this report to refer to both products and packages subject to EPR regulation.
3. In a very few places, local governments have enacted EPR legislation as described later in this report.
4. The notable exception is a recent study for the European Commission by Bio-Deloitte, *Development of Guidance on Extended Producer Responsibility* (Monier et al. 2014d). It is one of only a handful of efforts to study multiple EPR systems in a detailed and systematic manner. Nonetheless, the study examines only 56 of 169 EPR systems in the European Union and does not include systems in Asia, North America, or developing countries. The research prepared by Monier et al. provides one of the few bases for drawing conclusions across jurisdictions.
5. The term "EPR systems" is used in this document to refer to the laws, policies, governance structures, and the ensemble of organisations and operations involved in EPR in a given jurisdiction. It includes government agencies, and the EPR chain, i.e. collectors, processors, end use industries, producer responsibility organisations, and clearinghouses. "EPR schemes" is intended to be a narrow term, referring to entities in the EPR chain, especially producer responsibility organisations (PROs).
6. Governance is typically defined to include matters of how formal and informal authority are established and exercised. In this report, the focus is on the role of government and the organisational structure and administration of EPR systems.
7. Recyclers may refrain from accepting some EoL products because they do not want incur the cost or regulatory burden of depollution, i.e. removing problematic substances or components.

8. In addition, a variety of other non-environmental market failures can contribute to the inadequate level of recycling and/or the inefficacy of a get-the-prices-right strategy including transactions and search costs in secondary materials markets; information failure and uncertainty related to waste quality; consumption externalities related to products derived from secondary materials; technological externalities related to recovery and reuse of secondary materials and market power; and vertical integration in waste recovery (OECD, 2005b). Analysis has not established a connection, however, between the non-environmental market failures and the specific characteristics of EPR as a remedy for those failures with the exception of technological externalities (Nicolli et al., 2012).

9. Note that there are many anecdotal accounts of DfE in response to EPR. What is missing are systematic evaluations at the national or sectoral level using methodologies that can distinguish the impact of EPR on DfE from other factors that may also motivate design changes. For an example of potential design changes that can be tied to EPR, see Klausner et al. (1998) and Lindhqvist and Lifset (1998). Tojo (Tojo 2004) in a detailed study of Japanese EPR for home appliances argues the close relationship between producers and recyclers prompted by EPR has encouraged DfE. See Box 2.1 for a description of the Japanese Specified Home Appliance Law.

10. This characterisation leaves aside the fact that recycling is in fact an intermediate objective, pursued to achieve reductions in a variety of environmental impacts up and down the product chain. These reductions can in turn be seen as means of accomplishing other, broader societal goals, e.g., improved human health, conservation of habitats and ecosystems, protection of economic assets, etc.

11. A positive price for an EoL product or material does not necessarily mean that there is no market failure if the level of recycling that occurs is less than optimal. Analogously, a negative price does not necessarily indicate market failure if the appropriate level of recycling is occurring.

12. The CWIT report emphasizes that "there is a varying degree of compliance and criminality that spans across a continuum ranging from minor unintentional violations or non-compliance by individuals to deliberate illegal activities following a criminal business model." Using the statistics, "unreported" export thus includes both more benign and more problematic forms of export (Huisman et al., 2015, 24)

13. "Jurisdictions" is used in this report so as to include local governments, states, provinces, national governments and supra-national governments, because the use of EPR is not limited to the national level.

14. The term "producers" throughout this report encompasses distributors, retailers, and importers where EPR laws mandate their participation, unless otherwise noted.

15. Individual producer responsibility can include both systems operated by individual producers and systems where producers share operational responsibility but remain individually responsible on a financial basis (Dempsey et al. 2010; Mayers et al. 2013).

16. The term "shared responsibility" has varied meanings in the context of EPR. With regard to the role of municipalities, it indicates that they share some degree of administrative or operational responsibility as indicated above. In the United Kingdom, it refers to the division of responsibility among actors in the supply chain (see discussion below on tradable credit systems). In the US, "shared product responsibility" was advocated by industries opposed to the establishment of EPR obligations (Lindhqvist and Lifset 1997).

17. Beverage container deposit systems also constitute competition for packaging PROs, albeit competition for inputs rather than for customers.

18. Some describe the United Kingdom EPR system for WEEE as a tradable credit system but it differs in subtle, but in important ways from the United Kingdom system for packaging, in part with respect to the transparency of the pricing. For more detail, see (Smith 2014).

19. *www.gov.uk/government/policies/reducing-and-managing-waste/supporting-pages/packaging-waste-producer-responsibility-regimes.*

20. The term product stewardship is used in different ways across countries and stakeholder groups. In the US, in the 1980s and 1990s, it was often used to refer to programs in the chemical industry focused on the management of potentially toxic substances and products (e.g., the Responsible Care program), often not extending to consumer use and disposal. More recently, in the US, environmental advocacy groups involved in EPR, have defined product stewardship as a policy where producers bear primary responsibility for health and environmental impacts across the product life cycle. In this view, EPR is a subset of product stewardship focusing on producer responsibility for end-of-life management of product and packaging (Upstream et al. 2012). In Canada, product stewardship refers to programs that typically allocate responsibility to municipalities or provincial governments using legislated fees or public funding.

21. CONAI (Italy), EcoEmbes (Spain), Ecopack (Bulgaria), ECO-ROM Ambalaje (Romania), EKO-KOM (Czech Republic), Envipak (Slovakia), Fost Plus (Belgium), Green Dot Cyprus, Greenpak (Malta), Nedvang (the Netherlands) and Valorlux (Luxembourg).
22. The arguments against for-profit PROs were only one of several topics addressed in the Manifesto.
23. A detailed and useful discussion of for- versus non-profit PROs in the Irish context, however, is provided by Gorecki (2014).
24. Case C-41/90 Höfner & Elser v Macrotron [1991] ECR I-1979 ("Höfner & Elser"), paragraph 21, cited in United Kingdom Office of Fair Trading (2011).

    Although, for example non-profit entities may have differing objectives, at the margin they have an interest in exploiting any ability they may have to raise prices because they use the profits thereby generated to fund their objectives. (Philipson and Posner, 2009) A review of the literature on non-profit, privately-owned hospitals in the United States found that there was no economic theory to support the presumption that non-profits will not exercise market power to the detriment of total or consumer welfare. The empirical literature showed that nearly all of them exercised market power by raising prices. (Dravone et al., 2012)
25. For information on the e-waste take-back law in New York City, subsequently repealed, see Cardwell and DePalma (2008). For Alameda County, California's pharmaceutical take-back ordinance, see Almeda County Public Health (n. d.). For the pharmaceutical take-back ordinance in King Country, Washington, see King County and Seattle Public Health (2013).
26. For the purposes of this report, it is not important to distinguish between various forms of and terms for local government. Many labels for local governments refer to cities (e.g., municipalities, councils, local authorities) while others refer to a level of government between cities and states or provinces (e.g., county). The important distinction in this context is between levels of government that usually formulate or enforce policy (states/provinces and national governments) and those that deliver services (local governments).
27. While technically packaging in France is managed by two PROs, the main shareholder of the second PRO is the first PRO, so it may be viewed as a single-PRO system.
28. In those EPR systems that allocate responsibility on the basis of "return share," that is, the proportion of a producer's goods among those collected for recycling, the relevant data are collected by PROs or clearinghouses rather than registries. Other systems allocate responsibility on the basis of waste generated using a "put on the market" calculation where sales data are adjusted to reflect product life spans. This is especially the case for long-lived durables where sales do not correlate with discards on an annual basis. For a discussion of calculation methods for this last approach in the context of the Canadian province of Ontario, see Waste Diversion Ontario (2010).
29. While ARMA is a nonprofit organization and is run by a Board of Directors representing stakeholder groups, because it reports to the Ministry of Environment it is viewed in Canada as a government agency (Alberta Recycling Management Authority 2015; Lange 2015; Kerr 2015).
30. Producers have the option of not submitting a plan and instead may choose to follow specific requirements detailed in the regulations.
31. Stewardship plans are prepared and submitted by PROs on behalf of participating producers and thus the legal obligation falls on the producer and the wording reflects that.
32. Under the Home Appliance Recycling Law, Japanese consumers pay a fee which is intended to support collection and transport of home appliances. The amount of the fee is displayed at the time of purchase, but collected at the time of discard (Hotta et al., 2014).
33. Mass balancing involves calculating inputs and outputs (plus changes in stocks) across facilities in the EPR chain. If the totals do not match, this is an indication that there are errors or gaps in the data (Kleijn 1999).
34. The amount of free riding can be substantial: In Germany, it is estimated that about 25-30% of the packaging collected through the EPR system from households is not covered by license fees from producers (Monier et al., 2014i). In some instances the free-riding reflects or is exacerbated by ambiguities in the laws or regulations governing EPR.
35. At the time of the preparation of this report, the summary of the study, *Countering WWEEE Illegal Trade (CWIT)*, had just been issued, though the detailed substance of the report was still in the process of review. Because the CWIT study addressed WEEE in the European Union and because WEEE in the EU is managed through EPR, the study implicitly addresses enforcement related to

EPR. Information from the summary has been included here, but the full study is likely to contain additional information relevant to enforcement of EPR-related laws and regulations.

36. The notable exception is Dual System Deutschland (DSD) in Germany.

37. *www.rijksoverheid.nl/documenten-en-publicaties/rapporten/2013/09/02/basisdocument-monitoring-verpakkingen.html*.

38. Producers, for example, filed a law suit against the City of New York City when the city proposed door-to-door collection of WEEE under the City's newly enacted waste electronics law (Gronewold 2010).

39. It is important to note that there are 3 types of competition in EPR systems: among PROs, among providers of collection, sorting and treatment services, and among end markets for secondary materials. The absence of competition among PROs does not preclude competition in other parts of the EPR system (Gorecki 2014).

40. Data are mostly for 2010 but 2009 and 2011 data are also used.

41. Delegation implies a shift of decision making in addition to administrative functions. Outsourcing, most frequently used in business contexts, suggests hiring of an external entity on a contractual basis to perform activities in which the contractor has more expertise. The distinction in this context is not significant.

42. These are approximate prices in the US in 2014 (Dimino 2015). It should be noted that transportation distances have a significant impact on the market value of recyclables. For instance, in Chile, which spans 4 000 km from North to South and has most recycling facilities in the centre, transportation costs can be significant.

References

Agrawal, V.V., A. Atasu and K.V. Ittersum (2015), "Remanufacturing, third-party competition, and consumers' perceived value of new products", *Management Science*, Vol. 61, No. 1, pp. 60-72.

Akenji, L. et al. (2011), "EPR policies for electronics in developing Asia: An adapted phase-in approach", *Waste Management and Research*, Vol. 29, No. 9, pp. 919-930.

Alberta Recycling Management Authority (2015),*www.albertarecycling.ca* (accessed 9 September 2015).

Almeda County Public Health. n.d. Safe Drug Disposal (n.d.), *www.acgov.org/aceh/safedisposal/index.htm?utm_source=Breaking+News+-+Alameda+Co.+PhRMA+Ruling&utm_campaign=alameda+co+lawsuit+ruling&utm_medium=email* (accessed 21 September 2014).

Armstrong, R.P. (2014), "In the matter of an arbitration to determine the 2014 Steward Obligation for the Blue Box Program", Ontario, *www.amo.on.ca/AMO-PDFs/Waste-Management/Blue-Box/Blue-Box-Arbitration-Public-Private-Decision-2014.aspx*.

Arnold, D. (2015), "Personal Communication with Arnold, D., Chief, Special Programs Branch, FAS Office of Acquisition Management", US Department of General Services, 5 October 2015.

Atasu, A. and R. Subramanian (2012), "Extended producer responsibility for e-waste: Individual or collective producer responsibility? ", *Production and Operations Management*, Vol. 21, No. 6, pp. 1042-1059, *http://onlinelibrary.wiley.com/doi/10.1111/j.1937-5956.2012.01327.x/pdf*.

Barde, J.-P. (2004), "Discussant commentary on political economy of implementing EPR-based policy instruments", in *Economic Aspects of Extended Producer Responsibility*, OECD Publishing, Paris, *http://dx.doi.org/10.1787/9789264105270-en*.

Bell, V. (2013), "Better by design: evolving EPR fees call for better design choices", in *Recycling Today*, May 6, *www.recyclingtoday.com/article/rt0513-extended-producer-responsibility-programs*.

Bisschop, L. (2012), "Is it all going to waste? Illegal transports of e-waste in a European trade hub", *Crime, Law and Social Change*, Vol. 58, No. 3, pp. 221-249, *http://link.springer.com/article/10.1007%2Fs10611-012-9383-0*.

British Columbia Ministry of the Environment (2006), *Recycling Regulation Guide*, edited by B.C.M. o. t. Environment. British Columbia: British Columbia Ministry of the Environment.

Bury, D.R.W. (2015), Personal communication with Bury, D.R.W., Principal, Duncan Bury Consulting, 12 February 2015.

Cahill, R., S.M. Grimes and D.C. Wilson (2011), "Review article: Extended producer responsibility for packaging wastes and WEEE – A comparison of implementation and the role of local authorities across Europe", *Waste Management and Research*, Vol. 29, No. 5, pp. 455-479, *http://wmr.sagepub.com/content/29/5/455.abstract*.

California Paint Stewardship Law (2010), "In Chapter 420, Statutes of 2010 [Huffman, AB 1343] ", State of California, United States, *www.leginfo.ca.gov/pub/09-10/bill/asm/ab_1301-1350/ab_1343_bill_20100928_chaptered.html*.

CalRecycle (2015), "Five indicted in CRV recycling fraud scheme: USD 14 million in fraud uncovered in multi-state investigation", *www.calrecycle.ca.gov/NewsRoom/2015/05May/10.htm* (accessed 3 October 2015).

Cardwell, D. and A. DePalma (2008), "Two-part plan on recycling of electronic gear passes", New York Times, New York, *www.nytimes.com/2008/03/27/nyregion/27recycle.html?_r=0*.

CCME. (2009), *Canada-wide Action Plan for Extended Producer Responsibility*, Winnipeg, Manitoba, Canada.

CENELEC (2014), "Collection, logistics & treatment requirements for WEEE", *www.cenelec.eu/dyn/www/f?p=104:110:825655051826301::::FSP_ORG_ID,FSP_PROJECT,FSP_LANG_ID:1258637,55327,25* (accessed 30 September 2014).

Ching-Wen, L. (2004), *Exploring Determinant Factors for An Extended Producer Responsibility Program in Taiwan: A Case Study of IT Productsthesis*, LUMES, Lund University, Lund, Sweden.

CM Consulting (2013), *The WEEE Report: Waste Electrical and Electronic Equipment Reuse and Recycling in Canada -2013*, Peterborough, Ontario, Canada: CM Consulting.

Darnall, N. and S. Sides (2008), "Assessing the performance of voluntary environmental programs: Does certification matter? ", *Policy Studies Journal*, Vol. 36, No. 1, pp. 95-117, *http://papers.ssrn.com/sol3/papers.cfm?abstract_id=1030622*.

Dempsey M. et al. (2010), "Individual producer responsibility: a review of practical approaches for implementing individual producer responsibility for the WEEE Directive", *INSEAD Faculty and Research Working Paper*, *www.insead.edu/facultyresearch/research/doc.cfm?did=45054*.

Dimino, R. (2015), "Personal communication with Dimino, R., Director of Public Policy", *National Association for PET Container Resources*, 23 January 2015.

Dinan, T.M. (1992), "Implementation issues for marketable permits: A case study of newsprint", *Journal of Regulatory Economics*, (Historical Archive), Vol. 4, No. 1, pp. 71-87, *http://link.springer.com/article/10.1007/BF00134220*.

Dranove, D. et al. (2012), "Brief of *Amici Curiae* Economics Professors in Federal Trade Commission v. Phoebe Putney Health System (US Supreme Court)", *Health Management Policy and Innovation*, Vol. 1, *www.hmpi.org/pdf/HMPI-%20Town,%20Phoebe%20Putney.pdf; http://papers.ssrn.com/sol3/papers.cfm?abstract_id=2153979*.

Eichstädt, T. and W. Kalhlenborn (2002), "The implementation and technological impact of Packaging and Packaging Waste Directive in Germany", in *A Directive Made to Fit: The Packaging and Packaging Waste Directive in Germany, France and Finland*, edited by W. Kahlenborn and R. Kemp, Berlin: Ecologic.

Environment Canada (2013), "Extended Producer Responsibility", *www.ec.gc.ca/gdd mw/default.asp?lang=En&n=FB8E9973-1* (accessed 9 May 2015).

EPR Club (2013), *Manifesto for successful Extended Producer Responsibility for packaging*, Brussels, Belgium, EPR Club, *www.eprclub.eu/upload/public/documents%20(public)/EPR%20Manifesto%202013%2001%2018.pdf*.

EU DG Environment (2014), *Frequently asked questions on Directive 2012/19/EU on Waste Electrical and Electronic Equipment (WEEE)* , Brussels: European Commission, Environment Directorate.

Extended Producer Responsibility Alliance (2013), *Best practices for successful EPR for packaging*, Brussels, Belgium.

Fan, K.S., C.H. Lin and T.C. Chang (2005), "Management and performance of Taiwan's waste recycling fund", *Journal of the Air and Waste Management Association*, Vol. 55, No. 5, pp. 574-582.

Ferguson, M.E. and L.B. Toktay (2006), "The effect of competition on recovery strategies", *Production and Operations Management*, Vol. 15, No. 3, pp. 351-368.

Galeano, S.F. (1996), *Extended product stewardship (EPS), a rational alternative to producer's sole responsibility*. Paper presented, Orlando, FL, United States.

Gorecki, P.K. (2014), *Competition in Extended Producer Responsibilty Schemes (Redacted): Review of the Producer Responsibility Initiative Model in Ireland: Annex to the Main Report*, edited by Economic and Social Research Institute and Department of Economics and D. Trinity College. Dublin, Ireland: Department of Environment, Community and Local Government.

Gottberg, A. et al. (2006), "Producer responsibility, waste minimisation and the WEEE Directive: Case studies in eco-design from the European lighting sector", *Science of the Total Environment*, Vol. 359, No. 1-3, pp. 38-56.

Gronewold, N. (2010), "Court showdown looms for NYC electronics recycling law", The New York Times, [New York City].

Gui, L.Y. et al. (2013), "Implementing extended producer responsibility legislation, a multi-stakeholder case analysis", *Journal of Industrial Ecology*, Vol. 17, No. 2, pp. 262-276.

Harrison, K. (1998), "Talking with the donkey: Cooperative approaches to environmental protection", *Journal of Industrial Ecology*, Vol. 2, No. 3, pp. 51-72.

Hickle, G.T. (2014a), "An examination of governance within extended producer responsibility policy regimes in North America", *Resources, Conservation and Recycling*, Vol. 92, No. 0, pp. 55-65.

Hickle, G.T. (2014b), "Moving beyond the 'patchwork': A review of strategies to promote consistency for extended producer responsibility policy in the US ", *Journal of Cleaner Production*, Vol. 64, pp. 266-276.

Hotta, Y. (2013), "Recycling policy: The sound material cycle society and 3R concepts from Japan to developing Asia", in *Waste as a Resource*, The Royal Society of Chemistry.

Hotta, Y., A. Santo and T. Otsuka (2015), "Personal communication with Hotta, Y., A. Santo and T. Otsuka", Review comments on "The role of governance in extended producer responsibility", Paris.

Huisman, J. et al. (2015), *Countering WEEE Illegal Trade (CWIT) Summary Report*, Market Assessment, Legal Analysis, Crime Analysis and Recommendations Roadma. Lyon, France.

Huisman , J. et al. (2007), *2008 Review of Directive 2002/96 on Waste Electrical and Electronic Equipment (WEEE), Final Report*, Bonn, Germany: United Nations University.

Ino, H. (2007), "Extended producer responsibility in oligopoly", *Economics Bulletin*, Vol. 17, No. 6, pp. 1-9.

Kalimo, H. et al. (2015), "What roles for which stakeholders under extended producer responsibility? ", *Review of European, Comparative and International Environmental Law*, Vol. 24, No. 1, pp. 40-57.

Kalimo, H. et al. (2012), "Greening the economy through design incentives: Allocating extended producer responsibility", *European Energy and Environmental Law Review*, Vol. 21, No. 6, pp. 274-305.

Kerr, S. (2015), "Personal communication with Kerr, S.", Review comments on "The role of government in extended producer responsibility", Paris.

Khanna, M. (2001), "Non mandatory approaches to environmental protection", *Journal of Economic Surveys*, Vol. 15, No. 3, pp. 291-324.

King, A.A. and M.J. Lenox (2000), "Industry self-regulation without sanctions: The chemical industry's responsible care program", *Academy of Management Journal*, Vol. 43, No. 4, pp. 698-716.

King County and Seattle Public Health (2013), "Overview of proposed secure medicine return rule and regulation", *www.kingcounty.gov/depts/health/board-of-health/regulations/secure-medicine/overview.aspx* (accessed 19 September 2014).

Klausner, M., W.M. Grimm, and C.T. Hendrickson (1998), "Reuse of electric motors in consumer products: Design and analysis of an electronic data log", *Journal of Industrial Ecology*, Vol. 2, No. 2, pp. 89-102.

Kleijn, R. (1999), "IN = OUT: The trivial central paradigm of MFA? ", *Journal of Industrial Ecology*, Vol. 3, No. 2-3, pp. 8-9.

Kunz, N. et al. (2014), "Extended producer responsibility: Stakeholder concerns and future developments", *White Paper*, Fontainebleau: INSEAD Social Innovation Centre.

Lange, A. (2015), "Personal communication with Lange, A.", Review comments on "The role of government in extended producer responsibility", Paris.

Lifset, R. (2014), "Extended producer responsibility: Insights from the academic literature", presentation at Global Forum on Environment: Promoting Sustainable Materials Management through EPR. Tokyo, Japan, 17-19 June 2014, Organization for Economic Co-operation and Development (OECD).

Lifset, R. (1993), "Take it back: Extended producer responsibility as a form of incentive-based environmental policy", *The Journal of Resource Management and Technology*, Vol. 21, No, 4, pp. 163-175.

Lindhqvist, T. (2000), *Extended Producer Responsibility in Cleaner Production: Policy Principle to Promote Environmental Improvements of Product Systemsthesis*, The International Institute for Industrial Environmental Economics, Lund University, Lund, Sweden.

Lindhqvist, T. and R. Lifset (1998), "A glimmer of success: EPR and the electronic data log", *Journal of Industrial Ecology*, Vol. 2, No. 4, pp. 10-12.

Lindhqvist, T. and R. Lifset (1997), "What's in a name: Producer or product responsibility? ", *Journal of Industrial Ecology*, Vol. 1, No. 2, pp. 6-7.

Manhart, A. (2011), "International co-operation for metal recycling from waste electrical and electronic equipment", *Journal of Industrial Ecology*, Vol. 15, No. 1, pp. 13-30.

Manomaivibool, P. (2009), "Extended producer responsibility in a non-OECD context: The management of waste electrical and electronic equipment in India", *Resources Conservation and Recycling*, Vol. 53, No. 3, pp. 136-144.

Manomaivibool, P. and S. Vassanadumrongdee (2011), "Extended producer responsibility in Thailand.Prospects for policies on waste electrical and electronic equipment", *Journal of Industrial Ecology*, Vol. 15, No. 2, pp. 185-205.

Manomaivibool, P. and S. Vassanadumrongdee (2012), "Buying back household waste electrical and electronic equipment: Assessing Thailand's proposed policy in light of past disposal behavior and future preferences", *Resources, Conservation and Recycling*, Vol. 68, pp. 117-125.

Mayers, C.K. (2007), "Strategic, financial, and design implications of extended producer responsibility in Europe – A producer case study", *Journal of Industrial Ecology*, Vol. 11, No. 3, pp. 113-131.

Mayers, K. et al. (2013), "Implementing individual producer responsibility for waste electrical and electronic equipment through improved financing", *Journal of Industrial Ecology*, Vol. 17, No. 2, pp.186-198.

Metzner, T. (2015), "Personal communication with Metzner, T.", Connecticut Department of Energy and Environmental Protection, 7 April 2015.

Miller, T.R. et al. (2012), *Characterizing Transboundary Flows of Used Electronics: Summary Report*, Cambridge, MA: Material Systems Laboratory, Massachusetts Institute of Technology.

Monier, V., M. Hestin and J. Cavé (2014a), *WEEE in Denmark. in Development of Guidance on Extended Producer Responsibility (EPR)*, European Union, Environment Directorate, Brussels, Belgium.

Monier, V., M. Hestin and J. Cavé (2014b), "Batteries in Denmark" *in Development of Guidance on Extended Producer Responsibility (EPR)*, European Union, Environment Directorate, Brussels, Belgium.

Monier, V., M. Hestin and J. Cavé (2014c), "Case study on WEEE in the United Kingdom" *in Development of Guidance on Extended Producer Responsibility (EPR)*, European Union, Environment Directorate, Brussels, Belgium.

Monier, V., M. Hestin and J. Cavé (2014d), *Development of Guidance on Extended Producer Responsibility (EPR)*, European Union, Environment Directorate, Brussels, Belgium.

Monier, V., M. Hestin and J. Cavé (2014e), "WEEE in Ireland" *in Development of Guidance on Extended Producer Responsibility (EPR)*, European Union, Environment Directorate, Brussels, Belgium.

Monier, V., M. Hestin and J. Cavé (2014f), "Case study on packaging in Austria" *in Development of Guidance on Extended Producer Responsibility (EPR)*, European Union, Environment Directorate, Brussels, Belgium.

Monier, V., M. Hestin and J. Cavé (2014g), "Packaging waste in the Czech Republic" *in Development of Guidance on Extended Producer Responsibility (EPR)*, European Union, Environment Directorate, Brussels, Belgium.

Monier, V., M. Hestin and J. Cavé (2014h), "Case study on packaging in the United Kingdom" *in Development of Guidance on Extended Producer Responsibility (EPR)*, European Union, Environment Directorate, Brussels, Belgium.

Monier, V., M. Hestin and J. Cavé (2014i), "Packaging waste in Germany" *in Development of Guidance on Extended Producer Responsibility (EPR)*, European Union, Environment Directorate, Brussels, Belgium.

Monier, V., M. Hestin and J. Cavé (2014j), "Graphic paper in Finland" *in Development of Guidance on Extended Producer Responsibility (EPR)*, European Union, Environment Directorate, Brussels, Belgium.

Monier, V. et al. (2013), *Study on The Quantification of Waste of Electrical and Electronic Equipment (WEEE) in France: Household and Similar WEEE Arising and Destinations.*

Morgenstern, R.D. and W.A. Pizer (2007), "Reality check : The nature and performance of voluntary environmental programs in the United States, Europe, and Japan", *Resources for the Future*, Washington, DC.

Murray, F.E.S. and R.H.K. Vietor (1993), *Xerox: Design for Environment. In Business Management and the Natural Environment: Cases and Text*, edited by F.L. Reinhardt and R.H.K. Vietor, South-Western College Publishing, Cincinnati.

Nahman, A. (2010), "Extended producer responsibility for packaging waste in South Africa: Current approaches and lessons learned", *Resources Conservation and Recycling*, Vol. 54, No. 3, pp. 155-162.

Nash, J. and C. Bosso (2013), "Extended Producer Responsibility in the United States Full Speed Ahead?", *Journal of Industrial Ecology*, Vol. 17, No. 2, pp. 175-185.

Nicolli, F., N. Johnstone and P. Söderholm (2012), "Resolving failures in recycling markets: The role of technological innovation", *Environmental Economics and Policy Studies*, Vol. 14, No. 3, pp. 261-288.

OECD (2014a), *Innovative Economic Instruments for Sustainable Materials Management*, ENV/EPOC/WPRPW(2014)8.

OECD (2014b), "The State of Play on Extended Producer Responsibility: Opportunities and Challenges", *Issues Paper*, Global Forum on Environment, 17-19 June 2014, *www.oecd.org/environment/waste/gfenv-extendedproducerresponsibility-june2014.htm*.

OECD (2013a), *Waste Management Services: Germany*, DAF/COMP/WP2/WD(2013)47, *www.bundeskartellamt.de/SharedDocs/Publikation/EN/Diskussions_Hintergrundpapiere/OECD_2013.10.18-Waste_Management.pdf?__blob=publicationFile&v=3*.

OECD (2013b), *What have we learned about Extended Producer Responsibility in the last decade? A survey of the recent economic EPR literature*, ENV/EPOC/WPRPW(2013)7/FINAL, *http://spot.colorado.edu/~daka9342/OECD_EPR_KO.pdf*.

OECD (2011), *Horizontal Agreements in the Environmental Context: 2010*, DAF/COMP(2010)39, *www.oecd.org/competition/cartels/49139867.pdf*.

OECD (2006), *Germany in Environmental Regulation and Competition: Series Roundtables on Competition Policy*, Paris, OECD, *www.oecd.org/regreform/sectors/37981581.pdf*.

OECD (2005), *Improving Recycling Markets*, OECD Publishing, Paris, *http://dx.doi.org/10.1787/9789264029583-en*.

OECD (2003), *Voluntary Approaches for Environmental Policy: Effectiveness, Efficiency and Usage in Policy Mixes*, OECD Publishing, Paris, *http://dx.doi.org/10.1787/9789264101784-en*.

OECD (2001), *Extended Producer Responsibility: A Guidance Manual for Governments*, OECD Publishing, Paris, *http://dx.doi.org/10.1787/9789264189867-en*.

Paben, J. (2015), "California beverage container program budget cuts off the table", *http://resource-recycling.com/node/5876* (accessed 3 October 2015).

Parson, E.A. and E.L. Kravitz (2013), "Market instruments for the sustainability transition", *Annual Review of Environment and Resources*, Vol. 38, No. 1, pp. 415-440.

Philip Lee (2014), *Corporate Governance Report. In Review of the Producer Responsibility Initiative Model in Ireland*, Department of Environment, Community and Local Government, Dublin, Ireland.

Philipson T.J. and R.A. Posner (2009), "Antitrust in the not-for-profit sector," *Journal of Law and Economics*, 52:1, pp. 1-18.

Potoski, M. and A. Prakash (2013), "Green Clubs: Collective action and voluntary environmental programs", *Annual Review of Political Science*, Vol. 16, No. 1, pp. 399-419.

Province of British Columbia (2004), *Recycling Regulation. in Environmental Management Act*, edited by M. o. t. Environment, Queen's Printer, Victoria, British Columbia.

Quinn, L. and A.J. Sinclair (2006), "Policy challenges to implementing extended producer responsibility for packaging", *Canadian Public Administration-Administration Publique du Canada*, Vol. 49, No. 1, pp. 60-79.

Quoden, J. (2014), "Experience from packaging recovery organizations in Europe", presentation at Global Forum on Environment: Promoting Sustainable Materials Management through EPR. Tokyo, Japan, 17-19 June 2014, Organization for Economic Co-operation and Development (OECD).

Quoden, J. (2015), "Personal communication with Quoden, J.", Managing Director, Extended Producer Responsibility Alliance.

Reinhardt, F.L. (2000), *Down to Earth: Applying Business Principles to Environmental Management*, Harvard Business School Press, Boston, Mass.

Renckens, S. (2008), "Yes, we will! Voluntarism in US E-waste governance", *Review of European Community and International Environmental Law*, Vol. 17, No. 3, pp. 286-299.

Republic of Lithuania (2012), *Law on the Management of Packaging and Packaging Waste*, in No. IX-517. Vilnius: Republic of Lithuania.

Røine, K. and C.Y. Lee (2006), "With a little help from EPR? Technological change and innovation in the Norwegian plastic packaging and electronics sectors", *Journal of Industrial Ecology*, Vol. 10, No. 1-2, pp. 217-237.

RPS, ESRI, B.I. Service and Philip Lee Solicitors (2014), *Review of the Producer Responsibility Initiative Model in Ireland: Main Report (Redacted)*, edited by Trinity College Economic and Social Research Institute and Department of Economics, Department of Environment, Community and Local Government, Dublin, Ireland

SAIC Energy, E.I., LLC (2012a), *Evaluation of Extended Producer Responsibility for Consumer Packaging: Final Report*, Grocery Manufacturers Association.

SAIC Energy, E.I., LLC (2012b), *Extended Producer Responsibility Evaluation*, Product Management Alliance.

Salmons, R. (2002), "A new area for application of tradeable permits: Solid waste management", *Implementing Domestic Tradeable Permits: Recent Developments and Future Challenges*, OECD Publishing, Paris, *http://dx.doi.org/10.1787/9789264191983-en*.

Sander, K. et al. (2007), *Review of the WEEE-Directive: The Producer Responsibility Principle of Directive 2002/96/EC on Waste Electrical and Electronic Equipment (WEEE) DG ENV*, Study Contract N° 07010401/2006/449269/MAR/G4. Ökopol GmbH.

Scarlett, L. et al. (1997), *Packaging, Recycling, and Solid Waste*, Reason Foundation, Los Angeles, California, United States.

Sirena, H. (2013), "Is EPR harmonization sufficient/necessary? Why EPR guidelines at EU level are needed" in *EPR Revolution in the European Context: On the Future of Extended Producer Responsibility Systems in the Waste Management Field*, Brussels: ACR+.

Spasova, B. (2014), "Competition among PROs and Role of Municipalities in an EPR system: Case study of EPR for household packaging in Belgium", Germany and Austria thesis, International Institute for Industrial Environment Economics (IIIEE), Lund University, Lund, Sweden.

Stiftung Elektro-Altgeräte Register (2014), *www.stiftung-ear.de/en/*.

Tietenberg, T. (2007), "Tradable permits in principle and practice" in *Moving to Markets in Environmental Regulation: Lessons from Twenty Years of Experience*, edited by J. Freeman and C.D. Kolstad, Oxford University Press , New York..

Toffel, M.W. (2004), "Strategic management of product recovery", *California Management Review*, Vol. 46, No. 2, pp. 21.

Tojo, N. (2004), "Extended producer responsibility as a driver for design change: Utopia or reality?", Ph.D. thesis, Lunds Universitet (Sweden), Ann Arbor.

Tong, X. (2014), "Personal communication with Tong, X.", Associate Professor, Peking University, 26 December 2014.

Tong, X. and L. Yan (2013), "From legal transplants to sustainable transition: Extended producer responsibility in Chinese Waste Electrical and Electronic Equipment Management Tong and Yan EPR in Chinese WEEE Management", *Journal of Industrial Ecology*, Vol. 17, No. 2, pp. 199-212.

Upstream, Product Stewardship Institute, and California Product Stewardship Council (2012), *Product Stewardship and Extended Producer Responsibility: Definitions and Principles*, Athens, GA, United States: Upstream; Product Stewardship Institute; California Product Stewardship Council.

Valiante, U. (2015), "The relative economic efficiency of shared responsibility versus the principled application of extended producer responsibility for printed paper and packaging", *Policy Brief*, Corporate Policy Group.

Varga, O. (2015), *Existing and Upcoming E-waste Legislations in Latin American Countries a Challenge for Producers of IT Equipment*, 1cc Compliance Consulting.

Vergara, S.E., A. Damgaard and D. Gomez (2015), "The efficiency of informality: Quantifying greenhouse gas reductions from informal recycling in Bogotá, Colombia", *Journal of Industrial Ecology*: n/a-n/a.

Vermont Agency of Natural Resources (2014), *2014 E-Cycles Report to the Legislature*, Vermont Agency for Natural Resources.

Vermont Agency of Natural Resources (2013), *2013 E-Cycles Report to the Legislature*, Vermont Agency for Natural Resources.

Waste Diversion Ontario (2010), *Data Requirements for Monitoring Effectiveness and Efficiency of Waste Diversion Programs in Ontario: Program Targets and Reporting*, Waste Diversion Ontario, Ontario, Canada.

Waste Reduction and Management Division (2014), "Promoting Sustainable Materials Management Through Extended Producer Responsibility", in *Canadian Waste Electrical and Electronic Equipment (WEEE) Case Study*, Environment Canada.

WEEELABEX (2013), *Final Report: WEEELABEX*. LIFE Project Number: LIFE07 ENV/B/000041,WEEELABEX.

Wollman, H., Partner, Schönherr Rechtsanwälte gmbh (2015), "Personal communication with Wollman, H., Partner, Schönherr Rechtsanwälte gmbh", 13 October 2015.

Yu, J. et al. (2010a), "Managing e-waste in China: Policies, pilot projects and alternative approaches", *Resources, Conservation and Recycling*, Vol. 54, No. 11, pp. 991-999.

Yu, J. et al. (2010b), "Forecasting global generation of obsolete personal computers", *Environmental Science and Technology*, Vol. 44, No. 9, pp. 3232-3237.

Zarrehparvar, M. (2012), *California Architectural Paint Stewardship Program Plan*, PaintCare, Inc., Washington, DC.

Website:

www.weeeblackbox.ie/website/faq.html.

Extended Producer Responsibility
Updated Guidance for Efficient Waste Management

PART II

Chapter 4

Competition and extended producer responsibility

This chapter investigates the effect of EPR schemes on competition in markets. While consensus exists between different jurisdictions on how to assess these effects, there are also differences. Among other things, the chapter demonstrates widespread agreement that: i) EPR policies should be as pro-competition as possible, ii) monopoly should not be the default structure for producer responsibility organisations (PROs), iii) agreements among competitors to establish PROs should be assessed externally; iv) competition authorities should not distinguish between voluntary and government-sponsored agreements; v) waste collection, sorting and treatment services should be procured by transparent and competitive tender.

This chapter investigates the effect of EPR schemes on competition in markets. While consensus exists between different jurisdictions on how to assess these effects, there are also differences. Among other things, the chapter demonstrates widespread agreement that: i) EPR policies should be as pro-competition as possible, ii) monopoly should not be the default structure for producer responsibility organisations (PROs), iii) agreements among competitors to establish PROs should be assessed externally; iv) competition authorities should not distinguish between voluntary and government-sponsored agreements; v) waste collection, sorting and treatment services should be procured by transparent and competitive tender.

4.1. Introduction

The concept of "Extended Producer Responsibility" ("EPR") has become a widely-established principle of environmental policy towards certain products. EPR may be promoted through a range of tools. These tools may affect competition in the markets for the products themselves as well as markets for waste management. How can EPR schemes be designed both to achieve environmental objectives and to protect competition in markets?

EPR makes producers responsible for the cost of managing their products once they become waste. EPR policies have been adopted in many OECD countries for packaging waste, electrical and electronic waste, batteries, tires and end-of-life vehicles. Pharmaceuticals, furniture, and agricultural-veterinary chemical containers are other examples. A take-back requirement imposed on producers appears to be the most commonly used instrument, with advance disposal fees and deposit/refund schemes used less frequently (OECD, 2013).

Competition laws in OECD and many other countries typically aim to promote economic efficiency, often along with other objectives. The laws define and prohibit anticompetitive conduct. Some also prohibit distortion of competition by state subsidy or state grant of exclusive rights. EPR policies have been the subject of competition authorities' advocacy for more competition-friendly regulation. Companies and consortia engaged in the fulfilment of EPR have sometimes infringed competition law:

> "The provisions most commonly examined by competition authorities concern limitations on independent collection and recycling services, quotas allocating recycled product to users based on historical market shares, and exclusivity-type provisions that prevent participants from dealing with third parties, thus preventing the development of rival waste management and recycling schemes." (OECD 2010, p. 13).

The purpose of this chapter is to investigate the effects of EPR schemes on competition in markets. It follows up on the competition chapter of the 2001 OECD Guidance Manual for EPR.

The remainder of this chapter is structured as follows. The introduction describes the EPR measures, focussing on collective take-back systems. The second section briefly introduces the competition concepts that have been applied to EPR systems. The third section examines the competition issues that EPR schemes have or may generate in four markets – for the collective schemes, waste collection, waste recovery and disposal, as well as products. The last part identifies those areas where there appears to be consensus on how to address competition issues. It also identifies some areas where there are apparently differences in views. These differences can arise because of differences among competition laws, as well as among the products themselves.

4.1.1. *EPR instruments*

The competition concerns of collective take-back systems are the focus of this chapter. Two other EPR policies – advance disposal fees and deposit/refund schemes – are sometimes considered, too. A recent survey found take-back requirements to be the most common policy (72% of those EPR schemes surveyed) and applied to a wide variety of products. The other two instruments are used less frequently, 16% and 11%, respectively among those surveyed (OECD, 2013). A related concept, "product stewardship," encompasses not only systems where producers have responsibility, but also where municipalities retain their waste management responsibilities and extend them to include recycling and reuse. In this paper, the term "collective take-back systems" includes systems where municipalities perform some waste management against payment from a collective of producers subject to EPR. In general, illegal collection, trade, recovery, and disposal of waste is not a subject of this paper; illegal handling can, if less costly than complying with laws, undermine legal markets and legally compliant businesses.

A product take-back policy requires "producers" in a jurisdiction (a term which includes anyone who puts a product on the market in that country) to take back the product at the end of its life. Product take-back is often accompanied by regulations that impose targets for reuse, collection or recycling. Take-back may be organised in different ways: Producers may take back the products themselves, or organise a cooperative system for doing so, or purchase the service. An advance disposal fee is an additional fee imposed at the point of sale; funds are used for disposal costs. A deposit/refund policy entails the purchaser paying a fee at the point of sale and, if the product is subsequently brought to a collection point, the fee being refunded.

EPR policies are not mutually exclusive, e.g. producers may charge an advance disposal fee to cover the cost of a take-back obligation. For example, with respect to waste collected from households, it is not uncommon for different fractions to be subject to fees, quotas on recycling, and bans on landfilling. Indeed, a well-established result in economics is that at least as many policy instruments – such as fees, quotas and bans – are needed as there are policy objectives – such as shares of different waste to be recycled or re-used (Tinbergen, 1967). Consequently, it is neither surprising nor inherently inefficient to subject waste streams to multiple instruments.

4.1.2. *Markets and actors*

EPR schemes may affect competition in several markets. One of these is the market for the organisation of systems or solutions to fulfil EPR obligations, that is, the market for PROs. Producer responsibility organisations ("PRO") are frequently established to fulfil producers' product take-back obligations. In some cases, a PRO may be created by groups of producers or waste management companies, or it may be an independent and unrelated company. A PRO frequently procures the services of waste collectors, sorters, and treatment companies, as well as monitors the fulfilment of the contracts so as to provide proof of fulfilment of the EPR responsibilities. Thus a PRO has several groups of users:

producers, collectors, sorters, and treatment service providers. Alternatively, a PRO may procure some services, such as waste collection, from municipalities. The structure of markets for PROs differs; many are monopolies although some are oligopolies or competitive markets. PROs may or may not own the waste, a distinction that affects who gets the residual value of the waste. In some instances PROs themselves perform some waste management services, whereas in others they contract for the services. Differences between national legislation tend to make markets for PROs no larger than national.

A second set of markets are those for waste collection and sorting. These tend to be local up to national in scope, depending on what is collected and from whom. One option for collecting packaging waste subject to EPR from households is kerbside collection; a further option is to require households to deposit such waste at designated facilities. A third option is for informal but not illegal actors to collect recyclable waste. Kerbside collection of waste from households is usually a local natural monopoly in OECD countries. A natural monopoly is a market where the conditions of cost and demand imply it is cheaper for one entity, rather than two or more, to supply the market. Consequently, this kerbside collection service is often performed by a regulated private monopoly or a municipal monopoly. Collection from designated deposit facilities may exhibit different scale and density economies.

By contrast, the collection of recyclable waste from businesses tends to be subject to competition from a handful of rivals, that is, to be oligopolies. The geographic extent of these markets varies from local to at least national. E.g. markets for collecting waste lead batteries are national in Italy and Poland, local for end-of-life vehicles in the Netherlands, and provincial for recyclable materials from Dutch end-of-life vehicles. The geographic extent of markets for collection from businesses depends on a number of individual factors, including legal restrictions and transport costs.

Waste is sorted after it is collected from households. The sorting is done in relatively capital-intensive plants, which therefore enjoy scale economies. Experience suggests that the minimum efficient scale for sorting is larger than for collecting (see paragraph 106). That is, the scale at which the cost of sorting is minimised is larger than the scale at which the cost of collecting is minimised. Commercial packaging waste is usually sufficiently sorted at source and needs no further sorting. Sorted waste is often transported to a consolidation point, where the heterogeneous arriving loads are rearranged into homogenous loads and dispatched to be transported to specialised waste treatment facilities. A consolidation point enjoys economies of scope. This implies that a PRO established in one waste stream would find it easier to enter a new waste stream in the same geographic area than would a PRO without an appropriately located consolidation point.

A third set of markets are for waste recovery and disposal. The geographic extent of these markets may be national or even international, with *inter alia* legal restrictions, transport costs, and scale economies affecting their extent. For example, evidence from one case suggests that although international trade in end-of-life vehicles is restricted by legal barriers, spare car parts are increasingly traded internationally (European Commission decision No. 2002/204/EC (ARN) OJ L 68/18, points 17, 18, 72). Hazardous waste is subject to stricter international trade conditions than non-hazardous waste. The market in which secondary material is sold may also include primary raw materials: This appears to be the case for glass for containers and for lead.

The three markets account for very different shares of the cost of handling waste for recycling or reuse according to EPR. One estimate is that PRO services approximately account for 5% to 10% of the total cost of EPR, whereas collection and sorting account for 60-80%, and recovery and disposal for the rest.[1]

A fourth set of markets are the product markets, that is, the markets for tyres or cars or consumer goods within packaging. The geographic extent of markets for products such as tyres, cars and electronics are usually national if not global. Suppliers to a product market may be of vastly different sizes. Depending on the waste stream, a given PRO may serve suppliers on many different product markets.

The relationship between these markets is a focus of this chapter. Figure 4.1 places these markets in relation to each other, and to the types of businesses or entities that act in them. If monopoly is the most efficient way to organise collection, then when is competition among PROs efficient? Are monopoly PROs subject to incentives to maintain competition in the markets where they procure services? When is it efficient for companies engaged in the different activities to contract exclusively with a single trading partner? Can competition in product markets be harmed by the conduct of PROs?

Figure 4.1. **Relationship among markets and actors**

PROs with operational responsibility

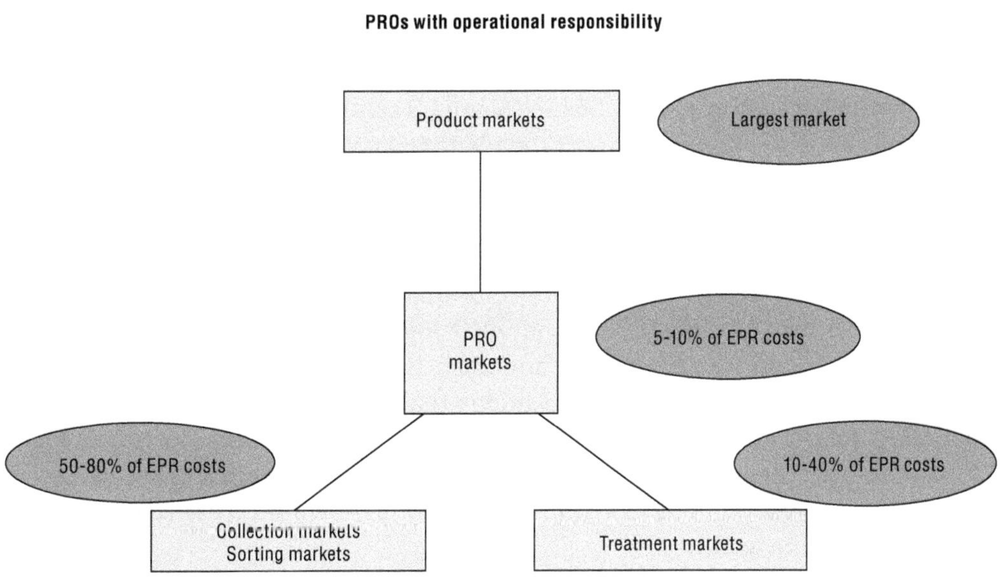

Note: In some countries collection, sorting and sometimes recycling is the responsibility of municipalities. As a consequence they decide with whom to contract for these services not the PROs.

4.1.3. *The 2001 OECD Guidance Manual*

The 2001 OECD Guidance Manual for EPR (OECD, 2001) identified a number of potential competition effects. Several of these concerned the effect on competition in product markets. First, to prevent a take-back obligation from serving as a barrier to entry into the product market, PROs need to be open to any producers of the products under a PRO's purview. Indeed, the Guidance Manual warned against not only denial of access but also discrimination because of the potential for discrimination or denial to disadvantage certain competitors in the product markets. Such disadvantage could distort the product markets. Second, the Manual identified the risk that producers could use their co-operation in the context of a PRO as a cover for collusion regarding product markets. Relatedly, producers could use the PRO as a means to pass on unnecessary costs of the EPR programme, or they could attribute excess costs to the PRO in order to raise the price of products and recoup the funds as excess dividends in their role as joint owners of the PRO.

Some of the other potential competition effects identified in the Manual concerned the market for PROs, themselves. The Manual urged governments to keep the regulatory barriers to entry into the PRO markets as low as possible, for example by refraining from giving official status to one particular PRO. As an additional measure to limit PROs' market power with respect to producers as well as with respect to sellers and buyers of collected materials, the Manual urged governments to allow for competition between PROs and producers' individual arrangements.

Another set of concerns in the Manual was the effect of PRO procurement on competition in collection services. It urged PROs to use open, competitive and fair procedures, to sign contracts that are not excessively long, and to not preference municipalities or incumbents.

The Manual also identified potential competition concerns with respect to secondary or recycled materials. One concern was the possibility that excess material may be sold at "below market values," harming competing recycled materials. A further concern was that mandates to preference local materials or that the specification to use particular materials could raise barriers to entry. The Manual also pointed out that requiring physical inspection of recycled materials can raise barriers to entry for recycled materials.

After more than a decade, the Guidance Manual's competition concerns continue to be relevant. Experience shows that some of the identified concerns did indeed come to pass, as did unanticipated competition concerns.

4.2. Brief introduction to competition concepts

Competition laws apply to the behaviour of undertakings. The laws restrict *inter alia* what agreements may be entered into and what information can be exchanged and, for undertakings in a dominant position in a market, the conduct they may engage in. The laws typically prohibit mergers and acquisitions that lead or may lead to a restriction of competition. Competition laws in the European Union, but in few other jurisdictions, have rules on state aid: These rules govern state subsidies as well as the grant of exclusive rights. The interaction of competition laws with other, e.g. environmental, laws is governed by national legal frameworks.

Most competition laws aim to prevent reductions in consumer welfare. Competition often leads to lower prices, increased variety or more innovation, each of which increases consumer welfare. Competition law aims to prohibit conduct that impedes competition, while subject to national frameworks as to how to take the objectives of other laws into account. National competition laws often have other objectives, too, and these vary from one jurisdiction to another. Although most competition assessments consider only economic costs and benefits, some competition laws include a general public interest objective, and in at least one jurisdiction this enables environmental costs and benefits to be taken into account.

Competition laws in many jurisdictions, including the European Union and United States, apply to undertakings regardless of their ultimate ownership – state, municipal or private – or profit/non-profit status. Under EU law, for example, any entity engaged in an "economic activity" is subject to competition law.[2]

The "relevant market(s)" is a fundamental concept in competition law and policy. It is a conceptual tool to structure the identification of the competitive constraints to which a firm or firms involved in an investigation are subject. When the competition concerns centre on a firm's supply – rather than purchasing – conduct, the competitive constraints include other firms that supply goods or services which customers consider to be good substitutes. Relevant markets are defined anew for each competition investigation. For the purposes of this chapter, a relevant market is a collection of products that are sufficiently close substitutes that customers switch among them and their suppliers compete. "Market" is used here, more loosely, to refer to collections of relevant markets that have something in common, such as physically similar services, even if those services are provided in different geographic areas. Also, "markets" are where transactions occur; they punctuate the physical process. Here, if it is relatively common for transactions to occur at a particular stage in the physical process, a "market" is referred to.

Certain terms recur in this chapter, and it may be useful to define them. A monopoly is a single supplier to a market. Compared with perfect competition, a monopoly will set higher prices, produce a lower output, and earn above-normal profits. It may also have less incentive to minimize costs or adopt new technology. A firm free-rides when it benefits from the actions of another without paying or sharing the costs. Free-riding could be inadvertent, e.g. when a consumer buys a product subject to EPR from a foreign supplier, but neither consumer nor supplier pays for the waste handling. One definition of a barrier to entry is a factor that prevents or deters the entry of new firms into a market despite the incumbents earning excess profits. (OECD, 1993, paragraphs 134, 91 and 14) A change in cost is with reference to an unchanging good or service; if the quality declines but nominal cost remains the same, the cost has risen.

The next three sub-sections briefly describe three aspects of competition law and policy: The assessment of agreements and of single-firm conduct, and, for European countries, state aid rules. Mergers are not addressed here since the EPR context does not seem to present unique issues and merger assessment is well described in guidelines issued by the various authorities. The final subsection briefly addresses how competition laws interact with other laws.

The third main section of this paper describes the competition concerns that have arisen or may arise in EPR schemes. Examples illustrate the application of several aspects of competition laws. They also illustrate the differences among assessments of agreements between competitors as well as in the trade-offs between environmental and competition objectives.

4.2.1. *Agreements*

Agreements are categorised by the relationship between the parties involved. Agreements among competitors – horizontal agreements – are generally considered a greater risk to competition than agreements between suppliers at different links along the supply chain – vertical agreements. Agreements between competitors about price, the allocation of markets or customers, and bid-rigging – collectively termed "hard-core cartels" – are generally presumed to be anticompetitive. The exchange of information that could help to form or monitor a cartel is also generally viewed as anticompetitive. However, an agreement among competitors may also generate economic benefits, and analytical frameworks have been developed to aid the assessment of effects on competition under these circumstances. In the competition law context, "agreement" has a broad meaning well beyond a signed document and, in some jurisdictions, includes tacit understandings.

An agreement among producers to establish a PRO to fulfil producers' EPR would usually be assessed as a joint venture to provide waste management and recycling. In contrast to hard-core cartel agreements, these agreements are assessed on a case-by-case basis, examining the facts of the individual situation to make an overall assessment. In many jurisdictions, these agreements are assessed under a two-step analytical framework, first to determine whether the agreement is anticompetitive, second to assess proponents' evidence that the benefits from the agreement outweigh the negative effects, and that these benefits could not be achieved by less anticompetitive means.[3]

An example of a two-part analysis is contained in one of the two central competition articles in the law of the European Union, Article 101 of the Treaty on the Functioning of the European Union ("TFEU"). The EU competition law is particularly relevant since many EPR agreements concern markets within the European Union. Article 101(1) prohibits anticompetitive agreements and decisions of associations of undertakings. If an agreement is found to violate Article 101(1), then Article 101(3) becomes relevant. Its purpose,

> "[I]s to determine the pro-competitive benefits produced by that agreement and to assess whether those pro-competitive effects outweigh the restrictive effects on competition. The balancing of restrictive and pro-competitive effects is conducted exclusively within the framework laid down by Article 101(3). If the pro-competitive effects do not outweigh a restriction of competition, Article 101(2) stipulates that the agreement shall be automatically void." (European Commission, 2011, para. 20, *footnotes omitted*).

Although many jurisdictions follow a two-step analysis, they differ in what benefits they take into account. For many OECD jurisdictions, the benefits must be economic, e.g. cost savings, better quality, greater variety or faster innovation, and they must accrue to the users of the good. For example, better waste management has been included in the concept of "technical or economic progress," one of the pro-competitive benefits listed in Article 101(3), in the VOTOB and DSD decisions, described below. By contrast, the competition laws of a few jurisdictions apply a public benefit standard that recognises non-economic benefits and detriments, such as environmental damage, that accrue to non-users of the good.[4] Where costs – such as restrictions of competition – are significant, this raises questions of how to measure non-economic detriments such as environmental damage, and whether the competition authority has the expertise to do so (OECD 2010).

Horizontal agreements in the United States

Two other large jurisdictions, the United States and Canada, follow different processes. An agreement among competitors is, under the United States antitrust laws, first assessed as to whether it falls into the category of agreements that are illegal *per se* (US FTC and DOJ, 2000). Agreements in this category always or almost always raise price or lower output, so do not warrant investigation. Agreements to fix prices or output, rig bids, or to allocate customers, suppliers, territories, or lines of business are examples of agreements in this category. All agreements falling outside the illegal *per se* category fall into the category of those assessed under a rule of reason. This is a factual inquiry into the agreement's overall effects on competition. The inquiry is flexible, depending on the nature of the agreement and the market circumstances.

Rule of reason analysis compares the state of competition with the agreement to the state of competition absent the agreement. The main question is: Does the agreement increase the ability or incentives profitably to raise price, lower output or quality, or delay innovation?

In the first stage of a rule-of-reason analysis, the absence of market power and the nature of the agreement can lead to the conclusion that the agreement is lawful. Market power for a seller is the ability to raise price above the competitive level for a significant period of time. Market power is unlikely if the cumulative market share of the parties to the agreement is small or if conditions of entry make it likely that a new entrant can compete effectively. Some agreements concern matters that have little effect on competition. By contrast, if harm to competition is evident from the nature of the agreement or if the agreement has already caused harm to competition, then the agreement is unlawful.

If the initial examination indicates there may be competition concerns, then the agreement is examined in greater detail. If this more-detailed examination finds that there is no potential for harm to competition, then the agreement is lawful. In the opposite case, then the question is whether the agreement is "reasonably necessary" to achieve "cognizable efficiencies." "Cognizable efficiencies" are efficiencies that have been verified by the [competition] Agencies, that do not arise from anticompetitive reductions in output or service, and that cannot be achieved through practical, significantly less restricted means" (US FTC and DOJ, 2000). "Reasonably necessary" does not imply "essential". Finally, to assess the overall competitive effect of an agreement, the magnitude and the likelihood of both the anticompetitive harms and the cognizable efficiencies are considered.

Environmental costs and benefits, like other non-competition public policy objectives, do not enter into the analysis.

Horizontal agreements in Canada

Canadian competition law divides agreements among competitors into two categories corresponding to, respectively, Sections 45 and 90.1 of the law (Canada Competition Bureau, 2009). In the first category are those that fix prices, allocate markets, or restrict output. These are illegal *per se*. However, an agreement in the first category may benefit from the ancillary restraints defence if it is directly related to, and reasonably necessary for giving effect to, a broader, lawful agreement. It need not be the least restrictive alternative to promote the objective of the broader agreement to qualify for this defence.

The second category of agreements consists of other forms of competitor collaborations. These are prohibited only if they are likely to substantially lessen or prevent competition. An agreement is assessed on the basis of a factual investigation. If it is anticompetitive, then it is illegal. However, if cost savings and other benefits from efficiency gains are "greater than and offset" any anticompetitive effects from the agreement, then the agreement is legal. Like in the United States, cost savings from reductions in output, service, quality or variety, or gains that are merely redistributive, or gains that will be attained if the agreement is prohibited or ordered to be modified, are excluded from consideration.

This brief description of how agreements among competitors are assessed under three different competition laws illustrates their subtle but important differences. Where, for example, competitors agree to charge a small fee to pay for recycling, these differences can generate different decisions on whether the agreement is legal.

Vertical agreements

Vertical agreements are assessed, under most competition laws, for their effects on consumers and competition on a case-by-case basis.[5] Vertical agreements usually facilitate better co-ordination by suppliers of complements, to the benefit of consumers. However, they may exclude, or significantly weaken competition from, rivals. In the context of PROs, a PRO may enter into a network of parallel vertical agreements with a number of other, e.g. collecting, companies. Some vertical agreements require one party to deal exclusively with the other. An exclusive agreement can give incentives for parties to make efficiency-enhancing investments. On the other hand, an exclusive agreement or network of such agreements can harm competition: This can occur if rivals need access to products that are unavailable for long periods due to the exclusive agreements. There is a greater risk of harm if a large share of the market is foreclosed, the exclusive agreements have long durations, and simultaneous entry into both markets is difficult (ICN, 2013).[6] Several cases have considered the competition effects of dominant PROs' networks of exclusive agreements.

To summarise, horizontal agreements to fix price, allocate markets or customers, or rig bids – collectively called "hard-core cartels" – are generally presumed to be anticompetitive in many competition laws. The exchange of information that can help to form or maintain a cartel is also anticompetitive. But a horizontal agreement, for example, to establish a PRO to provide waste collection and recycling that did not exist earlier, would generally be assessed on the basis of its overall effect. The case-by-case assessment of the specific facts of the agreement and its context would aim to identify the agreement's likely benefits and detriments. Jurisdictions differ in terms of which benefits and detriments can be considered. Vertical agreements are assessed on a case-by-case basis; although they usually generate cost savings to the benefit of consumers, they can nevertheless harm competition and consumers in some circumstances.

4.2.2. *Single-firm conduct*

The conduct engaged in by a single undertaking, acting alone, may violate competition laws. The terminology differs between laws e.g. "abuse of dominance" in European countries, "monopolisation" in the United States and "misuse of market power" in Australia, among others. The definitions also differ, although they all require at least that an undertaking have significant market power, which could be indicated by having a persistently high share in a market that is difficult to enter. An undertaking in a dominant position that abuses its dominance violates competition law. Whether conduct is abusive or not is determined case-by-case. Most often, abuse consists of conduct that excludes competitors or that increases the difficulty of market entry, although exploitation of customers, e.g. excessive pricing, constitutes an abuse under some competition laws. In order not to deter competition, "abuse" is not defined too broadly.

4.2.3. *State aid*

State aid, or government subsidy, as well as exclusive rights, may distort competition by allowing inefficient firms to remain in a market or to supply more to a market than they otherwise would.

Article 107(1) TFEU defines state aid as, "[A]ny aid granted by a Member State or through State resources in any form whatsoever which distorts or threatens to distort competition by favouring certain undertakings or the production of certain goods," with a requirement that it affect inter-State trade. In the context of EPR instruments, in particular advance disposal fees, a key question is whether the charge is compulsory (European Commission, 2012, para. 34-35). Even if funds are administered by a private consortium that is independent of public authorities, if the funds are financed through compulsory contributions and managed according to legislation, then they are considered to be State resources within the meaning of the state aid rules. This point is relevant in two decisions involving advance disposal fees for environmental purposes, one on meat and the other on new cars.[7]

The process by which the level of fees is determined could also affect whether an advance disposal fee constitutes state aid. The service for which compensation is received has to actually be performed, the compensation cannot exceed what is necessary to cover the costs, and the way the compensation is calculated has to be established in advance and be objective and transparent. The level of "necessary costs" may be determined through the use of a public procurement procedure or by an analysis of the costs of a well-run, typical undertaking, including reasonable profit (European Commission, 2012, para. 42-43).

A threshold, EUR 200 000 per undertaking over any three year period, can eliminate state aid concerns for very small schemes (European Commission, 2012, para. 41).

In summary, state aid rules aim to ensure that subsidies and legal monopolies do not distort competition. State aid is not generally a topic in competition laws outside the European Economic Area, which includes the EU, or EU candidate countries.

The next topic, however, arises in every jurisdiction with a competition law: How can other social objectives and competition objectives accommodate each other?

4.2.4. Competition as one of several laws

National legal frameworks govern how laws with different objectives, such as environmental and competition laws, interact. In European Union countries, the interaction with Union-level competition law is also relevant. Conflicts in a particular case can be resolved by pursuing, for example, the environmental objectives in the least anti-competitive way.

When examining an instance where, for example, environmental law affects the conduct of an undertaking, a key distinction is made between conduct that is allowed and conduct that is compelled by the environmental law. In many jurisdictions anticompetitive conduct can be shielded only if the "other" law requires the anticompetitive conduct, or precludes any other conduct.[8] Advocacy for competition during the preparation of legislation or other policy measures can influence the design of those measures to achieve the policy objectives at lower cost to competition.

An undertaking that performs a public service is, like other undertakings, subject to competition laws. Within the European Union, Article 106(2) TFEU addresses the situation of undertakings operating services of general economic interest or that have been granted a revenue-producing monopoly. Member States may not enact measures contrary to *inter alia* the competition rules. These undertakings are subject to the competition rules insofar as the application of the competition rules does not obstruct the performance of their assigned tasks. Legislation must define the obligations of the undertakings and of the authority granting the special task (European Commission, 2012, para. 51).

An Italian case illustrates the resolution of a conflict between competition and a public service law. Article 8, 2 of Law 287/90, says that the provisions of the Italian competition law,

> "...do not apply to undertakings which, by law, are entrusted with the operation of services of general economic interest or operate on the market in a monopoly situation, only insofar as this is indispensable to perform the specific tasks assigned to them."

The competition authority had found that COBAT, the consortium for collecting used lead batteries, had infringed the competition law through *inter alia* allocating quotas of used batteries to recycling companies. The First Instance Administrative Tribunal found that COBAT had been established to serve public interest objectives and thus its conduct fell under the provision cited above. Upon further appeal, the Council of State found that the restrictions on competition were not indispensable to the public interest objectives, and thus confirmed the infringement (OECD, 2010, pp. 64-5).[9]

The OECD Council Recommendation on Competition Assessment (2009) says, in the section on revision of public policies that unduly restrict competition, that "Governments should adopt the more pro-competitive alternative consistent with the public interest objectives pursued and taking into account the benefits and costs of implementation."

4.3. Experience with competition in EPR

EPR schemes can affect competition both in waste management markets and in markets for the products subject to EPR. This section describes and illustrates many of the competition issues that have, or may, arise. Following along the scheme outlined in section 4.1, markets for PROs are addressed first, followed by markets for waste collection and for waste recovery and disposal. The final section focusses on markets for products, e.g. tyres, cars, and batteries. Product markets can have very large volumes of trade, and harm to competition in these markets that raises price – or reduces quality or choice – by only a small amount can imply large losses of consumer welfare.[10]

Although not a complete inventory of EPR schemes, the extensive list compiled by Kaffine and O'Reilly (OECD, 2013) indicates that EPR schemes are used more commonly in European and North American countries than elsewhere. Of the 385 schemes listed, 167 are at a national level in Europe and 179 in Canada at the provincial or federal level or in one of states of the United States. Of the 284 take-back schemes, the corresponding figures are 144 and 115, respectively. Nevertheless, the experience with competition cases involving EPR schemes appears to be largely European, with some cases from other jurisdictions. The reason for this disparity is not clear. Reflecting this disparity, much of the discussion of experiences focuses on markets within Europe.

4.3.1. PRO *markets*

PROs fulfil producers' EPR by organising *inter alia* the necessary collection, sorting and treatment of the specified waste. A PRO may be a monopoly, or it may compete against other PROs. Producers may also organise the fulfilment of their own EPR, but this is uncommon in practice. Many PROs were originally organised as monopoly joint ventures, fulfilling EPR for all producers selling specified products into a given country. This section first addresses some of the factors that influence whether monopoly or competition are more efficient. Three arguments that are often put forward to support monopoly for PROs are that the activity enjoys economies of scale, that a monopoly makes it easier to control free-riding, and that a monopoly is easier for regulators to oversee. But monopolies are subject to diminished incentives for efficiency and, where a monopoly service is legally required, buyers who have no choice but to deal with the monopoly can be exploited. This review of the arguments for and against monopoly is based largely on theory; the empirical experience comparing competition and monopoly among PROs in similar markets is limited to the few instances where competition replaced monopoly. If competition provides better outcomes than monopoly, then high barriers to entry or high switching costs can protect a monopoly or dominant firm. These are the second and third topics of this section. If it is difficult for new PROs to enter into competition with the incumbent(s), then competition may not develop. If any type of user – producers, collectors, sorters or treatment providers – finds it difficult to switch to a different PRO from its current PRO, then competition is dampened directly and, by making entry more difficult, indirectly. Conduct that raises barriers to entry and switching has been subject to competition proceedings.

They key insights into the competition concerns that PRO markets have generated are the following:

- The first is linked to the question of whether and when a monopoly PRO is the most efficient way to organise the fulfilment of EPR. The cost of handling waste packaging in Germany fell significantly with the introduction of a number of changes including competitive bidding for collection and sorting, as well as competition among PROs and competition to supply services to PROs. Some argue that the quality fell as well. However, studies comparing PRO market structures in different countries and waste streams do not provide a clear answer, and the individual characteristics of markets likely lead to different answers. Key arguments concern scale economies, suppression of producer free-riding and regulatory oversight. A different argument – that temporary monopoly may be necessary to induce investment – may apply at the outset of an EPR scheme, particularly where future costs and revenues are very uncertain.
- Second, competition in PRO markets can be suppressed by difficult conditions of entry by rival PROs. Some of these may be structural, but others may be strategic, i.e. entry being made more difficult by the conduct of incumbent PROs. Competition investigations have identified long-term exclusive contracts with waste collectors as raising barriers to entry. Sharing of collection infrastructure has been identified as one way to make entry easier, including entry at national scale. Prohibiting contracts with collectors that are long-term or exclusive are another way.
- Third, difficulty in switching PROs can harm competition among PROs. The cost of switching can be influenced by *inter alia* vertical integration, the structure of fees, long-term exclusive vertical agreements, and non-portability of financial reserves.
- Finally, although the experience to date is limited, clearing rules can directly affect competition among PROs.

Monopoly

Whether EPR for a set of products is most efficiently implemented by a monopoly depends on a number of factors. Each case is different, but some factors to consider are whether the activity benefits from significant economies of scale, and how the number of PROs affects the cost of free-riding and cost and effectiveness of regulatory oversight. In general, the main argument against monopolies is that they are often economically inefficient: Being free of competitive pressure – the risk that users will switch to a better offer – monopolies tend to be slower to seek more user-friendly solutions or lower costs, and are under less pressure to pass cost savings on to customers. It is unfortunate that the question, under what conditions are monopoly PROs more efficient than competitive ones, cannot be answered empirically. Although data has been collected, too many important cost factors differ among waste streams and countries to answer the question.[11] In line with the Council Recommendation on Competition Assessment quoted above, many competition authorities take the view that any restriction on the establishment of multiple PROs or on new entry should be examined critically during the design phase, and if any restrictions are put into place, they should be phased out as soon as possible[12] (OECD, 2013, p. 126).

Where producers jointly establish a monopoly PRO to handle their EPR, the PRO can be viewed as a joint venture to produce an input. An input production joint venture ("IPJV") would normally be assessed under competition law on a case-by-case basis, as described above. Although there are potential cost savings, economic theory also points to potential negative effects: An IPJV may be used as a tool to enable the parent companies to charge cartel prices, or parent companies may free-ride on the efforts of others, resulting in inefficiency of the IPJV.[13]

This section finds that three arguments are often put forward to support monopoly in PRO markets, and a fourth argument in favour of temporary monopoly. These are: 1) the activity exhibits significant economies of scale compared with the size of market demand, 2) monopoly makes it cheaper to limit free-riding by producers, 3) monopoly makes regulatory oversight cheaper, and 4) temporary monopoly gives greater incentives to make risky investments.

Where there are significant scale economies, e.g. for some collection, this need not imply monopoly along the entire supply chain: Collection infrastructure may sometimes be shared, or allocated to different PROs at different periods. The amount of free-riding depends on both the incentives to engage in it and to suppress it. The benefits of free-riding would be lower if PRO fees were lower, which in theory is the case when PROs compete. Some of the methods used to suppress free-riding in PROs have been studied, but the studies did not allow a comparison of their effectiveness in monopoly versus competitive PRO markets. Regarding regulatory oversight, some direct costs may rise with more PROs, but so would the amount of information available to the regulator.

The arguments in favour of a temporary monopoly at the outset are somewhat different. If establishing a PRO incurs high sunk costs and there is uncertainty about the future costs and revenues of a PRO, then a temporary monopoly may be more efficient. Concentrating early-stage demand can reduce some of the uncertainty of a new venture. The consensus among competition authorities is that any restriction on the establishment of multiple PROs or on new entry should be examined critically during the design phase, and if any restrictions are put into place, they should be phased out as soon as possible (OECD, 2013, p. 126). In general, a monopoly not subject to economic regulation can exercise market power in the form of higher prices or lower quality, and is under less pressure to seek lower costs. These result in lower efficiency and lower consumer welfare.

Market power – monopoly and monopsony. The problem with monopolies stems from their not being subject to competitive pressure, that is, the risk that users will switch to a better offer. Even where monopolies are owned by their users, the absence of competitive pressure allows them to be inefficient (Ross and Szymanski, 2006). Where demand is obligatory, a monopolist's market power is even greater since users cannot choose to do without (see example in Box 4.1).

One consequence of weak competition is that suppliers do not pass on cost savings to buyers. Passing on any cost savings generated by the above-mentioned merger to waste-holders was considered "extremely doubtful." In addition, monopoly collecting-recycling

Box 4.1. **Bargaining position when buying obligatory service**

The weakness of waste holders' bargaining position vis-à-vis a monopoly was part of the analysis that led to the denial of a merger. The two lead collecting-and-recycling firms in Poland proposed to merge. The merger was found to be anticompetitive and denied. It was reckoned that the monopoly would have faced almost perfectly inelastic demand since the waste-holders had no alternative but to deal with the lead collector-recyclers at any price. Given this, the monopolist would have profitably raised price substantially (Acquisition of Baterpol Sp. z o.o. by Orzel Bialy S.A., 5 March 2009, cited in OECD, 2010, p. 81). Producers subject to EPR may have additional options not available to waste-holders in this market, so may be in a better bargaining position.

firms have little incentive for efficiency since they can recover any losses through higher fees. The Norwegian competition authority identified several examples where producer-importer owned collecting-recycling monopolies incurred excess costs and suggested this formed a pattern of inefficiency (OECD, 2006, p. 135). A further consequence of weak competition is that suppliers are under less pressure to adopt new, cost-saving technology. The adoption of better sorting techniques is attributed to opening of competition among German packaging PROs: "While new sorting techniques were already available during DSD's monopoly, they became widely established only after competition was introduced" (OECD, 2013b, p. 107). In practice, however, it is difficult to estimate whether technological change is inefficiently slow or rapid.

Monopsony, or single buyers (typically the case of single PROs that procure services), generates analogous efficiency problems. Compared to a situation where they could negotiate with a number of potential buyers, suppliers facing a single buyer must accept worse terms. In the extreme, monopsony low prices can lead to suppliers exiting the market.

If self-provision of EPR were a feasible alternative for producers, then the threat of self-compliance might restrict the exploitation of market power by PROs. The Swedish competition authority investigated whether an individual firm could realistically satisfy its EPR without joining a PRO. It noted that PROs tended to be controlled by "the major actors in the market." The investigation "found that the chances [of self-provision] tend to be either remote or non-existent" (OECD, 2006, p. 146). This finding suggests both that the self-compliance option places no meaningful restraint on the conduct by PROs, and that suppliers who compete against the major actors depend on their rivals for a necessary input.

In summary, a monopoly that is not subject to a real competitive threat – or regulation – can exercise market power by *inter alia* charging high prices and not tackling inefficiency. Such a monopoly is under less pressure to lower costs, to adopt cost saving technology, and to pass on any cost savings to users Similar arguments apply to single buyers, or monopsonists. While these have been more theoretical arguments, there is some empirical support. One study found that self-provision of collection-recycling was not a realistic threat (OECD, 2006, p. 146). Another study found that certain monopoly PROs incurred excess costs. (OECD, 2006, p. 135) A merger decision found that a monopoly provider of a legally required service would be able to raise prices to an extreme level (Baterpol Sp. Zoo by Orzel Bialy S.A., cited in OECD 2010, p. 81). Together, these suggest that, if a PRO is a monopoly, it will have and exercise market power. Where a monopoly PRO is owned and run by the obligated industry, then there is a risk that it be used to exercise market power by raising prices, and a risk that free-riding by individual firms within the obligated industry will reduce the PRO's efficiency.

Economies of scale. Monopoly may, however, be the least-cost structure to supply a market. If technology is stagnant and products are fairly homogeneous, and there are significant economies of scale, then the market might be supplied at least cost by a monopoly. Economies of scale means that average cost is minimised at a scale that is large compared with the size of the market. An example where scale economies might be significant would be the treatment of waste that requires highly specialised systems with onerous permitting procedures (both of which imply high fixed costs), but that arises in relatively small quantities. If the international movement of waste is restricted, then demand may be sufficiently low that these services are provided at lowest cost by a monopoly.

One potential source of economies of scale is high fixed costs. A key question is how large are the fixed costs incurred by a PRO from organising the collection, transport, and treatment of waste, and on-going monitoring costs. If these constitute a relatively small share of total costs, then this would not imply scale economies. Collection frequently enjoys substantial scale economies. Collection could therefore be a source of scale economies for a vertically integrated PRO. However, if shared access to the collection infrastructure is feasible, then the scale economies in that activity need not imply scale economies for integrated PRO services. The Swedish competition authority pointed out that separating the financing of producer responsibility from the collecting part would aid in assuring equal access to infrastructure (OECD, 2006, p. 146). Collection infrastructure has been shared in different ways.[14] Economies of scale at smaller scale can limit the feasibility of producers fulfilling their EPR obligations themselves, since individual producers usually generate smaller quantities of waste than the entire national market.

Free-riding. Reducing the cost of free-riding is another argument made in favour of monopoly PROs. Free-riding is where one firm benefits from the actions and efforts of another without paying or sharing the costs. It arises when it is difficult to exclude users who do not pay. Free-riding reduces the incentives to provide the good or service, and in the extreme may result in it being withdrawn. In the PRO context, the main concern is that producers may not pay for all of the collecting-recycling services they use. They may, for example, under-report quantities or not know the movement of their products, so not know which quantity to report in which market. In addition, producers may split their EPR among more than one PRO in a given market. Relevant factors in assessing this argument include how the structure of PRO markets affects the incentives to free-ride, the difficulty of detecting it, and the effectiveness of suppression tactics. A separate concern is that competitive PROs may shirk their obligations with respect to public information campaigns, or to inspect the quality of the sorted material. Regulatory oversight can detect the former type of free-riding, and independent inspection the latter.

Lower fees would lower incentives to free-ride, all else equal. In principle, more competition leads to lower costs and lower prices. However, as pointed out in the introduction, the effect of number of PROs on collection-recycling fees is not proven in the limited empirical literature. It is, however, argued that PROs in competitive markets have incentives to target free-riders, since these producers would incur no switching costs.

It is argued that detecting free-riding is easier if there is a single PRO. Where multiple PROs provide the same service in a market, a system is required to ensure that the total quantity of collected waste from individual producers does not exceed the total quantity reported to PROs as having been put on the market. National registers, operated by Member States of the EU, is one mechanism that has been put into place too recently to evaluate their effectiveness.

A study was carried out on the effectiveness of tactics to suppress free-riding on PRO services (Marbek, 2007). A four-case study on efforts to reduce producer free-riding in European packaging waste and WEEE take-back systems found a number of common elements: First, the programmes were mandatory, without free-riding by design: Even small producers had to pay something. Second, the programmes were operated by monopoly PROs. Third, government failure to enforce was identified as the key determinant of remaining free-riding. Closer reading of the case studies shows that members were made at least partly responsible for detecting free-riders, e.g. retailers agreed with the PRO to monitor whether suppliers had audited proof of having paid the recycling fee, and to charge them if not. Also, members were given incentives to monitor compliance, e.g. collectors were paid only for the weight of material for which a fee had been paid, and not for all material they might collect. In the case of bottle-collecting machines, they could detect whether a deposit has been paid, and did not pay out if it had not (Marbek, 2007). Whether the effectiveness of the tools mentioned – retailer monitoring of suppliers, monitoring of collected material, paying collectors only for fee-paid collected waste – would be reduced by multiple rather than monopoly PROs was not studied, since all the case studies concerned monopoly PROs.

Regulatory oversight. Ease of regulatory oversight is another argument made in favour of monopoly PROs. Regulators incur higher costs to award multiple licenses, monitor data from multiple sources, and hold multiple PROs to account, it is argued. Regulators may therefore prefer to deal with only a single entity. Experience shows, however, that regulators do not face insurmountable difficulties in licensing non-monopolies (RPS et al., 2014, pp. D-13, 14, 16). However, where the incremental cost of regulation is high, the consumer benefits from additional entry – e.g. lower costs or higher quality or quicker innovation – need to be weighed against those costs.

Regulatory oversight may be enhanced by having multiple regulated firms. In particular, the multiple sources of information represented by multiple firms (in this case PROs) may provide more information to the regulator. More information can allow better regulation.

In summary, the cost and effectiveness of regulatory oversight may be affected by the number of firms regulated. Costs of licensing and inspection may be lower with fewer firms, but the regulator would have access to more information, and perhaps different points of view, where there are more firms subject to its oversight. This can lead to better regulation.

Incentives to invest. An argument for a temporary monopoly at the outset of an EPR scheme is to give incentives to make risky, irrecoverable investments. Where there is uncertainty, the fact that sunk costs cannot be recovered if the venture turns out badly can make the initial investment uninviting. Guaranteed monopoly and, with legal obligations to buy the service, guaranteed demand reduce the risk of an unprofitable venture.

Monopoly may also be used to aggregate early-stage demand in order to exploit scale economies, which in turn gives incentives to make sunk investments. Although it does not concern a PRO, the Sydhavnens Sten and Grus decision illustrates these dynamic considerations for a new waste recovery facility. In the decision, the European Court of Justice considered whether the grant of exclusive rights to process non-hazardous building waste arising within Copenhagen violated European competition law (Case C-209/98, *Entreprenørforeningens Affalds/Miljøsektion (FFAD) v Københavns Kommune*, judgment of the ECJ of 23 May 2000, ECR [2000] I-3743). The court found that although the grant of an exclusive right led to a restriction of competition, it could be regarded as necessary for the performance of a task serving the general economic interest. (*ibid.*, point 81) The restriction enabled the chosen undertakings to receive a sufficiently large flow of waste for them to be interested in expanding the limited processing capacity. (*ibid.*, pt. 83, 79) In this case, the benefit of the restriction – the establishment of capacity to manage building waste which would not otherwise be built – was seen to outweigh the cost in terms of restricted competition for the duration.

A temporary monopoly may be efficient at the outset of an EPR programme. Relevant factors to compare costs and benefits include uncertainty, the size of sunk costs, and the existence of significant scale economies over the size of demand. Although it may be more efficient to establish a temporary monopoly, the basis for the decision should be reviewed critically at the design stage, and subsequently reviewed regularly to learn whether conditions have changed. Restrictions on competition should be removed as soon as possible.

Barriers to entry

The exercise of market power can be limited by new entrants. But if entry is not likely, or timely, or sufficient in size or scope, then the entry of new competitors cannot be relied upon to keep market prices low or quality high. Consequently, an examination of barriers to entry is very often part of a competition analysis and comments on barriers to entry are very often part of the assessment of the impact of regulation on competition in a market.

A barrier that simply slows entry, without preventing it entirely, can affect competition in a market. Entry barriers are often divided into structural and strategic. Structural entry barriers are cost and demand factors. They include economies of scale and, usually, the time and cost to meet legal requirements. A law that grants an exclusive right to serve a market is an entry barrier. Strategic barriers are barriers that have been deliberately created or enhanced by the market incumbent. They can include long-term exclusive agreements or certain pricing practices that give buyers incentives not to switch suppliers. It may be difficult to distinguish legitimate business conduct from conduct to raise barriers to rivals' entry. This subsection examines some of the conditions in PRO markets that may delay or even prevent entry.[15]

Structural barriers to entry. Sunk costs are investments that cannot be recovered once committed. Costs to meet legal requirements are an example. The riskiness of entry depends on the interaction between sunk costs and uncertainty about future market conditions. If sunk costs are high, then more uncertainty about future profits in the market makes entry less attractive.

An obligation to enter a market nationwide, increases sunk costs if the best entry strategy absent the obligation would be to enter at small scale in a limited area. A universal service obligation is often imposed to prevent new entrants from "cherry picking" the most profitable areas. Substantial experience with ensuring universal service in, e.g. postal delivery and tele-communications, shows that these obligations are often unnecessary and may protect inefficient incumbents. Changing the service obligation can considerably shrink the sunk costs of entering, which improves the prospects for entry. Among the questions to be considered in assessing a universal service obligation is whether the service, as well as the price, need be the same in high- and low-cost areas (OECD, 2004b). An example where sharing certain waste collection infrastructure lowered the cost of nationwide coverage is provided in Box 4.2.

Box 4.2. **Shared collection infrastructure**

Access to the incumbent's collection infrastructure was at issue in Sweden. Regulation required waste packaging PROs to serve the entire country. An entrant was unable to duplicate the incumbent's collection infrastructure: In rural areas, it was very costly, and in urban areas the incumbent used municipal sites that could not be duplicated. The incumbent was accused of denying access. The entrant complained to the competition authority, accusing the incumbent of abuse of dominance. After consultations with the competition authority, the parties entered into commercial negotiations that lead to a solution whereby the two firms shared the collection infrastructure at issue, and shared the costs. This enabled both PROs to offer nationwide service (Nordic Competition Authorities, 2010, pp. 51-2; Plastkretsen/FTI dnr 152/2008, decided 10 July 2009)

Subsidy to enter a market reduces the level of sunk cost. In some instances in Norway, the first PROs established in a waste stream received direct economic support or other services from the state (OECD, 2006, p. 122).

Two other structural barriers to entry that could be relevant for PROs are economies of scale and network effects. The former are entry barriers because entrants typically operate at smaller scale than incumbents, so have higher average costs than incumbents. Network effects arise when the value a user gets from a product depends not only on his own consumption but also on the number of other users who also consume it. In the case of different types of users, A and B, then the value a user of type A gets depends on the number of users of type B. An example is newspaper readers and advertisers. An entrant would need to attract not just users of type A, as it must in markets for products without network effects, but also simultaneously attract users of type B, e.g. both producers and collectors are necessary to the success of a PRO.

Strategic barriers to entry. Strategic barriers, in contrast to structural barriers, are deliberately created or enhanced by the market incumbents.

Denying entrants access to "essential facilities" or strategically increasing users' switching costs are strategic entry barriers that have featured in competition cases in the PRO market. Although the definition of "essential facilities" differs somewhat between jurisdictions, the basic idea is that there is something to which access is necessary to compete in a market, it cannot be feasibly duplicated, it can be feasibly shared, and it is controlled by a monopolist or a dominant firm. If a competitor is denied access, then the monopolist may be ordered to grant access under reasonable conditions – itself hard to define. The recognition that ordering access can diminish incentives for private investment in such facilities has limited the frequency with which mandatory access is ordered under competition laws.

Collection infrastructure has several times been found to be an essential facility for packaging PROs. Leading examples are the European Commission's decision on DSD's agreements, as well as its decisions granting exemptions for the French Eco-Emballages and for the Austrian Altstoff Recycling Austria AG ("ARA"). These decisions essentially prohibit long-term exclusive contracts between PROs and waste collectors. An exclusive contract means that one or both parties agree to deal, in a certain product, only with the other party. The decisions restrict the duration of contracts (EC DG Competition, 2005, point 81). And they prohibit any requirement that collectors send all their waste to a single PRO (Commission Decision No. 2001/837/EC *DSD* 2001 OJ L 319/1, confirmed by Case T-289/01 DSD, judgment of 24 May, 2007; Commission Decision No. 2004/208/EC *ARA, ARGEV, ARO* 2004 OJ L 75/59; Commission Decision No. 2001/663/EC Eco-Emballages, OJ 2001 L 233/37). However, the Commission's decision on DSD explicitly accepts the necessity of a countrywide network of long-term exclusive contracts for collection and sorting in order to incentivize the investments for the first-ever extensive take-back system (Commission Decision No. 2001/837/EC *DSD* 2001 OJ L 319/1, point 156).

More recently, the statement of objections sent to ARA by the European Commission in 2013 concerns *inter alia* an alleged refusal by ARA to grant access to its household collection infrastructure. This infrastructure consists of containers and bags, as well as the contracts with waste collectors and municipalities. Austrian law requires PROs to offer collection nationwide, but duplicating the infrastructure is impossible. Thus, any competitor would depend on access to ARA's infrastructure. If proven, denial of access to an essential facility would constitute an abuse of dominance (European Commission, 2013). At writing, proceedings are ongoing. In Austria, a new law would allow the nationwide coverage requirement to be met by a combination of own and shared collection infrastructure (OECD, 2010, p. 72).

Conditions of entry are important determinants of competition in markets. Some barriers erected by incumbents, such as denial of access to essential facilities, can be addressed under competition law. In some instances, this implies sharing of infrastructure or limiting the extent and duration of exclusive contracts (see Box 4.2). Regulation may inadvertently make entry more costly or time consuming or, if imposing numerical restrictions, impossible.

Switching costs

Switching costs can be a barrier to entry and directly dampen competition. Although some switching costs are inevitable, a dominant firm can make entry more difficult if it can increase the costs that suppliers of a necessary input incur when they switch customers. High switching costs can also dampen competition directly: If, for example, it is costly to switch suppliers, then other suppliers are viewed as less substitutable for a buyer's current supplier.

Several elements that raise switching costs are reviewed here. One is vertical integration, i.e. ownership links between producers, collectors or treatment providers and a PRO. Second, the structure of fees charged by a PRO has been found to raise producer switching costs. Third, operating in the same way are requirements that a producer channel all its EPR through a single PRO. Fourth, the non-portability of a PRO's financial reserves can raise the cost of producers switching. Long-term agreements that make one PRO the exclusive trading partner of waste collectors, reviewed above as a strategic barrier to entry, can also be viewed as raising collectors' switching costs.

This section finds that vertical integration can make entry by new PROs more difficult: Ownership links between producers, collectors or treatment providers and a PRO discourage them from switching their custom to a rival PRO. Also, the structure of the fees charged by a PRO, non-portability of financial reserves built up in a PRO, and requirements contracts can raise the cost of producers switching. Long-term exclusive agreements between a PRO and waste collectors can raise collectors' switching costs.

On the other hand, these ownership arrangements and agreements can have positive effects, such as encouraging investments that pay off only if the relationship is maintained. Therefore, as indicated in in the following, these types of arrangements and conduct need to be assessed case-by-case.

Vertical integration. Vertical integration between users and PROs may hinder users' switching to rival PROs. For example, the acquisition from producers by a financial investor of the German packaging PRO, DSD, was seen as freeing producers to choose the best waste solution on a purely economic basis. This promoted entry by new providers of waste solutions, as well as made DSD no longer subject to interference in favour of shareholders. Earlier, waste management companies had ended their ownership of DSD. As a consequence of the sale, the German competition authority ended a proceeding against DSD (OECD, 2006, p. 105). The Norwegian competition authority, also, has expressed the view that "splitting the waste-management value chain and differentiating the various recycling sub-markets can resolve the problems of today's waste recycling systems" (OECD, 2006, p.138).

On the other hand, producers on whom the EPR are imposed may find that joint ownership of a PRO is the best way to ensure that the organisation and infrastructure will be in place when those responsibilities have to be executed, and that producers would have the best incentives to ensure that the EPR are fulfilled in the most efficient manner. Producers may also be concerned that an independently owned PRO may exercise market power against them in the form of inefficiency and higher fees.

Whether vertical integration results in a more or less efficient provision of EPR is an empirical question. One relevant comparison is the liquidity of the market for the shares of a producer-owned PRO and the liquidity of the market for PRO services: If the capital market is less liquid, then it impedes producer switching. Not only is its effect on switching costs and therefore competition among PROs relevant, but also other effects on competition in product markets. These are addressed in a later section.

Fee structures. The structure of fees can raise switching costs. A loyalty discount or rebate means that the price depends on the quantity or proportion of purchases from the given supplier in a way that discourages buying from a different supplier. The economic incentives from a loyalty discount or rebate bind the buyer to the supplier, having the same effect as an exclusive contract.

Requirements contracts. A requirement that a producer use a single PRO within a waste stream rather than splitting the service among several PROs can make entry into the PRO market more difficult. A newly-entered PRO may be unable to provide the range of services required by a given producer. The European Commission has regarded the practice as "necessary to encourage vital investment in...collection and recycling infrastructure," but it would no longer regard it with such leniency if recovery and recycling targets had been reached (DG Competition, 2005, points 72-5).

Box 4.3. **Fee structure raised switching costs**

The structure of fees initially charged by the German packaging PRO, DSD, was found *inter alia* to raise producer switching costs to exclude rivals in the PRO market. More detail is provided in Box 4.14. The European Commission found that DSD had abused its dominant position by charging customers according to the volume of packaging bearing the Green Dot™ trademark rather than according to the volume of packaging for which DSD provided the take-back and recycling service. This discouraged producers from switching PROs or self-complying since such actions would not reduce the fee owed DSD and would increase costs, e.g., fees paid to a different PRO. DSD was ordered to modify its pricing formula so that fee were payable only on packaging benefiting from the PRO services (European Commission Decision No. 2001/463/EC (*DSD*) 2001 OJ L 166/1, points 114-116, 154).

Financial reserves. Non-portability of financial reserves has been identified by some as an impediment to competition among PROs for producers. In some jurisdictions, producers who switch PROs cannot take with them the share of the contingency reserves which they have paid in, although they do take their EPR responsibility with them to the PRO which they newly join. This imposes a high switching cost on producers. A report commissioned by the Irish environment department recommends that a switching code be developed for Ireland "to facilitate the transfer of the producer's contribution to the contingency fund from one PRO to another." It points out that, although potentially complex, other sectors such as pension funds manage such transfers. The proposal anticipates payments made by a producer who withdraws from the market would remain in the PRO to pay for treatment of the orphaned products. And it recognises that a PRO's financial reserves need to be sufficient to cover the cost of collecting and treating waste in the event the PRO ceases operation (RPS et al., 2014, pp. 61-6).

The Norwegian competition authority takes a similar position on the portability of financial reserves. The authority recommended that two features of national environmental regulations for PROs be changed: The absence of an upper limit on the size of financial reserves a PRO may accumulate, and members of a PRO having no right to take "their" share of the reserves with them when they switch PROs. Non-portability of PRO financial reserves has arisen in Norway in the case of end-of-life vehicles and WEEE (Konkurransetilsynet 2008c; Nordic Competition Authorities, 2010, p. 51).

Non-portability of financial reserves is not universally viewed as an impediment to producers' switching. An Oslo district court heard the complaint of a large WEEE producer who had switched to a new PRO without taking a share of the previous PROs' financial reserves. The producer later asked for a share of the reserves to be paid out and was denied. The court noted that the producer had indeed switched to a new PRO, and that several smaller producers had earlier departed the PROs without asking for a share of the financial reserves. The court ruled in favour of the PROs.[16]

Other competition issues

A few other competition issues have arisen in PRO markets. One is the effect of clearing rules on competition. Second is an agreement for two PROs to specialise in different areas, and not to compete to serve the same waste stream. A third issue is the possibility of predatory pricing: In this abusive strategy, prices are initially low to encourage exit but later, after rivals have exited, prices are exploitatively high.

Clearing. Where there are multiple PROs in a market, "clearing" is needed to ensure that the legal threshold for taken-back waste has been met (Bio, 2014, p. 105). Clearinghouses collect and aggregate data from the various PROs to ensure the data is fair and accurate, liaise with public enforcement authorities, and may allocate costs for, e.g. reimbursement of local authorities for help desks and provision and maintenance of areas for waste collection containers. The design of the clearing system should take into account the incentives of firms to create and exploit market power; and it should avoid becoming a tool for cartelisation through information exchange. One of the ways regulation can increase market power is for the penalty for non-compliance to be high, so that firms have little real alternative than to deal with an exploitative monopoly.

Although it has now been changed, and appears to be an isolated example, the former system of WEEE clearance in the United Kingdom illustrates how monopoly can be created in an apparently competitive PRO market. Under the former regulation, the treatment of every kilogram of WEEE collected from households had to be financed by a PRO. However, the actual obligation of each PRO was revealed only after the end of each compliance period (one year). At the end of a compliance period, the different PROs would settle up, with those in surplus selling "evidence" of compliance to those in deficit. The regulation guaranteed that there would be demand for 100% of the obligated WEEE: A PRO who did not fulfil its obligation would be subject to criminal sanctions. Thus, it was advantageous for a PRO to have access to WEEE that exceeded its forecast obligation, since it could predict with certainty that another PRO would require that surplus in order to meet its obligations. At this point, the PRO in deficit was subject to exploitative pricing by the PRO in surplus. In addition, this system dis-incentivised PROs from attracting producers/importers from their rivals (United Kingdom Department for Business, Innovation and Skills, 2013, points 21, 43, 38).[17] All in all, these regulations provided little incentive to reduce the cost of collecting and treating WEEE. The new regulations make a number of changes. First, they reduce the penalty for not meeting the obligation: A PRO in deficit at the end of a compliance year must pay a compliance fee, rather than be subject to criminal sanctions as before. Second, producer/importers may not withdraw from a PRO during the course of a compliance year (United Kingdom Department for Business, Innovation and Skills, 2014, p. 13). The first change reduces the market power of PROs who over-comply, and the second change removes disincentives to compete for producers.

Specialisation agreement. An agreement between competitors to share markets is, in general, presumed to be anticompetitive. However, an agreement between competitors to concentrate on different parts of a market may in some circumstances be viewed as a specialisation agreement that is assessed under the two-step general framework for horizontal agreements falling outside the hard-core cartel category (see Box 4.4). If an agreement generates economic efficiencies, they would result from the parties bringing together complementary skills and assets. If there are few complementary skills or assets, these efficiencies are unlikely to arise. Similarly, if there is not effective competition, for example from third parties active in the market, then any cost savings are unlikely to be passed onto consumers.

Predatory pricing. Predatory pricing is a strategy that involves charging low prices for a period in order to force small rivals from the market, and charging higher prices after the rivals exit. The definition of predatory conduct – price is just one tool that may be used – differs

Box 4.4. **Specialisation agreement**

A market sharing, or reciprocal specialisation, agreement between two WEEE PROs was authorised by the Swiss competition authority in 2005. The two PROs agreed that each would specialise in the collection and treatment of specific categories of equipment. For example, one withdrew from handling electrical appliances and the other office equipment. The authority found that this agreement may be authorised on the basis that it increased economic efficiency. In particular, it reduced transactions costs and allowed the firms to realise economies of scale. Potential competition – the threat of entry by new suppliers – was viewed as sufficient to ensure that the agreement did not suppress effective competition. Under the legislation, producers could self-comply (OECD, 2006, pp. 149-150; *Swico/Sens* DPC 2005/2, p. 251).

between jurisdictions and over time, but the essential elements are that the predator engages in some conduct – e.g. price low, expand capacity – that encourages rivals to exit or to behave meekly, and that the predator can gain extra profits – e.g. later or in other markets – to recover at least what it lost through predation. In many jurisdictions, pricing below the alleged predator's average variable cost is suspect, but the "right" test to use in a given circumstance is often debated. And barriers to re-entry or poorly-funded, easy-to-expel rivals can make recoupment of the profits more likely. The competition analysis is case-by-case and quickly becomes complex. The recognition that lower prices and more capacity benefit consumers, and that over-enforcement against predatory conduct can discourage competition means that allegations of predatory conduct must exceed a high bar in many jurisdictions. An example of predatory pricing is provided in Box 4.5.

Box 4.5. **Allegation of predatory pricing**

Predatory pricing by a PRO was alleged in the market for the organisation of collection and handling of WEEE in Norway. Ragn-Sells alleged that a rival, Elretur, engaged in predatory pricing. Elretur had built up financial reserves, and decided in autumn 2005 to reduce the reserves by temporarily cutting the fees it charged by 75%, a level it maintained for 18 months. The competition authority recognised that similar circumstances had arisen in other PRO markets. The authority pointed out to the pollution control authority that its regulations, including those related to financial reserves, could restrict competition, but took no action in the individual cases (Konkurransetilsynet, 2008a, 2008c).

4.3.2. *Waste collection services markets*

Markets for waste collection and waste sorting range from geographically small up to national in scope. Collection is done differently for different waste streams in different places. Recyclable waste from households may be collected from the kerbside by a formal operator or by informal but not illegal actors, or it may be deposited by households at designated facilities. Collecting involves visiting pick-up points and transport to a sorting facility, or if sufficiently sorted at collection, to a consolidation point. Thus the density of the pick-up points and transport costs are important in determining the geographic extent of these markets. Legal restrictions on trade in waste can prevent these markets from extending across national borders.

Some collecting markets are local natural monopolies. A natural monopoly is a market where the conditions of cost and demand mean it is cheaper for one, rather than two or more, entity to supply the market. In particular, the kerbside collection of waste from households is usually a local natural monopoly in OECD countries. The economic characteristics that tend to make it so are economies of population density and economies of scale. Economies of population density relates to the fall in average cost as population density changes. Higher population density allows shorter travel between pick-up points. Economies of scale means that average cost falls as quantity of output increases, over a relevant range of output. Kerbside collection of household waste is often performed by a regulated private monopoly or municipal monopoly in OECD countries. Although the economics of the collection of household packaging waste has not been studied as much as the collection of all household waste has been, the two activities have a similar number and density of collection points and the quantity per pick-up point is smaller for packaging. These characteristics suggest that kerbside collection of household packaging waste is, also, a local natural monopoly. In any case, it, too, is often done by a single entity, indeed often by the same entity as collects residual household waste.

Waste that is collected in larger quantities or requiring special handling does not exhibit as strong economics of population density. This is illustrated in Sweden where the collection of recyclable fractions of solid waste from multi-family apartment buildings is subject to competition, but collection from individual households – which involves smaller quantities at each pick-up point – is done by a municipal monopoly.

Markets for the collection of waste from businesses are generally subject to competition from a handful of rivals. This waste is usually collected from fewer collection points so does not enjoy significant economies of population density. While recyclable waste is similar to residual waste in terms of location and pick-up frequency, the former has specific reporting rules whereas the latter does not (EC, 2005, points 38-9). If anything, more reporting requirements would increase fixed costs and therefore tend to reduce the number of competitors.

Two key insights that have been gained from recent experience are:

- Economic characteristics determine whether a market for waste collection is, or could be competitive, or whether it is a natural monopoly. Economies of population density and economies of scale imply that some collection markets, e.g. kerbside collection of waste from households, are usually local natural monopolies. In this situation, these markets are most efficiently served by a single entity. But many other collection and sorting markets tend not to have these characteristics and therefore tend not to be natural monopolies. The geographic extent of collection and sorting markets depends *inter alia* on transport costs and on legal restrictions on trade in waste. They vary from local to national.
- Many PROs procure waste collection and sorting services. Procurement by fair and competitive tenders can result in provision by the most efficient provider at a cost that reflects no excess profits. There is evidence that the use of competitive tenders significantly reduces collection costs. However, discriminatory tenders, tenders with inappropriate duration, or tenders that do not attract enough qualified bidders, do not have these efficient results. Even weak bidders can strengthen competition in tenders. Changes in the tender rules and procedures can attract more potential bidders. Consequently, the tender rules and procedures used by PROs can have an important impact on the cost of services they procure.[18]

Markets can be local, national or international, depending on the waste type

The geographic extent of markets for the collection of recyclable waste from businesses varies. Markets to collect commercial packaging waste were thought likely to be regional or national (EC, 2005, point 45). Markets for the collection of waste lead batteries were found to be national in both Italy and Poland (OECD, 2010, pp. 64, 81). Evidence from a Dutch case suggests that markets for the collection of end-of-life vehicles are local, but for the collection of recyclable materials from car wreck dismantlers are provincial.[19] Others suggest that WEEE markets are regional or national, limited by the cost of transport to consolidation points.

Sorting seems to operate at larger scale than collection

Waste is sorted after it is collected from households. The plants exhibit economies of scale, and the costs of getting planning permission further increase scale economies (United Kingdom Office of Fair Trading, 2006, p. 58; EC 2005, points 39, 40). The experience of the German DSD in procuring packaging waste sorting and collection both bundled and separately, in 2003 and 2004 respectively, suggests that the minimum efficient scale – the scale at which average unit cost is lowest – for sorting is larger than for collecting. Commercial packaging waste is often pre-sorted at source, so need not be further sorted (DG Competition, 2005, point 40). Transport costs influence the geographic extent of markets. The trade-off between transport costs and scale economies plays an important role in determining the number of competitors.

Competition can be induced through competitive tendering of concessions

Even where it is a monopoly, waste collection may be subject to competition "for" the market. That is, there may be competition for the right to be the monopoly supplier for a specific period of time. Often, the competition takes the form of a formal bidding contest, where different firms submit bids and the winner is the one who makes the best offer. In the best case scenario, the temporary monopoly is won by the firm that can provide the service at least cost and consumers do not pay excessive prices. Competition "for" the market may lead to efficient provision if several conditions hold. Among these are that the procurement be fair and competitive, which means *inter alia* that incumbents are not advantaged, potential bidders receive the same information at the same time, with sufficient time to prepare their bids, and that a sufficient number of potential bidders submit bids.[20] The extent and duration of the contract are other factors that affect efficiency. Of course, the efficiency of awarding a temporary monopoly by competitive bidding is undermined by bid-rigging, or if excessive formation of bidding consortia reduces the number of independent, qualified bidders.

A PRO may procure waste collection services by holding a tender. ARN, the Dutch end-of-life vehicles PRO, held such tenders and contracted with a single collector per province. In choosing the winner, ARN took into account not only the collection price bid but also the technical quality and organisational suitability of the bidder. Bidders had to qualify to enter the tender, for example, they had to hold an environmental transport license (European Commission decision No. 2002/204/EC (ARN) OJ L 68/18, point 17).

Competition for the temporary monopolies can lead to significant cost savings. The introduction of competitive tendering is one change to which the dramatic fall in the cost of collecting and recycling packaging waste in Germany is attributed. Previously, DSD had procured collection and sorting services in a bundle and without competitive tenders. In 2003, DSD began to procure collection and sorting in separate competitive tenders (OECD, 2013, p. 104). The amount it paid for certain collection and sorting fell immediately by more than 20% (DG Competition, 2005, point 81). Presently, each collection area has a single collector for a three-year term which is chosen by competitive tender. Since there are now several packaging PROs in Germany, the waste is shared amongst the PROs in accordance with the quotas for which they have been contracted by producers (OECD, 2013, p. 104).

Alternatively, municipalities often have a monopoly to collect or collect and sort recyclable waste from households. As a consequence, PROs can then not hold competitive tenders, but must instead negotiate with municipalities regarding the terms for collection and sorting. To limit municipalities' exercise of monopsony power, the price at which municipalities perform this service is sometimes regulated. One form of regulation used in some markets is the use of "reference cost," a variation of "yardstick competition".[21] Belgium uses a reference cost system for waste batteries and oils. (Bio Intelligence Service, 2014, pp. 92-3) A reference cost is established by examining the cost of collection in municipalities across the country and then applying this cost to the collection contracts that PROs have with municipalities. Yardstick competition can provide incentives for efficiency since cost reductions yield greater revenues. However, heterogeneity among regulated entities can be difficult to take into account, and it can be difficult to prevent reductions in quality that are not measured or not measurable.

Discrimination in the tender process can distort competition

Discrimination in the tender process may distort competition to provide waste collection services to a PRO. The Spanish competition authority authorised ECOVIDRIO, the glass packaging PRO, in 2005 subject to it abiding by certain conditions. One of these conditions was that it must procure collection, sorting, recovery and disposal services through competitive bidding, and apply objective, transparent and non-discriminatory conditions. In 2010, the authority found that ECOVIDRIO had faked the competitive bids for collection and treatment services, favouring associated undertakings and managing to exclude at least one competitor from the glass packaging collecting market (OECD, 2010, p. 85).

The incentive to discriminate among waste collectors can be eliminated by excluding such firms from owning a PRO where they engage in collecting. In Germany, for example, after the competition authority announced it would no longer tolerate the restrictive agreements within the packaging system, DSD, waste management companies withdrew from partial ownership in 2003. DSD was sold to a financial owner a few years later (OECD, 2010, p. 53).

Non-discrimination between private and publicly-owned firms, or "competitive neutrality," is important where firms of the two ownership types compete against one another (OECD 2009b, pp. 35-42). There is a risk that a less-efficient publicly-owned firm may outbid a more-efficient privately-owned firm. A publicly-owned company has a lower cost of capital – since lenders realise it cannot be declared bankrupt – and it can cover any revenue shortfall from general tax funds. In the case of collection services, it may also have been granted a legal monopoly from which, if oversight is poor, it may cover revenue shortfalls arising from a competitive market. These advantages allow a publicly-owned firm to bid lower and thus can discourage equally efficient private companies from bidding against a publicly-owned firm.

The duration of contracts plays an important role

The duration of contracts between waste collectors and PROs can affect competition in both markets. The length of time required to recover sunk costs should affect the duration of collection contracts. Too short, and prices are higher to enable the costs to be recovered more quickly. Too long, and some of the benefits of competition, e.g. the adoption of more efficient technology, are lost. Further, future competition is harmed if those waste collectors who do not win a PRO contract in one year find it difficult to "survive" and be a viable bidder in the next procurement opportunity. In its decisions on DSD and ARA, the European Commission decided that a duration of three years for contracts between waste packaging collectors and PROs was indispensable (DG Competition, 2005, points 80, 81).

The number of bidders affects competition for concessions

Attracting a sufficient number of bidders is important to the success of a competitive tender. Where there are only a handful of potential bidders, then the loss of a single one has a substantial effect on the outcome of a competitive tender, even where the excluded bidder is a weak one (OECD, 2006b, pp. 32-4). Bidders can be discouraged if two separable services are bundled together or if contracts are large. An example can be bundling collection and sorting. If they are procured together, rather than separately, then only bidders who can perform the two services – or perform one and buy the other – may bid. If one service has significantly larger minimum efficient scale, then there would in general be more potential bidders if the two services were tendered separately (see example in Box 4.6).

Box 4.6. **Separate procurement of collection and sorting yielded more competition**

An example of how changes in tenders can attract more bidders is provided by the German packaging PRO, DSD. Its first call for tenders in 2003 did not result in much real competition in many contracting areas. Indeed, in around half the contracting areas, only one bid was received. In contracting areas where a single bid was received, prices averaged 70% higher than the lowest price offered in contract areas where two or more bids were submitted (Bundeskartellamt, 2003). DSD modified the tender conditions to attract bids from more small- and medium-sized companies, and in 2004 it succeeded. By 2005, DSD had reduced its costs of collecting and sorting by more than 20% as compared with 2003 (OECD, 2007b, p. 98). More recently, the German competition authority has expressed the view that "ensuring the separate tendering of collection services is particularly important. This safeguards that competition in sorting are not distorted" (OECD 2013, p. 107).

4.3.3. *Waste recovery and disposal markets*

Competition concerns have arisen in the markets for waste recovery and disposal services. There can be concentration in the markets, making it profitable to charge high prices. In addition, the practice of allocating waste according to historical product market share can harm competition in product markets. Purchasing cartels distort the market for waste by lowering the price below the competitive level. Both of these latter phenomena can be linked to the closed nature of some collection-and-recycling systems, where firms at the different levels of collection, sorting and treatment are obliged to deal only within the PRO. A further competition concern arose when increased collection for recycling to constituent materials could reduce the input available for producers of "recycled" or "re-conditioned" products, where the post-consumer product is re-used; re-filled printer cartridges might be an example. These competition issues are illustrated by cases that all, it turns out, concern waste with a positive market value. Concern has also been expressed that, in markets where EPR is introduced, existing small-scale waste treatment providers will be forced to exit the market if they cannot offer treatment on the scale required by PROs.

A number of insights can be drawn from recent experience. The first is that concentrated treatment markets may result in high costs of recovery and disposal, or alternatively low prices for waste to be treated. For example, a purchasing cartel is easier to organise if there are fewer competing treatment companies. Concentration combined with vertical integration into a PRO can also lead to discrimination. Charging high prices for treatment, or paying low prices for waste, reduces incentives to recycle waste beyond the legal minimum. Competitive tendering can reduce the market power of treatment companies.

The second concerns the importance of allowing trade outside the closed circle of a PRO. If collectors, for example, can deal directly with recovery firms, then those recovery firms who value the waste the highest, perhaps because they are more efficient, will gain access to more waste than under a central allocation scheme. Removing restrictions on trade in waste outside members of a PRO can be key to promoting entry by competing PROs – the new entrants thereby gain access to collected waste and to firms that can treat it. However, vertical dis-integration may require additional monitoring to ensure that the environmental targets are not undermined.

The third concern is the way waste is allocated within a PRO. Where the allocation is at below-market prices, then an administrative system based on historical market shares has the effect of retarding shifts in product market share. This dampens competition in the product market. Instead, regular competitive tenders to provide recycling services are one means to provide incentives for this service to be provided efficiently (Netherlands, OECD 2010, p. 76).

The fourth concern focuses on the competition between "new" raw materials and secondary or recycled materials. Where secondary materials are cheaper or cheaper to use, then competition from new raw materials limits the market power that can be exercised by recovery firms. But where competition occurs between re-conditioned or re-used products and new products, then the re-used products often limit the exercise of market power by new product producers. In these cases, the collection of waste to be recycled into secondary materials can soften the competitive pressure exerted by suppliers of re-used products on suppliers of new products.

Market power

Waste treatment markets can be concentrated. In Spain, for example, the competition authority concluded that "the main concern for competition stems from the ownership and access to treatment infrastructure – mainly waste recovery and disposal plants." Owners of treatment plants can charge abusive prices to undertakings that are legally obliged to manage their waste but have no access to alternative treatment facilities. Where the treatment facility owners form collective systems, they may discriminate against undertakings who choose to fulfil their EPRs through the systems (OECD, 2010, p. 87).

The use of competitive tenders can reduce the market power of waste treatment firms. A 2006 review of WEEE PROs found that those who used multiple recyclers and transport firms, chosen by competitive tender, reported that they had managed to reduce costs substantially. This contrasts with those that had chosen a single supplier (Savage et al., 2006, p. 38). The introduction of competitive tendering reportedly also contributed to the development of new recycling technologies, suggesting that large scale guarantee of demand helped to encourage investments (Veerman in OECD, 2004, p. 145).

Where small scale waste treatment providers were established prior to the introduction of EPR, its introduction can stimulate demand for large-scale, nationwide treatment solutions. PROs may see a trade-off between the lower costs resulting from greater competition in un-concentrated markets and higher costs due to non-realisation of scale economies and the requirement to contract with more providers.

Anticompetitive quotas

The allocation mechanisms used by PROs for waste glass and waste lead have been found to be anticompetitive in several cases. In these instances, valuable waste was allocated according to producers' historical market share at prices below the market value (see example in Box 4.7). The effect is to protect incumbent producers, distorting competition in the product market, as well as to hinder the development of independent collection-recycling systems (OECD, 2010, p. 13).

Box 4.7. **Anti-competitive quotas for allocation of waste for recovery and disposal**

The method employed by COBAT (Consorzio Obbligatorio Batterie al Piombo Esauste e Rifiuti Piombosi), the exclusive consortium for the collection of used lead batteries in Italy, to allocate used lead batteries among recycling companies was found to be anticompetitive. The batteries were allocated according to the productive capacity of each producer company, thereby maintaining historical market shares. If a recycling company got batteries directly from a collector, then COBAT would reduce its quota by the same amount. The Italian competition authority found that the practice restricted competition both by discouraging the development of new collection systems alternative to COBAT and by blocking recycling independent of COBAT's system (Case 1697 COBAT cited in OECD, 2013, p. 127).

Collusion in the recycling of used lead batteries was also uncovered in the COBAT case. The recycling companies regularly exchanged information about the quantities of used batteries they received from COBAT and reached an agreement about the allocation of used batteries. They also hindered attempts to develop recycling independent of COBAT's system, which deprived battery manufacturers of a less costly alternative.

Quotas assigning waste packaging containing cellulose by the consortium COMIECO, was the subject of an Italian competition authority case. Waste paper was assigned to each paper manufacturer in proportion to the amount it had put on the market in the previous year. This tended to maintain shares in the paper markets. After the investigation began, COMIECO presented commitments to assign 40% of its waste paper by competitive auction, and retain the old system for the remaining 60% so as to fulfil the consortium's public interest objectives. The auctions revealed that waste paper had a market value, which led many producers to leave the consortium to seek better prices in the market. This, in turn, led to the consortium lowering its participation fee by more than two-thirds (1730 Gestione dei rifiuti cartacei COMIECO, case closed 16 March 2011, cited in OECD, 2013, p. 128).

An earlier liberalisation of trade within the German packaging PRO, DSD, led to the entry of new PROs. Initially, recyclers received the sorted material from DSD at no cost. Under the modified system, DSD charged recyclers when the market price was positive and allowed the sale of recyclable materials outside the DSD system, provided rebates were paid to DSD. The vertical separation of collection and recycling opened up the market for competing packaging PROs (see example in Box 4.8).

Box 4.8. **Purchasing cartel**

The joint purchase of all waste glass collected in Germany was found to constitute a buying cartel and pronounced illegal. German glass container producers established Gesellschaft für Glasrecycling und Abfallvermeidung ("GGA") in 1993 to jointly purchase all waste glass collected from households from waste management companies. It organised the delivery of the waste glass to recycling plants, from which the producers received a secondary material that was cheaper, and cheaper to use, than primary raw material. The German competition authority found that the purchasing cartel eliminated competition over a substantial part of the waste glass market, and that the cartel was not necessary to the achievement of the environmental goals – the recycling target was easily overachieved – but rather they could be achieved by a less-anticompetitive means. Thus, the cartel was found to be an infringement (OECD 2013, p. 105, citing OECD 2010).

The combination of prohibitions on trade with collectors and recovery firms outside a PRO, as well as price-fixing on trade within the PRO, was found to be unnecessary to the achievement of the environmental goals for which the PRO was established in Turkey (see Box 4.9).

Box 4.9. **Exclusivity agreements and price-fixing**

In Turkey, two rival associations for collecting and recycling lead from batteries and accumulators were established. One has members who are producers accounting for about 90% of accumulator sales. The other has members who are importers. Agreements within the larger association prohibited sales of accumulators collected by its collectors to others, nor could recovery firms take accumulators from collectors from outside the association. The prices at which waste accumulators were traded within the large association were set by the association. The smaller association had no such exclusivity or price control restrictions. The Turkish competition authority found that the founders of the large association, large producers, were using the association "to secure the supply of waste accumulators at price determined by themselves." Further, the exclusivity provisions put importers at a competitive disadvantage in the accumulator market because it hindered importers in fulfilling their extended producer responsibilities. The price-fixing and exclusivity provisions were found not indispensable to the achievement of the environmental goals for which the associations were established (Accumulator decision 08-34/456-161 of 20.05.2008, OECD, 2010, pp. 90-4).

The above cases illustrate the importance of exclusivity provisions in a PRO. Restricting trade between collecting and recovery firms to those who are "in" a PRO can disadvantage non-members who cannot achieve minimum efficient scale in fulfilling their legal responsibilities, as in the Turkish accumulator case. Agreements that limit dealing with third parties may also prevent the development of alternative collection and recovery systems, whether in the form of a PRO or in the form of markets. Where trade is restricted, it does not necessarily take place at market prices. The Italian waste paper case illustrates the use of markets to "discover" the true value of waste. And the German waste glass case illustrates how trade restrictions may yield a cartel outcome. The next case raises the issue of whether take-back systems, by removing some of their input, could erode competition from manufacturers of re-conditioned products.

Impact on product market competition from reconditioned products

Box 4.10. **Competition from reconditioned products**

The Japan Fair Trade Commission engaged in a confidential consultation concerning a proposed joint collection of containers by five information equipment manufacturers. The firms make and sell consumables B for their information equipment A. The B of one company is not compatible with the A of the other companies. Third parties collect and manufacture recycled B products. The five firms proposed to place containers for the used B in post offices, sort them according to manufacturer, and return them to the respective plants for recycling into materials. The fee charged would be less than 1% of B's sales price. The firms were free to decide whether and how much of the fee to pass on. This was considered not to affect competition. A second issue was whether competition from manufacturers of recycled products would be harmed. The conclusion was that collection by third parties, which was largely done at mass retailers, would not be significantly affected (OECD, 2013, pp. 136-7).

4.3.4. *Product markets*

The formation and operation of PROs may have spill-over effects that harm competition in product markets. Product markets are large, so a small increase in inefficiency due to a lessening of competition can have large effects on economic welfare. Producers may use PROs for collusion, either as a cover for collusive meetings or by using excess fees to raise the price of products. Discrimination by PROs can distort competition in product markets. Specific spill-over effects are discussed below. First addressed is, under what circumstances is it anticompetitive for producers to agree to charge a "visible" fee for waste management?

The protection of competition in the product market is an important focus of competition policy towards EPR schemes. Cases involving PROs have identified collusion, exploitation, and discrimination/exclusion from the product market as potential or actual anticompetitive effects. Agreements to form and operate a PRO are often assessed under a general competition framework for assessing horizontal agreements that may have pro- as well as anticompetitive effects. These frameworks differ, as described later. The application of these frameworks to instances where producers agree to charge a small "visible" fee has yielded different outcomes. In some instances, the small "visible" fees are found to be necessary and not to cause a discernible harm to competition. In others, they are viewed as not necessary to the broader co-operation agreement and anticompetitive.

Other concerns regarding the effect of PROs on competition in product markets are collusion, exploitation, reduced variety, discrimination, and allocation of secondary materials that freezes changes in market share. The exchange of competitively-sensitive information within a PRO may facilitate price-fixing or market allocation. Mechanisms to avoid this have included the use of reliable independent firms to collect information and then release to each firm only the information it requires to fulfil its own obligations. If a monopoly PRO's fees are large, perhaps because the recycling of certain waste is very costly, then the scope for competition in the product market can be reduced. In other circumstances, producers can be incentivised to gravitate towards common designs. Although this could dampen competition, it may not: Consumers may be indifferent to the dimension along which variety is falling. Excess recycling fees can be used to help to raise consumer prices charged by all suppliers when they otherwise would not be able to do so. If a monopoly PRO discriminates among producers, for example in the structure of fees it charges, it may distort competition in the product market or even force some competitors to exit. The allocation of valuable post-consumer material can also distort product market competition.

A second set of competition concerns arises from the power of the state to subsidise or to grant exclusive rights. Key factors in determining whether competition is distorted by an advance disposal fee are whether it is voluntary, whether the scheme favours certain undertakings or certain goods, and whether the payment exceeds the cost of disposal.

Agreement to charge a "visible" fee

A number of factors determine whether an agreement among competing producers in a collective collection-recycling scheme to pass-on a fee violates the applicable competition law. First, does legislation specifically require passing-on a "visible" fee? If not, then agreements to pass-on a fee would be seen as illegal price fixing in some jurisdictions. In other jurisdictions, the entire collection-recycling agreement, of which the fee agreement forms a part, would be assessed within the general national frameworks outlined above. Whether an agreed fee is necessary or small has been relevant in these assessments.

An agreement to pass onto consumers the fee charged by a PRO is typically viewed as illegal price-fixing. This is the case even if making a fee "visible" is perceived as necessary to induce consumers to change their behaviour. An agreement to pass on the PRO fee reduces the scope for competition: Absent the agreement, competitors would decide individually what fraction of the PRO fee to pass on to customers.[22] However, if an agreement to pass on fees forms just one part of an agreement concerning a collection-and-recycling scheme, then the entire agreement is likely to be assessed for its overall economic effect under the general frameworks described above. From the decisions, it appears that a very small agreed fee is often viewed as foreclosing competition to a lesser degree than a large agreed fee. In some instances, other legislation requires producers be allowed to pass-on an agreed, "visible" fee. This is illustrated in an example concerning Dutch WEEE (Box 4.11).

Despite the typical view, agreements to charge a "visible" fee have been exempted in some jurisdictions where the fee was very small compared with the total price of the product. Examples are a EUR 45 fee on new cars and a five-JPY fee on plastic shopping bags, described in Box 4.12. As very small fractions of the total price, these fees were viewed as unable to discernibly affect competition in the product market.

Australian competition law allows the competition authority to authorise an anticompetitive agreement – but not an exclusionary one – if it results in a public benefit and the public benefit outweighs the public detriment from the lessening of competition.

Box 4.11. Unnecessary fee agreements denied

Batteries in the Netherlands

Provisions to pass-on the fee for battery recycling as a separate invoice item were found to be neither necessary nor beneficial to consumers by the Dutch competition authority. The obligation to recycle batteries prompted the association of Dutch battery makers and importers to establish Stibat to handle their collection and recycling obligations. Battery producers and importers paid fees to Stibat to cover the system's estimated costs. However, the proposed plan also included an obligation to pass on the fee to the next distribution stage and to show the fee on the invoice. These latter provisions were found to be unnecessary to the scheme and not beneficial to consumers (Batteries, case 51 and case 3142, cited in OECD, 2010 p. 76; OECD 1998, pp. 4-5).

Plastic in Japan

An agreement on a levy on raw materials to fund recycling would have been found to be anticompetitive under Japanese competition law. A Japanese trade association of raw material producers, processing companies, and moulding equipment manufacturers established a recycling centre. The scheme as proposed in the consultation with the Japan Fair Trade Commission was considered not to affect competition. However, according to the competition authority, if the trade association had decided to add a certain mark-up to the selling price of raw materials, then it would have constituted a price restriction under the Antimonopoly Act (OECD, 2006, p. 112).

Box 4.12. Fee agreements found to be necessary and small

Dutch end-of-life vehicles

An agreement to pass on a "visible" disposal fee of EUR 45 for end-of-life vehicles was found not to violate the Dutch competition law. Automobile companies established a collective system to recycle end-of-life vehicles before the adoption of the European directive on end-of-life vehicles. At the time, it was less costly to shred and dump end-of-life vehicles than to recycle them. The new recycling system would reverse the hierarchy, thus resulting in environmental gains. According to the assessment, the agreement would lead to the creation of a market for recycling end-of-life vehicles, and the individual parties were not able individually to create such a market. In addition, the car wreck disposal fee of EUR45 charged consumers was found to be such a small harmonisation of a cost component as not to increase co-ordination in market conduct. For these reasons, the competition authority found the agreement did not infringe the Dutch Competition Act (OECD, 2010, p. 76; European Commission 2001c, 2002).

Shopping bags in a Japanese city

An agreement to fix and charge a "visible" fee on plastic shopping bags was found not to infringe Japanese competition law. A committee composed of nearly all the retailers in a city and of resident groups agreed that consumers would be charged five-JPY for each plastic shopping bag in order to reduce their use. The city government adopted an act recommending the fee's introduction. Fixing a minimum fee was deemed necessary to prevent participating retailers from charging a lower or no fee, as had occurred after an earlier initiative. Proponents consulted the Japan Fair Trade Commission, which found that a fixed fee would be effective in attaining the goal of reducing the use of plastic bags, that an agreement was necessary to ensure the fee was sufficiently high to reach the goal, and that the fee was not unacceptably high. Also, the fee cannot be considered a burden: Plastic bags are not indispensable – shoppers may bring their own – and the resident groups agreed to it (OECD, 2013, pp. 137-8; OECD, 2010, p. 149).

In Australia, the test has been interpreted to allow the authority to take into account all detriments and benefits likely to result from the relevant conduct. The case in Box 4.13 illustrates the competition authority taking environmental benefits into account directly in a decision on whether to re-authorise a collecting and recycling scheme. This contrasts with the approach more typical in other jurisdictions, where the assessment concerns economic benefits and detriments.

Box 4.13. **Public benefit from agreement including a "visible" fee**

Farmers, suppliers of agricultural and veterinary ("agvet") chemicals, and others in Australia asked the competition authority to re-authorise a recycling scheme. The scheme imposed a levy of four cents per litre/kilogram of agvet chemicals on manufactures and suppliers, to be passed onto end-users, to finance the collection and recycling of empty agvet chemical containers. The competition authority found the levy would likely raise the price to end users of agvet chemicals, but that the levy was reasonable compared with the cost of collection and recycling. The authority compared the future with and without the scheme and found that the scheme would generate significant environmental benefits. Without the scheme, it figured that agvet chemical users would need to organise individual solutions at higher cost than the present scheme, and that the high cost may result in environmentally damaging disposal such as on-farm storage, burning and burial. The vast majority of interested parties submitting comments supported the re-authorisation of the scheme. The authority found that the public benefits "in the form of significant environmental benefits" likely outweighed the public detriment, and granted a five year authorisation (Australian Competition and Consumer Commission, 2009).

Other considerations – voluntariness and requirements of other laws. The Japan Fair Trade Commission's 2001 "Guidelines Concerning Joint Activities for Recycling under the Antimonopoly Act" points out that agreeing a specific fee for recycling, added onto the existing product price, could become a problem under the Antimonopoly Act (OECD, 2006, p. 111). A voluntary fee would not be seen as an infringement. In one instance, a Japanese trade association of office equipment manufacturers set a voluntary standard according to which members should charge a fee when collecting used products. When consulted, the competition authority viewed the standard as not infringing the competition law because the standard was voluntary (OECD, 2006, p. 112).

In contrast with the above examples, sometimes other legislation requires producers be allowed to pass-on an agreed, "visible" fee. In one instance, the Dutch competition authority exempted from the national competition law an agreement to pass on a "visible" fee for WEEE because of the 2003 European Directive on WEEE. This Directive required Member States of the European Union to allow, during a transition period, producers to show purchasers the cost of environmentally responsible disposal[23] (OECD, 2010, p.76).

However, competition laws vary, and not all can be described here. For example, the Turkish competition authority authorised the fixing and separate invoicing to consumers of a fee to cover the cost of a PRO to collect and recover end-of-life tyres. The PRO's founders together accounting for about 60% of the tyre market in Turkey. The fee on new tyres amounted to 0,6% to 0,7% of the price. The authority accepted that the fee should be invoiced separately so as avoid any collusion to increase the price of new tyres (OECD, 2013, p. 218).

Exchange of competitively-sensitive information that aids collusion

The key message in this section is that PROs and national registers require accurate information to be able to operate. In line with the general framework for assessment of agreements, the design of information processes should take into account what is necessary for the scheme to efficiently achieve its objectives as well as the possible detrimental effect on competition, such as cartelisation in the product markets.

A PRO operated jointly by producers may harm competition in the product market by facilitating the exchange of competitively-sensitive information. This exchange can facilitate collusion. The joint operation of a PRO can entail regular meetings between competitors where topics may not be restricted to the PRO but also veer into product market topics. In addition, to ensure a fair allocation of costs, a PRO may collect more detailed or timely sales information than would otherwise be available to competitors.[24]

Competition agencies consider a number of factors in assessing information exchanges within the context of a broader competitor collaboration. First, how concentrated are the markets, and what will be the extent of the information exchange? Greater market concentration and greater coverage by the information exchange mechanism are more problematic. Second, what is the nature of the information? More recent, more disaggregated information is more problematic. Third, what is the modality of the information exchange? Private information exchange is generally viewed more sceptically (OECD, 2010b).

Co-operation in a PRO may be used as a cover for collusion in the product market. The Swedish competition authority warned of the potential in 2003 (OECD, 2003b, p. 12), and a few years later cited two instances where companies in the same industries had received an exemption to co-operate on environmental issues, but also took the occasion to engage in anticompetitive co-operation (OECD, 2006, p. 147). The German competition authority found that the corporate structure of DSD, the German packaging PRO, permitted trade association representatives who set the license fees paid by trade and industry and waste disposal firms in DSD's supervisory board to co-ordinate the disposal fees (Bundeskartellamt, 2005, p. 38).

An institutional solution applied by some registers, e.g. the WEEE Register Society in Ireland, is to use a credible, independent entity to collect, collate, and communicate confidential data. Each PRO sends the required information, and gets information only about its own obligations. In this way, the information necessary to the achievement of the national register function is made available, but not more. In particular, company-specific information is not available to other companies. PROs may also use a similar institutional fix to maintain confidentiality between individual producers.

Increasing cost commonality

If a PRO significantly increases the commonality of producers' costs, this may harm competition in the product market. If a PRO is a monopoly or dominant, and if the fees constitute a substantial part of producers' variable costs, then the intensity of competition in the product market could be reduced. This potential concern was mentioned in connection to certain types of light bulbs (DG-Competition 2005, point 158 citing ELCF 2003, p. 12).[25] Where common costs account for a substantial part of variable costs, then the product market would be examined for whether it has other characteristics that make collusion feasible, "such as, for example, transparency, stability and the level of concentration" (EC 2011, point 187). If collusion appears feasible, then any advantages and detriments of a monopoly PRO would need to be weighed against the harm to competition in the product markets.

Alternatively, a PRO may harm competition in the product market by giving competitors incentives to increase the commonality of their designs. This may reduce variety – itself a dimension of competition – and raise barriers to entry. It also potentially makes collusion easier, though this risk "will largely depend on the level of homogeneity of the product in question" (Kienapfel and Miersch, 2006, p. 54). The risk to competition in variety would seem to depend on whether any harmonisation with respect to waste treatment implies harmonisation with respect to characteristics that influence consumer choice. For example, packaging harmonisation would seem not to affect competition in markets where products are chosen only on the basis of their technical characteristics. Where packaging affects buyers' choices, harmonisation with respect to waste treatment need not imply harmonisation of design. Packaging may be more or less attractive yet have the same recycling costs.

In principle, increased harmonisation of cost or design can diminish competition in the product markets. A key factor is whether common costs constitute a high share of producers' variable costs.

Excessive fees for recycling

Excess recycling fees may be used as a tool to raise prices to consumers and increase suppliers' profits. If market demand is fairly inelastic, then it is profitable to increase price if rivals can be blocked from undercutting the price. If product suppliers pass on the recycling fee to consumers in the form of higher prices, then consumers experience a price increase. In several instances in Norway, a producer-owned PRO has made lump-sum payments to owners, typically after the PRO has built up sufficient reserves. The reserves were built up because the recycling fees exceeded the costs (OECD, 2006, p. 123). Whether this conduct in fact constituted exploitative abuse of dominance was not established.

Discrimination among producers

Competition in the product market may be harmed by discriminating among producers. If some producers receive advantageous terms from a PRO, then their competitors may be weakened or even forced out of the product market. The structure of PRO fees may be discriminatory, for example against foreign suppliers or small suppliers. If the fixed costs of participating in a PRO are high, and self-compliance is infeasible, then small producers may exit and entrants be discouraged (see Box 4.14). If incumbents form the unique PRO, and the PRO imposes unfair terms that exclude competitors, then the PRO's terms may be assessed as to whether they deny access to an essential facility (OECD, 2006, p. 118).

Box 4.14. **Potential for discrimination to harm market competition**

Mineral water bottle pool not harmful

The Chambre Syndicale des Eaux de Source ou Minérales (Belgium) complained to the European Commission about conduct by the German Cooperative for Spring Water (Genossenschaft Deutscher Brunnen, "GDB"). The GDB had established a standardised packaging pool for returnable bottles. The complainant was unable to show limited or restricted access to the German market. The Commission established that German distributors did not in general object to mineral water in containers different from those in the GDB pool. It found that the cartel did not significantly harm third parties or trade between Member States, so rejected the complaint (Decision of 22 December 1987, unpublished; Commission of the European Communities, 1987, point 75).

Box 4.14. **Potential for discrimination to harm market competition** *(cont.)*

Packaging fee structure discriminatory

The initial fee structure for Duales System Deutschland AG ("DSD"), the first German packaging PRO, was found to discriminate among producers. DSD had abused its dominant position by charging customers according to the volume of packaging bearing the Green Dot trademark rather than according to the volume of packaging for which DSD provided the take-back and recycling service. The European Commission felt that this compelled producer/importers who wished to not use DSD for some of their packaging to separate their packaging and distribution channels. This would be impractical for some and disproportionate for others, and was found to be an exploitative abuse of dominance. The fee structure also harmed competition in the PRO market. DSD was ordered to modify its pricing formulae so that fees were payable only on packaging benefitting from the PRO services. Note that PROs in other countries do usually not use a trademark on packaging to indicate that a fee had been paid.

Anticompetitive allocation of secondary materials

The allocation by quota of valuable secondary material to producers at prices below the market price can distort competition in the product market. This effect was described above in connection with waste recovery and disposal markets.

State aid

State aid, or government subsidy, may also distort competition. Advance disposal fees, one of the tools of EPR, have been the subject of at least two state aid decisions where the concern centred on competition in the product market. The first is a European Commission Decision on *inter alia* the state aid aspects of the agreement, discussed above, to impose a EUR 45 disposal fee on new cars in the Netherlands. The second is a preliminary ruling from the Court of Justice on a disposal tax imposed on meat in France (Box 4.15). The decisions provide a clear structure for determining whether state aid exists.

Box 4.15. **State aid and advance disposal fees**

The European Commission investigated the 2001 agreement to pass on a EUR 45 disposal fee for end-of-life vehicles in the Netherlands to ascertain whether the disposal fee might overcompensate for the cost of collecting, dismantling and recycling. The fee was calculated to equal the average cost of dismantling a car wreck. Key points of the assessment included:

- Was the fee obligatory? Manufacturers or importers could establish individual systems or use alternative systems to fulfil their EPR, and these were real alternatives, at least for manufacturers and large importers. Therefore, the fee was voluntary or at least optional. Consequently, the proceeds of the fee were not state resources.
- Even if the proceeds of the fee had constituted state resources, the system would not have been considered as state aid because it did not favour certain undertakings nor the production of certain goods. Car makers and importers were not favoured since the charge and the payment relate to the same good, albeit at different times. Further, the system ensures that car makers and importers pay a significant part of the collection costs, consistent with the EU Directive on End-of-Life Vehicles.

Box 4.15. State aid and advance disposal fees *(cont.)*

- Although not part of the conclusion, the Commission also assessed whether the fee was higher than the minimum needed to handle the wrecks. The answer was negative, both because of the process (the system manager used competitive tenders to get the lowest possible prices for the services) and the outcome (only one dismantling company received payments exceeding its costs, and the excess was small).

The Commission concluded there was no state aid because no state resources were involved and, even if there had been, the system favoured no undertakings nor certain products (Commission Decision No. 2002/204/EC (ARN) OJ L 68/18; European Commission 2001c).

Fee on meat to fund disposal of unfit meat

The second state aid matter concerned a tax on meat to finance environmentally-appropriate disposal of dead animals. A French law imposed a tax on the purchases of meat and products, including animal feed made from meat or offal, by persons who engage in retail sales of meat and products. The purpose of the tax was to finance the collection and disposal of animal carcasses and of slaughterhouse waste that was unfit for human or animal consumption. These services are provided free of charge to the slaughterhouses and farmers. Holders of animal carcasses exceeding 40 kilograms are legally obliged to use a carcass disposal service. While it is not immediately obvious that this is an instance of an advance disposal fee, the unfit-to-consume carcasses and waste are a co-product of meat, and meat is the product that has a value available for taxation. From an economic perspective, if the waste and meat are produced in constant proportion, a tax on one is equivalent to a tax on the other.

The European Court was asked to issue a preliminary ruling on whether the scheme constituted state aid.

- Was the state's intervention to relieve holders of carcasses of the cost of disposal an economic advantage liable to distort competition? Yes, since the scheme relieved farmers and slaughterhouses of an inherent expense of their economic activities.
- Did the state measure favour "certain undertakings or the production of certain goods"? Yes, since Law No. 96-1139 essentially benefits farmers and slaughterhouses.
- Was intra-Community trade affected? Yes, since the scheme advantages French meat producers over others, who must normally pay carcass disposal costs. And foreign-reared meat sold in France was subject to the tax.

Therefore, "[A] system such as that at issue in the main proceedings, which provides farmers and slaughterhouses with the free collection and disposal of animal carcasses and slaughterhouse waste, must be classified as State aid." (Ministre de l'économie, des finances et de l'industrie and GEMO SA, Case C-126/01, 2003, ECR I-13769; see also the Opinion of Advocate General Jacobs).

Whether the fee was voluntary was an important question in the analyses. In the car wreck case, the fact that there were real alternatives to paying the fee, making it voluntary, immediately brought the scheme out of the state aid box. The second main question is whether the scheme favoured certain undertakings or the production of certain goods. In the car wreck case, the fee was imposed on the same objects as received the service, albeit at different points in time. In the meat case, the tax was placed on the sale to French retailers of meat for human or animal consumption but the proceeds were paid to French farmers and slaughterhouses for the disposal services of something different, carcasses and waste unfit for human or animal consumption. Because there was international trade in meat, the meat and the unfit-carcasses were not necessarily co-products: Foreign-reared meat was taxed when sold in France, and exported French-reared meat escaped the tax but benefited from the disposal services. This non-coincidence of tax and benefit is a key distinction from the car wreck case, and contributed to a different finding. A further key point in a state aid assessment is whether the proceeds exceed the minimum necessary to pay for the assigned tasks.

Although not all competition laws have state aid provisions, the potential for government subsidy or exclusive rights to distort competition make them topics for competition advocacy.

4.4. Key measures to address competition issues

The OECD Council has, on several occasions, addressed the relationship among different economic policies, including between competition and other policies. In addition, OECD competition authorities have discussed the application of competition law and policy to environmental agreements and more specifically to PROs, both in principle and in specific cases. Indeed, many of the cases here were presented at those discussions. It is evident that there is a degree of consensus on how to address many competition issues that have arisen, but also that there are some relevant differences between competition laws and policies. Based on these and other materials, points about which there appears to be consensus include:

- EPR policies should be as pro-competitive as possible while achieving their environmental policy goals. This is in line with the OECD Council Recommendation on Competition Assessment (2009). EPR policies should be subject to competition impact assessments, a process consistent with the Council Recommendation on Regulatory Policy and Governance (2005).[26] Competition authorities should help to formulate less anticompetitive policies. EPR policies that, *inter alia* limit the number of market participants or their incentives to compete, can reduce the intensity of competition, which imposes costs on society.
- Monopoly should not be the default market structure. While there may be a case for a monopoly PRO at the outset, the arguments for a single system should be assessed critically at the design stage. Once underway, restrictions that prevent new entry should be phased out as soon as possible. A temporary initial monopoly may be sensible if sunk costs are high and there is uncertainty about the future costs and revenues. Concentrating early-stage demand can reduce some of the uncertainty.
- Agreements among competitors to establish a PRO should be assessed under the jurisdiction's general framework for assessing horizontal agreements that may have pro- and anti-competitive effects.[27]
- Competition authorities should not, in their analyses, distinguish between purely voluntary agreements and government-sponsored agreements.[28]
- The procurement of services such as waste collection, sorting, and treatment, should be performed by transparent, non-discriminatory and competitive tenders. Relevant factors include contract duration and scale to provide incentives to invest, and scale and level of aggregation to facilitate bidding by all qualified firms. Where ownership links exist between service providers and PROs, providing an incentive for discrimination, special scrutiny may be required to ensure fair competition in the tenders. At least one

competition authority goes further, finding that separate tending of collection is particularly important to insulate competition in sorting and processing from distortion.

- Contracts between service providers and PROs should be assessed on a case-by-case basis within the general framework for assessing vertical agreements. These assessments include the benefits as well as the competition detriments. For example, the duration and recovery of sunk costs are related. One guideline,[29] for example, says that contracts between packaging waste collectors and dominant PROs exceeding three years duration are not indispensable, and that collectors and recyclers should not be obliged to contract exclusively with one PRO.
- Post-consumer materials should not be allocated in a way that raises barriers to entry or expansion in the product market. An example of an allocation system that restricts competition is one where material is allocated at below market prices according to historical product market share.[30]
- PROs, national registers or other clearing houses should avoid sharing confidential market information with users that may aid cartelisation.
- Competition authorities should issue up-to-date guidelines or provide other easily accessible guidance. National competition laws are heterogeneous and change over time. Up-to-date guidelines can help both private actors and government officials to design environmental schemes that are compatible with competition law and avoid inadvertent infringements.

There is no consensus on a few issues where there have been a sufficient number of decisions or comments to detect differences.

The general frameworks for assessing horizontal co-operation agreements differ. Many limit consideration to those benefits and detriments that are economic, and that occur within the same market as the competition harm occurs, or at least to the same set of consumers. Others consider also non-economic benefits and detriments, for example environmental ones. And they may consider benefits and detriments that occur more widely than the consumers who suffer from harmed competition.

Competition laws differ with respect to the legality of agreements fixing a small, "visible" fee for fulfilling EPR. The practice in a number of jurisdictions appears to be that fees that are both necessary and very small compared with the total price of a product would not be prosecuted. The detrimental effect on competition of agreement on a small component of total price is viewed in these analyses as small as compared with the benefit of the collection-recycling system that is made possible. By contrast, in other jurisdictions such agreements would be illegal price-fixing.[31]

Other issues have an insufficient number of decisions or comments to conclude whether or not there is consensus.

- The vertical separation of producers, waste collectors and waste treatment firms has been promoted in several jurisdictions. The experience in Germany, where first waste management companies and then trade and industry companies sold their holdings in DSD, is viewed positively by *inter alia* the German competition authority. The French competition authority has expressed its opposition to vertical integration by PROs into waste collection, sorting, recovery and disposal. The Norwegian competition authority is of the opinion that vertical dis-integration in the waste-management value chain would promote efficiency.

- An obligation to provide collection services nationwide hinders entry by new PROs. Duplication of infrastructure in areas that generate little recyclable waste is wasteful or impossible. In these instances, the incumbent may be required to grant access to its infrastructure. This was a solution in Sweden, for example, and a mixture of own infrastructure and shared infrastructure would be allowed under a new Austrian law. Where sharing is not feasible, then lifting the universal service obligation would offer a way for competition in parts of the market.
- The non-portability of financial reserves has been identified by some as an impediment to PRO competition for producers. Financial reserves of a certain level are necessary to ensure that the obliged services can be paid for. Some observers see portability as feasible and non-portability as raising the cost of producers switching to a different PRO. Others see non-portability as not in practice limiting producers switching PROs, so not anticompetitive.

Notes

1. Estimate from OECD Secretariat, based on discussions and expert judgements.
2. Case C-41/90 Höfner and Elser v Macrotron [1991] ECR I-1979 ("Höfner and Elser"), paragraph 21, cited in United Kingdom Office of Fair Trading (2011). Although, for example non-profit entities may have differing objectives, at the margin they have an interest in exploiting any ability they may have to raise prices because they use the profits thereby generated to fund their objectives. (Philipson and Posner, 2009) A review of the literature on non-profit, privately-owned hospitals in the United States found that there was no economic theory to support the presumption that non-profits will not exercise market power to the detriment of total or consumer welfare. The empirical literature showed that nearly all of them exercised market power by raising prices. (Dravone et al., 2012) The economic theory cited is not specific to hospitals, but applies equally to any non-profits that may use profits they generate to fund their activities.
3. Two guidelines that illustrate this process in Europe are the European Commission's 2011 Guidelines on Horizontal Co-operation Agreements and, with more examples specific to environmental objectives, the Dutch Authority for Consumers & Markets' 2014 Vision Document on Competition & Sustainability.
4. An Australian authorisation decision, below, illustrates a public benefit standard in use. The treatments of non-economic benefits and costs, and of indirect – not in the same market – benefits and costs, have changed over time in European competition law. The United Kingdom contribution in OECD (2010) cites several older Commission decisions where non-economic and indirect benefits were cognizable, and more recent Court of First Instance (now the General Court) decisions where indirect benefits were recognised. Compare, too, the recognition of reduced harmful emissions as a benefit in a 2000 Commission decision involving an agreement among washing machine manufacturers with the indirect but economic benefits cited in the lightly camouflaged example in the 2011 EC Horizontal Co-operation Agreements Guidelines, para. 329 (Commission Decision No. 2000/475/EC (CECED) 2000 OJ L 187/47, para. 56; European Commission, 2011).
5. This broad statement is only roughly accurate. The European Union, for example, has a block exemption for vertical agreements meeting certain criteria. The criteria were designed to relieve agreements unlikely to harm competition from a costly case-by-case assessment.
6. European Commission Notice "Guidelines on Vertical Restraints" (2010) provides guidance on how vertical agreements are assessed. Sections VI.2.2 and 2.6 concern exclusive distribution and exclusive supply agreements. The guidelines relate to Commission Regulation (EU) No 330/2010 on the application of Article 101(3) of the Treaty on the Functioning of the European Union to categories of vertical agreements and concerted practices, OJ L 102/1.
7. Despite applying to the waste sector, a 2011 study found that no Member State had notified to the Commission any state aid to waste-related services of general economic interest since 2005 (Commission Staff, 2011, pp. 17-8).
8. "It is only if anti-competitive conduct is required of companies by national legislation, or if the latter creates a legal framework which precludes all scope for competitive activity on their part,

that Article 101 does not apply. In such a situation, the restriction of competition is not attributable, as Article 101 implicitly requires, to the autonomous conduct of the companies and they are shielded from all the consequences of an infringement of that article." (European Commission, 2011, para. 22, footnotes omitted). In the United States, the so-called "state action doctrine" guides the relationship between federal antitrust laws and state regulatory schemes.

9. An introduction to the resolution of conflicts between competition and other laws is in ICN 2004.

10. Loss of consumer welfare in a market can often be approximated by the volume of commerce times the increase in unit price.

11. Considering only cost factors, "[F]ees vary because of differences in the collection system, the target, the proportion of the costs of collection, sorting and recovery accounted for by the fees and the types of collection, sorting and recovery channels covered (e.g., household, industrial and commercial). Second, collection, sorting and recovery costs are likely to be a function of landfill costs, incineration charges and so on, which are likely to vary by Member State." (RPS et al., p. D-20).

12. The Executive summary in OECD 2010 observes that, "Experience suggests that competition among recycling schemes produces substantial benefits and leads to higher recycling rates at lower costs, as competitive pressure forces the schemes to improve their efficiency and to pass on benefits to consumers....There might be a better case for encouraging a single collection and recycling scheme at the outset in order to get recycling efforts off the ground. But any arguments in favour of a single system should be critically reviewed to examine whether competition would be the superior alternative from the start, and if a single system is accepted care must be taken that any restrictions that may prevent new entry are phased out as soon as possible."

13. Arrangements by which independent firms organize the joint production of an input can have both positive and negative economic effects. Joint production may reduce costs, for example, as compared with individual production. However, input production joint ventures ("IPJVs") may allow the downstream firms to charge monopoly prices (Chen and Ross, 2003). This result obtains even when the IPJV is operated optimally, without disagreement among the parent firms. If the IPJV were granted independent decision-making powers, then the downstream prices would be yet higher and the parent firms' profits lower. Shapiro and Willig (1990) anticipated the monopoly pricing result, while also positing that parent firms may free-ride on the efforts of other parents, thus reducing the efficiency of the IPJV. Sports leagues, also, can be viewed as production IPJVs for some inputs. In its 2010 decision in American Needle, Inc. v. National Football League, the US Supreme Court emphasized that the teams were independent businesses with separate economic interests, and they compete with one another for fans, gate receipts and personnel. They also had common interests, e.g., promoting the NFL brand, and the need to cooperate, "provides a perfectly sensible justification for making a host of collective decisions." Hovenkamp (2010), commenting on the decision, pointed out that although the individual teams have incentives to compete with one another, the organization may have incentives to maximize joint profits by behaving as a cartel. "This conclusion also has a flip side: The individual members may have an incentive to free ride on the investments of other members, while the organization has an interest that each member do its part." With both positive and negative effects on efficiency, it is not surprising that input production joint ventures are assessed under competition law on a fact-specific, case-by-case basis. Using the terminology of this chapter, the potential negative effects of a collective take-back and recycling scheme are that it may serve to raise the prices consumers pay for products subject to EPR, or the scheme may operate less efficiently than desired by the producer-owners.

14. In Sweden, the competitors share the cost of jointly used infrastructure. In Germany, the individual packaging PROs are assigned monopoly collecting areas in proportion to their share of the producers who have purchased PRO services nationally. In other markets, WEEE PROs are assigned individual types of collecting infrastructure in defined areas, e.g., retailer collection points in an area for a time period.

15. Many of the categorisations and definitions of this section follow OECD 2005.

16. Elkjøp Norge Grossist AS v Elretur AS and Hvitevareretur AS, Oslo Tingrett (Oslo District Court), 13 March 2011, p. 21.

17. Switching costs were high because the "new" PRO would need to procure the amount of WEEE evidence corresponding to the amount of products put on the market by the producer/importer. Since the cost to procure "evidence" in the secondary market was often higher than the cost of organising the collection and treatment of WEEE, and was uncertain, then PROs were not incentivised to attract new producer/importers (United Kingdom Department for Business, Innovation and Skills, 2013, Points 38, 59).

18. Note that collection services generally require less investment than recovery services and typically serve a smaller area. Collection services are therefore often local or regional, while this s not the case for recovery services. There are usually numerous tenders for collection services for different parts of a country, while tenders for recovery services are frequently regional or nation-wide.

19. The Dutch end-of-life vehicle PRO, ARN, had 267 affiliated car dismantling companies. ARN organised tenders to award contracts to collect car wrecks. It awarded one contract per province (European Commission decision No. 2002/204/EC (ARN) OJ L 68/18, points 12, 17). The latter suggests that ARN felt that a local monopoly for car wreck collection was most efficient.

20. Further guidance on the design of auctions is provided in OECD (2014) and publications referenced therein.

21. Yardstick competition is a regulatory tool to give incentives for efficiency to undertakings that are not subject to competition. There are different forms of yardstick competition depending on the market failure it is intended to address. Yardstick competition establishes a "shadow competitor"; the more efficient the regulated undertaking is compared with its "shadow competitor", the greater its profits (Netherlands Bureau for Economic Policy Analysis, 2000). In the reference cost scheme, the collection costs of various municipalities are used to establish the maximum price that a given municipality may charge for collection.

22. Competition authorities have long opposed agreement on "visible" fees. In 1990, an association of independent operators of storage tanks for third parties, Vereniging van Onafhankelijke Tankopslag Bedrijven (VOTOB) decided to increase their prices by a uniform, fixed amount, an "environmental charge," to partially cover the investment costs to reduce vapour emissions from the tanks. The charge would be a separate invoice item. The European Commission's view was negative on the bases that the charge was fixed, uniform, and invoiced as a separate item. The fixing was seen as eliminating competition on that price element, as well as reducing incentives on members to meet the objectives with least cost. The uniformity meant that differences in the level and timing of costs would not be reflected. The separate itemisation was seen as suggesting the change was imposed by the government. In the event, the association changed its policies and the Commission agreed to suspend proceedings (European Commission, 1993, paragraphs 177-186).

23. The 2003 WEEE Directive says, "Member States shall ensure that for a transitional period… producers are allowed to show purchasers, at the time of sale of new products, the costs of collection, treatment and disposal in an environmentally sound way." (Article 8, point 3, Directive 2002/96/EC of the European Parliament and of the Council on Waste Electrical and Electronic Equipment (WEEE), OJ L 37/24). The 2012 WEEE Directive says that Member States may require producers to show purchasers, at the time of sale of new products, the information on costs (Article 14, point 1, Directive 2012/19/EU of the European Parliament and of the Council on Waste Electrical and Electronic Equipment [WEEE], OJ L 197/38).

24. Similar issues may arise with respect to national registers. Member States of the European Union are required to establish national registers by Article 16 of Directive 2012/19/EU, the recast directive on waste electrical and electronic equipment. Producers must report the weight of EEE placed on the market, by category, for each reporting period. (OJ L 197/38) The categories are fairly broad, and weight may not be useful for monitoring EEE sales.

25. At the time, the ELCF estimated the cost of recycling a WEEE lamp at 60% or even 80% of the retail price of the lamp.

26. The Recommendation of the Council on Competition Assessment (2009) identifies policies that limit the number or range of market participants and their incentives to behave in a competitive manner, among others, as policies to which particular attention should be given. The Recommendation of the Council on Regulatory Policy and Governance (2005) recommends Members to "Integrate Regulatory Impact Assessment (RIA) into the early stages of the policy process for the formulation of new regulatory proposals. Clearly identify policy goals, and evaluate if regulation is necessary and how it can be most effective and efficient in achieving those goals. Consider means other than regulation and identify the tradeoffs of the different approaches analysed to identify the best approach."

27. Executive summary in OECD 2010 observes that, "Competition authorities in most member and observer countries must examine agreements that pursue environmental goals under the framework that is generally applicable to all competition law analysis. Accordingly, if an agreement is found to potentially restrict competition, a competition authority will consider as justifications only direct economic benefits that are typically recognised in competition law analysis, such as cost savings, innovation, improved quality, and other efficiencies. Non-economic benefits and more remote economic benefits that do not accrue to the users of the products or services covered by the

agreement will not be taken into account when determining whether an environmental agreement violates a jurisdiction's competition laws."

28. Summary of discussion in OECD 2010.
29. European Commission, DG-Competition 2005.
30. Executive summary in OECD 2010 observes that, "A]llocation systems [in which recycled product was allocated to manufacturers based on their historical market shares] were considered anticompetitive as they protected the position of incumbent market players by guaranteeing them cheaper access to a valuable input; they distort competition for the recycled product and make it more difficult for new producers to obtain."
31. Executive summary in OECD 2010 explains that, "Because pricing restraints conflict with the core goals of competition laws, competition authorities will typically challenge environmental agreements among competitors that directly affect the price at which they sell their products to customers. For example, agreements to pass on environmental charges to customers would almost invariably be considered unlawful even if it could be argued that such a pass-on might motivate customer conduct consistent with environmental policy goals."

References

Abbott, A. and J. Wright (2010), "Antitrust analysis of tying arrangements and exclusive dealing," in *Antitrust Law and Economics* (K. Hylton, ed.), Edward Elgar Publishing.

Akerlof, G. (1970), "The market for 'lemons': Quality uncertainty and the market mechanism," *Quarterly Journal of Economics* 84: 353-374.

Australian Competition and Consumer Commission (2009), *Final Determination: Application for authorisation* A91105, AgStewardship Australia Limited.

Bio Intelligence Service (2014), *Development of Guidance on Extended Producer Responsibility*, Final Report for European Commission DG Environment, 25 March 2014.

Bundeskartellamt (2003), "Searches in the waste management sector", Press release, 11 September, *www.bundeskartellamt.de.*

Bundeskartellamt (2005), "Activity report 2003/2004, Abbreviated version", *www.bundeskartellamt.de.*

Canada Competition Bureau (2009), *Competitor Collaboration Guidelines*, Ottawa, *www.competitionbureau.gc.ca/eic/site/cb-bc.nsf/eng/03177.html.*

Canada Competition Bureau (2011), *Merger Enforcement Guidelines*, Ottawa, *www.competitionbureau.gc.ca/eic/site/cb-bc.nsf/vwapj/cb-meg-2011-e.pdf/$FILE/cb-meg-2011-e.pdf.*

Chen, Z. and T.W. Ross (2003), "Co-operating upstream while competing downstream: A theory of input joint ventures", *International Journal of Industrial Organisation*, 21: 381-397.

Commission Staff Working Paper (2011), *The Application of EU State Aid Rules on Services of General Economic Interest Since 2005 and the Outcome of the Public Consultation*, SEC(2011) 397 of 23 March 2011.

Dravone, D. et al. (2012), "Brief of *Amici Curiae* Economics Professors in Federal Trade Commission v. Phoebe Putney Health System (US Supreme Court)," *Health Management Policy and Innovation*, Vol. 1.

European Commission (2013), Press release, IP/13/711 of 18 July.

European Commission (2012), "Commission communication on the application of the European Union State aid rules to compensation granted for the provision of services of general economic interest", OJ C 8/4.

European Commission (2011), "Commission Communication on Guidelines on the Applicability of Article 101 of the Treaty on the Functioning of the European Union to Horizontal Co-Operation Agreements", OJ C 11/1.

European Commission (2010), "Commission Notice on Guidelines on Vertical Restraints", OJ C 130/1.

European Commission (2005), *DG Competition Paper concerning issues of competition in waste management systems*, Directorate General for Competition, *http://ec.europa.eu/competition/sectors/energy/waste_management.pdf*, accessed 11 December 2014.

European Commission (2002), "Decision No. 2002/204/EC on the waste disposal system for end-of-life vehicles implemented by the Netherlands", OJ L 86/18.

European Commission (2001), "The Commission investigates aid elements in Dutch disposal system for end-of-life vehicles", Press release IP/01/281, 28 February 2001.

European Commission (2001b), "The Commission defines the principles of competition for the packaging-waste disposal market", Press release IP/01/850, 15 June 2001.

European Commission (2001c), "The Commission concludes that Dutch disposal system for end-of-life vehicles does not constitute State aid", Press release IP/01/1518, 30 October 2001.

European Commission (1993), *Twenty-second Report on Competition Policy 1992.*

European Commission (1988), *Seventeenth Report on Competition Policy 1987.*

European Lighting Companies' Federation (2003), "Let there be light", position paper, *www.elcfed.org/documents/let_there_be_light_brochure.pdf*, 27 May 2003, accessed 11 December 2014.

Evans, D. and R. Schmalensee (2014) "The antitrust analysis of multi-sided platform businesses", Chapter 18 in R. Blair and D. Sokol, eds. *Oxford Handbook on International Antitrust Economics*, Oxford University Press.

France Autorité de la concurrence (2012), "*Avis no. 12-A-17 du juillet 2012 concernant le secteur de la gestion des déchets couvert par le principe de la responsabilité élargie du producteur.*"

Heyes, A. (2009), *Is Environmental Regulation Bad for Competition? A Survey*, J. Regul Econ, 36:1-28.

Hovenkamp, H. (2010), "American needle: The Sherman Act, conspiracy, and exclusion", *The CPI Antitrust Journal*, June.

ICN International Competition Network (2004), "Antitrust Enforcement in Regulated Sectors Working Group, Subgroup 1: Limits and constraints facing antitrust authorities intervening in regulated sectors", Report to the Third ICN Annual Conference Seoul, April 2004.

ICN (2013), "Exclusive dealing", *Unilateral Conduct Workbook*, Chapter 5.

Japan Fair Trade Commission (2001), *Guidelines Concerning Joint Activities for Recycling under the Antimonopoly Act, www.jftc.go.jp.*

Kienapfel, P. and G. Miersch (2006), "Competition issues in waste management systems", *Competition Policy Newsletter*, No. 1, pp. 52-56.

Konkurransetilsynet (2008a), *A2008-20 Ragn-Sells AS – konkurranseloven §12, jf. §§10 og 11 – avslag på anmodning om å gripe inn mot Elretur AS* (Ragn-Sells AS – Competition law §12, jf. §§10 and 11 – rejects the request to intervene against Elretur AS), 2 July.

Konkurransetilsynet (2008b), *A2008-22 Veolia Miljø Metall AS – konkurranseloven §12, jf. §§10 og 11 – avslag på anmodning om å gripe inn mot Autoretur AS* (Veolia Miljø Metall AS – Competition law §12, jf. §§10 and 11 – rejects the request to intervene against Autoretur AS), 2 July.

Konkurransetilsynet (2008c), *Påpekning – konkurranseloven § 9 første ledd e) – ordninger for produktgjenvinning* (Pointing Out – Competition Act § 9 first paragraph e) – arrangements for product recovery), 2 July.

Marbek Resource Consultants (2007), *Analysis of the Free-Rider Issue in Extended Producer Responsibility Programs, www.ccme.ca/files/Resources/waste/extended/free_riders_1.0_1380_e.pdf.*

Netherlands Bureau for Economic Policy Analysis (2000), "Yardstick competition theory, design, and practice", *Working Paper* No. 133, December.

Netherlands Authority for Consumers and Markets (2014), "Vision Document: Competition and Sustainability", *www.acm.nl.*

Nordic Competition Authorities (2010), "Competition Policy and Green Growth", *Joint Report* No. 1/2010.

OECD (2014), *Hearing on the Use of Tenders and Auctions: Issues Note by the Secretariat*, DAF/COMP/WP2(2014)15, *www.oecd.org/officialdocuments/publicdisplaydocumentpdf/?cote=daf/comp/wp2(2014)15&doclanguage=en.*

OECD (2013), *What Have We Learned About Extended Producer Responsibility in the Last Decade? A Survey of the Recent Economic EPR Literature*, OECD Paper ENV/EPOC/WPRPW(2013)7/FINAL available at *http://spot.colorado.edu/~daka9342/OECD_EPR_KO.pdf.*

OECD (2013b), "Waste management services", *Competition Policy Roundtables*, OECD, *www.oecd.org/daf/competition/Waste-management-services-2013.pdf.*

OECD (2010), "Horizontal agreements in the environmental context", *Competition Policy Roundtables*, OECD, *www.oecd.org/competition/cartels/49139867.pdf.*

OECD (2010b), "Information exchanges between competitors under competition law", *Competition Policy Roundtables*, OECD, *www.oecd.org/competition/cartels/48379006.pdf*.

OECD (2010c), "Standard setting", *Competition Policy Roundtables*, OECD, *www.oecd.org/daf/competition/47381304.pdf*.

OECD (2009a), "Two-Sided Markets", *Competition Policy Roundtables*, OECD, *www.oecd.org/daf/competition/44445730.pdf*.

OECD (2009b), "State-owned enterprises and the principle of competitive neutrality", *Competition Policy Roundtables*, OECD, *www.oecd.org/daf/competition/46734249.pdf*.

OECD (2007), *Improving Recycling Markets*, OECD Publishing, Paris, *http://dx.doi.org/10.1787/9789264029583-en*.

OECD (2007b), "Public procurement", *Competition Policy Roundtables*, OECD, *www.oecd.org/competition/cartels/39891049.pdf*.

OECD (2006), "Environmental regulation and competition", *Competition Policy Roundtables*, OECD, *www.oecd.org/regreform/sectors/37981581.pdf*.

OECD (2006b), "Competition in bidding markets", *Competition Policy Roundtables*, OECD, *www.oecd.org/competition/cartels/38773965.pdf*.

OECD (2005), "Barriers to entry", *Competition Policy Roundtables*, OECD, *www.oecd.org/competition/abuse/36344429.pdf*.

OECD (2005b), "Predatory foreclosure", *Competition Policy Roundtables*, OECD, *www.oecd.org/competition/abuse/34646189.pdf*.

OECD (2005c), *Analytical Framework for Evaluating the Costs and Benefits of Extended Producer Responsibility Programmes*, OECD, *http://dx.doi.org/10.1787/oecd_papers-v5-art13-en*.

OECD (2004), *Economic Aspects of Extended Producer Responsibility*, OECD Publishing, *http://dx.doi.org/10.1787/9789264105270-en*.

OECD (2004b), "Universal service obligations", *Competition Policy Roundtables*, OECD, *www.oecd.org/regreform/sectors/45036202.pdf*.

OECD (2003), *The objectives of competition law and policy*, CCNM/GF/COMP(2003)3, *www.oecd.org/daf/competition/2486329.pdf*.

OECD (2003b), *Annual Report on Competition Policy Developments in Sweden 2003*, *www.oecd.org/sweden/34722334.pdf*

OECD (2001), *Extended Producer Responsibility: A Guidance Manual for Governments*, OECD Publishing, Paris, *http://dx.doi.org/10.1787/9789264189867-en*.

OECD (1998), *Annual Report on Competition Law and Policy Developments* - The Netherlands. *www.oecd.org/netherlands/1820951.pdf*.

OECD (1993), *Glossary of Industrial Organisation Economics and Competition Law*, *www.oecd.org/regreform/sectors/2376087.pdf*.

OECD (1992), *Objectives of competition policy*, DAFFE/CLP(92)2/REV1.

Philipson T.J. and R.A. Posner (2009), "Antitrust in the not-for-profit sector", *Journal of Law and Economics*, 52:1, pp. 1-18.

Ross, S. and S. Szymanski (2006), "Antitrust and inefficient joint ventures: Why sports leagues should look more like McDonalds and less like the United Nations", *Marquette Sports Law Review*, 16:2.

RPS, ESRI, Bio Intelligence Service and Philip Lee (2014), *Review of the Producer Responsibility Initiative Model in Ireland (Main Report, Redacted)*, Report for Ireland Department of Environment, Community and Local Government, July.

Savage, M. et al. (2006), "Implementation of the Waste Electric and Electronic Equipment Directive in the EU", *Technical Report Series*, IPTS, European Commission Joint Research Centre.

Shapiro, C. and R.D. Willig (1990), "On the antitrust treatment of production joint ventures", *Journal of Economic Perspectives*, 4:3, 113-130.

Tinbergen, J. (1967), *Economic Policy: Principles and Design*, Amsterdam: North-Holland.

United Kingdom Office of Fair Trading (2006), "More competition, less waste: Public procurement and competition in the municipal waste management sector", *Discussion Paper* No. 841.

United Kingdom Department for Business, Innovation and Skills (2013), *Waste Electrical and Electronic Equipment (WEEE) System, Impact Assessment*, 11 October.

United Kingdom Department for Business, Innovation and Skills (2014), *WEEE Regulations 2013: Government Guidance Notes*, March.

United Kingdom Office of Fair Trading (2011), *Public Bodies and Competition Law: A Guide to the Application of the Competition Act 1998*, December, OFT1389.

United States Federal Trade Commission and Department of Justice (2000), *Antitrust Guidelines for Collaborations among Competitors.*

Walls, M. (2004) "EPR policy goals and policy choices: What does economics tell us?" Chapter 1 in OECD (2004), *Economic Aspects of Extended Producer Responsibility*, OECD Publishing, *http://dx.doi.org/10.1787/9789264105270-en.*

WEEE Register Society (2014), FAQs, *www.weeeregister.ie*, accessed 18 December 2014.

Extended Producer Responsibility
Updated Guidance for Efficient Waste Management

PART II

Chapter 5

Incentives for eco-design in extended producer responsibility

One of the key objectives of extended producer responsibility is to shift responsibility for end-of-life management to producers and therewith incentivise them to invest in design-for-environment (DfE) in order to minimise waste management costs. However, due to the fact that most EPR systems have been implemented in the form of collective producer responsibility, DfE incentives have often been found to be very weak. This chapter discusses some of the ways that can help to strengthen these incentives, such as through the implementation of full cost recovery from producer fees, the use of variable rather than fixed producer fees, as well as modulated fees that take account of product design features.

5.1. Introduction

EPR contributes to sustainable materials management by stimulating reuse, recycling and energy recovery of waste (PRO, 2010; Rotter, 2011; Bio IS, 2012). For example, the Korean EPR schemes have augmented recycling rates for products such as packaging and tires to respectively 74% and 62% (see Annex J). Similarly, the European EPR directives on packaging (2004/12/EC) and cars (2000/53/EC) have been the main drivers to increase average recycling rates to respectively 64 % and 84% (EU27 – 2011, Eurostat).

According to the OECD manual (2001), EPR has two specific objectives. The first objective is to shift the responsibility of waste management from municipalities to producers. Considering that producers will integrate EPR costs in the price of their products, the cost of waste management will be (partially) internalised by consumers that purchase waste-generating goods. Consequently, EPR implements the "Polluter Pays Principle" in an indirect way. The second objective is to give an implicit incentive to producers to invest in Design for Environment (DfE), also commonly referred to as Eco-design.

DfE is a broad design concept that aims to reduce the environmental impact generated over the lifetime of a product, from the production until the end-of-life phase. A commonly used method to adopt DfE is the Lifecycle Design Strategies (LiDS) wheel shown in Figure 5.1 (Brezet and Van Hemel 1997). With respect to EPR, the four most relevant dimensions of the LiDS wheel are "Selection of low impact materials" (e.g. avoiding hazardous substances or using recycled resources), reduction of material (e.g. development of lightweight products or reducing packaging), optimisation of initial lifetime (e.g. increasing durability or re-use of components) and optimisation of end-of-life system (e.g. design for disassembly or producing mono-material goods).

Figure 5.1. **The Lifecycle Design Strategies (LiDS) wheel illustrates the different dimensions of Design for Environment (DfE)**

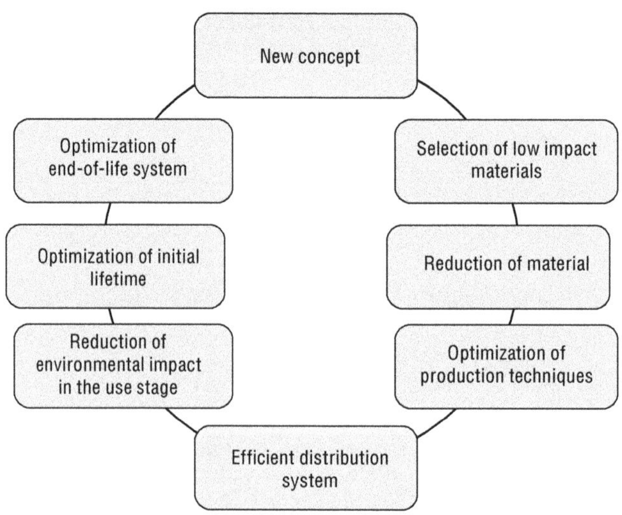

Numerous EPR policies stress the necessity of stimulating DfE. E.g. article 8 of the European Framework Directive (2008/98/EC) mentions explicitly that EPR should "*encourage the design of products in order to reduce their environmental impacts and the generation of waste in the course of the production and subsequent use of products*". Similarly, the Washington law on electronic product recycling (RCW 70.95N) states that "*the system must encourage the design of electronic products that are less toxic and more recyclable*".

The effectiveness of EPR with respect to DfE is, however, disputed. Many academic contributions and policy reviews criticise the actual structure of EPR schemes and state that EPR incentives for DfE are currently (too) weak (Lindhqvist and Lifset, 2003; Sachs, 2006; Gottberg et al., 2006; Roine and Lee, 2006; Walls, 2006; Rotter, 2011; Gui et al., 2013; Tong and Yan, 2013). The contrast between policy expectations and realisations, therefore, seems stark.

Owing to the global disparity of waste policies and the wide range of products that are concerned, there is a high diversity of existing EPR schemes. Accordingly, the questions to address are: "Which type of EPR stimulates DfE most?", "Are the DfE incentives from EPR the same for all products?", "How can DfE incentives of EPR schemes be improved?". To answer these questions, this chapter reviews the literature on EPR incentives for DfE and uses case studies to illustrate the analysis. The chapter especially draws on insights from the OECD Forum "Promoting sustainable materials management through Extended Producer Responsibility", 17-19 June 2014, Tokyo. Since product design is typically firm-specific, the chapter focuses on the financial incentives given to individual producers.

The chapter is structured in four sections. Section 5.2 discusses the challenge to assess the extent to which EPR fosters DfE. Section 5.3 groups EPR schemes in three types depending on the financial incentives for individual producers. Section 5.4 further highlights the impact of six key features of collective EPR schemes. Section 5.5 provides a conclusion.

5.2. Assessing the impact of EPR on Design for Environment

Assessing the impact of EPR on DfE is challenging for three reasons. First, although the LiDS wheel allows relative comparison of products within DfE dimensions, the environmental assessment cannot be aggregated easily to a cross-dimensional comparison. In addition, design changes often affect different stages of the life cycle simultaneously. On the one hand, positive interrelated effects occur. For example, by limiting material use and facilitating material recovery, DfE does not only reduce downstream environmental impacts from waste treatment, but also prevents the upstream environmental impacts from resource extraction and refining (Damgaard et al., 2009; Dubois, 2013; Acuff and Kaffine, 2013). On the other hand, negative interrelated effects can exist. For example, a car completely made out of steel can be easily recycled at end-of-life, but the energy consumption during the use phase of a steel car will be much higher compared to a car made from carbon fibres (Duflou et al., 2009). In consequence, the implementation of DfE requires making trade-offs between environmental impacts along the life cycle, whereas different stakeholders may have distinct design priorities.

Second, companies have diverse drivers to implement DfE: customer demands for green products, stakeholder pressure, corporate social responsibility, personal values of managers and designers, increasing resource prices, legal obligations such as RoHS[1] (Restriction of Hazardous Substances – 2002/95/EC) and financial incentives of EPR schemes or other economic policy instruments (van Hemel and Cramer, 2002). These drivers interact and can reinforce each other, including some design trends, such as miniaturisation, suit consumer demands, but also reduce material use and waste management costs. Further, limiting packaging weight reduces EPR induced costs, but also optimises transport. Moreover, incorporating recycled resources into new products can be driven by higher prices due to resource scarcity but can also be inspired by corporate social responsibility. In other words, it is challenging to determine the contribution of an individual driver, such as EPR, for "DfE successes".

Third, some of the impact of EPR is difficult to measure. Anecdotal evidence indicates that EPR contributes to DfE by fostering communication between producers and recyclers (Van Rossem et al. 2006; Dempsey et al., 2010). In Japan companies that operate both production and recycling plants for consumer electronics, such as Sharp, Sony and Mitsubishi Electric, offer trainings, plant visits and guidance on recycling practices for designers. In a similar way, EPR has triggered active communication between car manufacturers and recyclers in Sweden (OECD, 2004). The design guidelines and testing procedures for PET bottles put forward by producers of beverage packaging in order to improve PET recycling are a further example of the strength of improved communication between the actors in the supply chain. Communication between producers can also facilitate standardisation and harmonised labelling that streamlines guidelines for consumers. For example, in Japan, standardised labels on rechargeable batteries help to sort the batteries into different types: lead-acid, nickel-cadmium, nickel-metal hydride or lithium-ion (see Annex G). Furthermore, the framing of materials as "environmentally damaging" or "disruptive for recycling", such as is the case for ceramic caps for glass bottles in France, incentivizes producers to use other materials to avoid reputational damage. Although these examples highlight the importance for DfE of intangible aspects, such as improved communication or brand image, it is difficult to quantify the impact.

Due to barriers for quantitative assessment, some studies have investigated the impact of EPR incentives on DfE in a qualitative way. According to Tojo (2004) and Van Rossem et al. (2006), manufacturers of consumer electronics stress the importance of EPR as a driver for DfE. Yeo (2005) also highlights the DfE improvements in consumer electronics since the start of the Korean EPR program. However, as pointed out by Walls (2006), most design changes cited were already well under way before implementation of EPR for e-waste. Gottberg et al. (2006) and Roine and Lee (2006) conclude, on the basis of several case studies, that EPR is one of the contributing factors for DfE, but seldom the triggering factor.

In contrast to intangible effects, such as improved awareness for waste management challenges by designers and industrial decision makers, financial incentives can be readily observed. Considering that the financial incentives from EPR schemes depend strongly on product characteristics and operational management, the next section will discuss different EPR types.

5.3. EPR types

From the perspective of DfE, the financial incentives from EPR schemes for individual producers can be grouped into three types (OECD, 2005): Individual Producer Responsibility (IPR), variable-fee Collective Producer Responsibility (CPR), and fixed-fee CPR. The following subsections discuss the three types.

5.3.1. *Individual Producer Responsibility (IPR)*

Under IPR, each individual producer is responsible for the collection and recycling of waste originating from his own products. As illustrated in Figure 5.2, the producer can either take back the product or pay a third party that collects and recycles his products. The

Figure 5.2. **Physical and financial flows for Individual Producer Responsibility (IPR)**

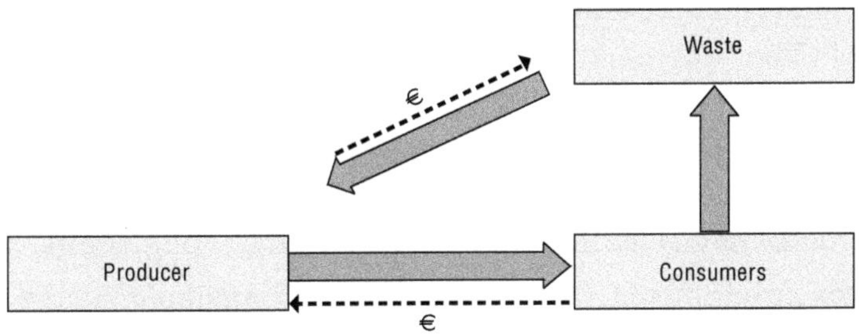

distinctive feature of IPR is that the cost of waste management of an individual producer is not affected by the behaviour of competitors. The economic benefits of DfE are therefore fully internalised, i.e. firm-specific waste management costs will lower if the producer succeeds in making products more lightweight, more durable or easier to recycle.

IPR is the purest form of 'producer responsibility' and is advocated by many academic papers and stakeholders owing to its positive effect on DfE (OECD, 2004; Dempsey et al., 2010). IPR especially has the potential to stimulate remanufacturing and dismantling for re-use of components, because end-of-life products can return to the original producer (Atasu and Subramanian, 2012). Companies such as Electrolux, Fujitsu, HP, Sony or Samsung promote IPR because it rewards individual firms for DfE investments (McKerlie et al., 2006). Other stakeholders in favour of IPR, are environmental NGOs (Van Rossem et al., 2006).

In contrast with the environmental benefits, IPR implementation may worsen market distortions due to monopoly power. IPR schemes may reduce competition in product markets due to the entry barriers created for new-comers and the drawbacks for small enterprises (Short, 2004; Didier and Sittler, 2014). The main reasons are the economies of scale present in waste management, which enables incumbent firms with large market shares to further strengthen their competitive position.

Despite the potential environmental advantages and the support of stakeholders, few IPR schemes exist (Lindhqvist and Lifset, 2003; Sachs, 2006; Mayers and Butler, 2013). In most countries, producers set up collective industry-wide EPR schemes that will be discussed in the following sections.

5.3.2. *Collective Producer Responsibility (CPR)*

There are several advantages to setting up industry-wide CPR schemes that collect and recycle consumer waste jointly:

- Since waste management is typically characterised by economies of scale (or density), joint collective schemes are more cost efficient. For example, Khetriwal et al. (2009) illustrate in a case study of Swiss e-waste management that transport costs are significantly lower if all brands are collected jointly, rather than individually.
- A joint collection scheme creates administrative ease for several actors. Consumers are mostly unwilling to bring end-of-life goods to different dedicated locations (see Annex G and Hickle, 2013). In addition, municipal collection points and multi-brand retailers that accept end-of-life goods, denounce the time needed to sort the waste stream into individual brands (see Annex G and Hotta et al., 2014). Furthermore, CPR schemes simplify administration for producers, especially for companies operating in different countries. With the help of a case study, Mayers (2007) highlights the complexity for

global firms to deal with disparate national EPR schemes. Accordingly, rather than setting up an individual collection scheme per country, companies typically prefer the convenience of CPR.

- At the start of an EPR scheme, overall costs are difficult to estimate due to uncertainty about amounts of waste to be collected, recycling processes and prices of secondary resources (Mayers and Butler, 2013; OECD, 2014). Therefore, producers typically opt for a collective approach that, at least, shares the risks among competitors.
- Fourth, also for governments CPR has benefits. For example, at the start of an EPR scheme for durable goods, as the treatment of 'orphan goods that have been put on the market by firms that are no longer present, will normally be financed by public money. Therefore, budget-constrained policy makers may prefer CPR schemes that include all products regardless of their brand. Moreover, individual producers can decide to retreat from local markets or can go bankrupt. Since it is unlikely that all producers retreat from the market simultaneously, most governments consider CPR as a suitable insurance for future waste management costs (OECD, 2004; Van den Abeele, 2006; Séguin, 2014; Grgulovà, 2014).

Most CPR schemes function with a Producer Responsibility Organization (PRO) as illustrated in Figure 5.3 (OECD, 2013). When governments impose take-back (i.e. collection or recycling) targets, producers commonly set up and finance a collective PRO to meet EPR obligations. The PRO will then negotiate contracts with independent collection and recycling companies. Considering that the PRO typically works with a balanced budget, all costs are charged to the producers. A PRO, therefore, serves as a collective "service-provider" that optimises and facilitates EPR implementation. Differences between products, markets and policies have induced CPR schemes with distinct characteristics.

Figure 5.3. **Physical and financial flows for Collective Producer Responsibility (CPR)**

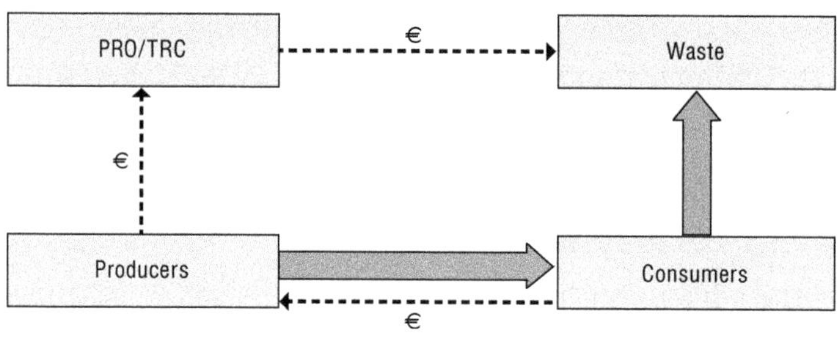

In some cases, such as for batteries in Belgium, industry has opted for one PRO per product stream, while in other cases, such as for consumer electronics in Japan, competitive firms have decided to establish multiple PROs (Bouteligier et al., 2014; Hotta et al., 2014).

CPR schemes such as the United Kingdom Packaging Recovery Notes use Tradable Recycling Credits (TRC). On the one hand, the scheme obliges retailers and producers to buy credits at a trading platform in proportion to the packaging volume put on the market. On the other hand, packaging recyclers can sell credits generated in proportion to the amount of waste recycled (Matsueda and Nagase, 2012). Hence, the trading platform for credits serves as a financial substitute for a PRO.

Some governments limit producer involvement to the payment of an Advance Disposal Fee (ADF), i.e. a tax that has to be paid at the purchase of a waste generating good. In such a CPR, no PRO or TRC is needed. Nonetheless, ADF is often categorised as an EPR policy because the financial incentives of ADF are similar to the incentives induced by a fee paid to the PRO (OECD, 2013). If ADF is combined with recycling subsidies such as in the Chinese E-waste Fund (see Annex D), the policy resembles the deposit-refund scheme that is typically modelled in economic literature (Fullerton and Wu, 1998; Calcott and Walls, 2000, 2005).

CPR schemes can have visible fees, i.e. the CPR fee is explicitly mentioned on the bill. On the basis of bounded rationality of consumers and producers, it is sometimes reasoned that the visibility and the handling of the fee would affect DfE incentives. For example, Quebec has forbidden visible fees in order to stimulate DfE. Since producers receive the money but have to pass it on to the PRO, they may be more aware of the financial benefits from DfE (Bury, 2013). However, currently there is no consensus on the impact of such visible fees.

Although disparity in CPR implementation is considerable, all CPR schemes impose a cost on producers. With respect to the financial incentives for DfE, the key element is cost allocation between producers. The following two sections discuss the implications of cost allocation via variable-fee CPR and fixed-fee CPR. Owing to its popularity, the following sections focus on CPR with one or more PROs that aim to achieve the take-back targets imposed by policy makers.

Variable-fee Collective Producer Responsibility (CPR)

A variable-fee CPR scheme is mainly used for mono-material products with limited durability, such as packaging or graphic paper. Importers, producers or retailers pay a fee in proportion with product weight, as illustrated with an example of the fees from Fost Plus, the Belgian PRO for consumer packaging, in Table 5.1.

Table 5.1. **Weight-based fees per material in Belgian PRO for consumer packaging, Fost Plus**

Packaging material	Weight-based fee (EUR/kg)
Glass	0.0227
Paper – cardboard	0.0167
Aluminium	0.0371
PET	0.1064
Mixed – can be valorised	0.2593
Mixed – to be disposed of	0.4084

Source: Fost Plus (2014), Tarifs, *www.fostplus.be/fr/entreprises/votre-declaration/tarifs.*

In order to optimise waste management, Fost Plus has significantly higher rates for multi-material packaging than for mono-material packaging. In addition, the fee for multi-material packaging that will be disposed of after consumption is higher than the fee for multi-material packaging that still has some valorisation potential. In a similar way, the French PRO for consumer packaging, Eco-Emballages, raises its fee by 50% if elements that complicate recycling are present, e.g. glass bottles with ceramic caps are financially penalised because ceramics impede glass recycling (Didier and Sittler, 2014; FME, 2014).

A weight-based fee gives financial incentives to make products more lightweight. In line with these incentives, many "light-weighting successes" for packaging can be found. The Belgian organisation "Preventpack" keeps a public database with more than hundred DfE examples for packaging. As shown in Table 5.2, PRO Europe, the federation of European packaging PROs, publishes some overall European trends: the weights of PET bottles, aluminium cans, glass bottles, tin cans, plastic bags and cardboard boxes have lowered significantly in the last decade. Similarly, the weight of PET bottles (after correction for changes in size) has decreased between 0.2% and 19% in Japan from 2004 to 2010 depending on the different types (size and purpose) of bottles (see Annex I). Moreover, the French PRO for graphic paper, Ecofolio, reports that the average weight of marketing flyers has been reduced by 15% since the start of the EPR scheme in 2007 (Didier and Sittler). As discussed in section 5.2, these results are achieved by a combination of drivers. It is, therefore, not possible to quantify the contribution of financial EPR incentives.

Table 5.2. **Changes in food packaging weight in Europe, 2006-10**

Packaging and product	Weight 2 000 (g)	Weight 2 010 (g)	% change
PET bottle of 1.5 L still water	31.8	28	-12
Aluminium can of 330 ml for soft drinks	15.8	0.0131	-17
Glass bottle of 250 ml for olive oil	223.6	200.2	-10
Tin can of 125 gr for fish	34.3	31.9	-7
Plastic bag for 1 kg of pasta	9.03	7.85	-13
Cardboard box of 500 gr for dry food	13.88	11.32	-18

Source: http://proeurope4prevention.org/packagings-trends.

Since variable-fee CPR schemes typically focus on weight or material type, incentives for other DfE aspects are weak. E.g. even though bulky waste items makes collection more expensive, PRO fees are rarely differentiated based on the product volume. Similarly, recycled content is not often taken into account for the determination of PRO fees. However, the Japanese example of PET bottles for green tea illustrates that EPR can contribute to DfE aspects other than weight or material type. Because these bottles used to be green, a separate collection scheme was needed to avoid colour degradation of the standard transparent PET. Thanks to consultations between producers, green PET bottles were replaced by transparent PET bottles with a green coloured label (Hosoda, 2004). This change resulted in lower collection costs and reduced impurities in secondary resources. Another example comes from Eco-Emballages in France. The PRO incentivises DfE aspects other than waste by lowering the fee for paper and cardboard packaging by 10% if the recycled content is more than 50% (Didier and Sittler, 2014).

Fixed-fee Collective Producer Responsibility (CPR)

A fixed-fee CPR typically applies for multi-material or durable goods such as electronic equipment, cars or furniture. The PRO fee is the same for all competitive products put on the market. For example, in the Netherlands, the PRO fee for recycling end-of-live vehicles is EUR 45 per new car regardless of brand or type (ARN, 2012). Similarly, the fixed fee charged to companies in the Chinese E-waste Disposal Fund amounts to 13 CNY per television and 7 CNY per washing machine (see Annex D). It should be noted, however, that some EPR schemes have weight-based PRO-fees for consumer electronics, such as computers and printers (Séguin, 2014).

Fixed-fee CPR is often applied for multi-material goods due to the challenging trade-offs for DfE aspects. For example, washing machines can either use a large metal or concrete component as a stabiliser. Although metal-containing models are often heavier which may negatively affect environmental impacts of transport, the value of the metal components makes them attractive to recycle. Moreover, for consumer electronics, such as mobile phones, the multitude of metals and the diversity in designs, complicate the comparison of environmental impacts between models.

Fixed fee CPR is also typically applied for durable goods due to the uncertainty about future waste collection costs, recycling processes that will be applied and future value of secondary resources. Therefore, many PROs have fixed fees regardless of the brand and apply a "pay-as-you-go" principle, i.e. producers do not pay for waste handling of products that are currently sold but pay for current waste management (OECD, 2004; Van Den Abeele, 2006). Hence, most PROs finance today's expenses with today's income. For example, producers pay a PRO fee per refrigerator that is currently sold in order to finance present-day waste management, not waste management in ten years when the currently sold refrigerator will become obsolete. Evidently, if PRO fees are identical for all products and unrelated to future waste costs, the incentives for DfE will be weak (Van Rossem et al., 2006; Atasu and Subramanian, 2012).

5.3.3. *IPR or CPR?*

Some EPR schemes for durable goods, apply brand recognition to allocate EPR costs among member producers. For example, in Japan collected e-waste is separated into brands and is physically sent to one of the two PRO schemes (Dempsey et al., 2010; Hotta et al., 2014). In Washington, e-waste is periodically sampled to allocate costs between brands (Hickle, 2013; Gui et al., 2013). In Maine, collected e-waste is registered per brand and producers have the choice between picking-up the e-waste and having it treated in the collective scheme (see Annex K and Dempsey et al., 2010; Gui et al., 2013). Such brand recognition should not be confused with IPR because brand separation in Japan is more like a fixed-fee CPR with two competitive PROs (Atasu and Subramanian, 2012). Periodic sampling and brand registration such as applied in Washington and Maine, may improve cost allocation but as long as the unit treatment cost is the same for all brands, it remains a collective scheme.

In the absence of EPR legislation, individual companies have started taking back end-of-life products in a perspective of corporate social responsibility. E.g. Best Buy and Dell accept end-of-life computers without any charge, all over the US regardless of legal obligations; Hyundai has voluntarily contributed to the set-up of a recycling scheme for end-of-life vehicles in Mongolia; producers of consumer electronics have set up take-back programs in Brazil (Neto and Van Wassenhove, 2013). Considering that these voluntary initiatives foster recycling and shift costs away from municipalities, they should be lauded. However, the incentives for DfE remain weak because most schemes accept end-of-life goods of all brands and typically only collect a limited fraction of own-brand models put on the market. Consequently, from the perspective of DfE, these schemes should be qualified as CPR not as IPR.

Interest is increasing in leasing and product-service systems (PSS) for durable goods (Isaksson et al., 2009; Subramanian et al., 2014). These alternative contracts imply that consumers pay for services while the producer (or leasing company) remains the proprietor of the product used to deliver the service. E.g. the "power-by-the-hour" service of Rolls Royce where flight operators pay for flight hours rather than for the airplane engine or maintenance; the "pay per lux" service of Philips where companies pay for light hours rather than for lamps, electricity or maintenance; car-sharing services where consumers in urban areas rent cars by the hour and pay per kilometre. If the producer remains the owner all along the life cycle of the product, incentives for DfE from PSS are similar to incentives from IPR. However, the drivers for leasing and PSS are typically financial, not environmental (Lifset and Lindhqvist, 2000; Plambeck and Wang, 2009; Agrawal et al., 2012). Therefore, PSS can induce DfE, but the incentives will depend on the specific contract conditions (Subramanian et al., 2009; Pangburn and Stavrulaki, 2014).

5.4. Features of Collective Producer Responsibility schemes

Since most EPR schemes are organised in an industry-wide way, the following section elaborates on six key features of CPR that determine the financial incentives for DfE: the use of uniform fees, the magnitude of financial incentives, the controversy on full cost internalisation, the scope of CPR schemes, the international disparity of EPR schemes and the role of technological innovation.

5.4.1. *Modulated fees*

If uniform PRO fees are applied for all brands and models of a specific product category, DfE incentives for individual companies are weak. To strengthen incentives, modulated fees can be introduced, i.e. differentiating the fee per product in function of a specific design feature (Gui et al., 2013). In this chapter, modulated fees refer to fees differentiated in function of a DfE aspect other than weight. For example, the Belgian PRO fee for optical smoke detectors is EUR 2.7 while the PRO fee for an ionic smoke detector is as high as EUR 30 (tariff 2013). This price difference discourages use of ionic smoke detectors due to their potential radioactivity in the waste stage. As discussed earlier, most EPR schemes apply uniform fees. However, the examples in Table 5.3 highlight that France successfully applies modulated fees for packaging, electronics and graphic paper.

Implementation and monitoring of modulated fees induces an additional administrative cost. Therefore, modulated fees should only be implemented when the environmental or financial gains are significant. Peeters et al. (2014) show that the disassembly time of LCD TVs can be reduced by 70% if equipped with specific elastomer fasteners, which equals a disassembly cost reduction between EUR 0.07 and EUR 0.91 per product depending on the labour wages. Their analysis also indicates that differentiating PRO fees with less than EUR 0.1 per LCD TV can already strengthen DfE incentives in regions with a labour cost of more than EUR 7/h. Mayers et al. (2013) elaborate a PRO scheme with modulated fees for CFC vs Pentane refrigerators and mercury containing LCD vs mercury free LCD displays. These propositions illustrate that further implementation of modulated fees can improve DfE incentives from fixed-fee CPR.

Since "dirty" products are penalised, modulated fees will affect competition. Producers of "dirty" models may see their sales reduced whereas producers of "eco-friendly" products may experience a competitive advantage. In consequence, achieving the consensus needed to introduce such modulated fees may be hard to achieve, when considering that producers are the founding members of the PRO (Didier and Sittler, 2014). Conversely, allowing producers to independently determine modulated fees, may foster collusion or abuse of market power (Fleckinger and Glachant, 2010). Therefore, policy makers should promote and monitor modulated fees.

Table 5.3. **Modulated PRO fees in France, 2015 – Packaging, electronic and electrical appliances and graphic paper**

	Packaging: Eco-Emballages – rates 2015
Fee reduction of 10%	Paper/cardboard packaging with more than 50% of recycled content
Fee increase of 50%	Glass packaging with ceramic cap Paper/cardboard Packaging for liquids with less than 50% of fibres Reinforced paper/cardboard packaging PET bottles that contain PVC or silicones with density more than 1
Fee increase of 100%	Non-recyclable packaging materials, e.g. ceramics
Fee reduction of 8%	Packaging that carries specific sorting instructions for consumers Applies to producers that organize additional prevention campaigns
Fee reduction of 8%	Weight reduction of more than 2% without changing the packaging material or function Volume reduction of more than 2% without changing the packaging material or function Applies to companies that develop rechargeable packaging
	Electronic and electrical appliances: Eco-Systèmes – rates 2015
Fee increase of 20%	Vacuum cleaners with brominated flame retardants (EUR 0.5 instead of EUR 0.42 per unit) Portable computers with brominated flame retardants (EUR 0.3 instead of EUR 0.25 per unit) TV with brominated flame retardants (depending on the unit weight: > 25 kg: EUR 8 instead of EUR 6.67 > 13 kg: EUR 4 instead of EUR 3.33 > 7 kg: EUR 2 instead of EUR 1.67 < 7 kg: EUR 1 instead of EUR 0.83)
Fee increase of 100%	Mobile phones without a standardised charger (EUR 0.01 instead of EUR 0.02 per unit)
	Electronic and electrical appliances: Récylum – rates 2015
Fee reduction of 20%	LED Lightning (EUR 0.12 instead of EUR 0.15 for conventional lightning bulbs) owing to the absence of mercury and the long life cycle
	Graphic paper: Ecofolio – rates 2015
Fee reduction of 10%	Publications with more than 50% of recycled content (EUR 47 instead of EUR 52 per tonne)
Fee increase of 5%	Use of primary fibres from forests without eco-management labels The presence of elements that hamper the recyclability: fibre colour, inks, glue and non-fibrous elements such as varnish or plastics (cumulative presence of inhibitors can lead to a 15% fee increase, i.e. EUR 60 instead of EUR 52 per tonne)

Source: www.ecoemballages.fr/; www.ecofolio.fr/adherents/le-bareme; www.eco-systemes.fr/documents/Bareme.pdf; http://reduction.ecoemballages.fr/catalogue/; www.recylum.com/enjeux/eco-conception-des-equipements/.

5.4.2. *Magnitude of the financial incentives*

The magnitude of the PRO fee (or TRC price/ADF) is as important as the allocation of the fee. Evidently, high PRO fees will drive DfE investments more than low PRO fees. For example, Didier and Sittler (2014) indicate that packaging PRO fees in France may be equivalent to 4 % of the sales revenues of packaging producers. Mayers (2007) cites an example where the German PRO fee for a plastic blister pack of an electronic toy accounts for 8 % of the packaging cost. Table 5.4 illustrates that the Belgian PRO for consumer electronics, Recupel, charges a fee of EUR 10 for a refrigerator. These individual cases illustrate that CPR incentives may be sufficient to stimulate design changes. Conversely, Table 5.4 also highlights that that PRO fees are often insignificant compared to the price of the product. Indeed, few computer producers will be driven to invest in DfE by a fee of only EUR 0.05.

Several effects contribute to the downwards evolution of PRO fees between 2002 and 2013 that can be observed in Table 5.4. An important element is that Recupel has changed its financial strategy. Initially it built up a provision to finance future costs of waste management for durable products that were put on the market today. In the last decade, Recupel has opted for a "pay as you go" where current expenditure for waste management is paid with current PRO fees. The consumption of the initial provision has evidently pushed PRO fees down. Secondly, the value of secondary materials has increased. Clearly, this reduces the overall

Table 5.4. **PRO fees of the Belgian PRO for consumer electronics Recupel**

Recupel fee (Belgium, incl. VAT)	2002	2013
Refrigerator	EUR 20	EUR 10
Washing machine	10	1
Iron	1	0.05
Laptop	2	0.05
TV	11	1
Drill	2	0.35

Source: Recupel (2013), *List of Appliances* (translated from: Apparatenlijst), Brussels.

costs of waste management. A third important element relates to the learning effects that are often achieved after start-up of an EPR scheme. Since its origin in 2001, Recupel has reduced PRO fees thanks to efficiency gains and new recycling techniques. In many other EPR schemes similar cost gains can be observed. Evidently, such efficiency gains should be lauded and show that initial cost estimations typically overrate long-term costs. Unfortunately, from the perspective of DfE, this cost reduction also weakens DfE incentives.

5.4.3. *Full cost internalisation*

Full cost internalisation is a controversial topic. The principle means that all costs related to waste management of products should be internalised via EPR schemes (Mc Kerlie et al., 2006; OECD, 2014). As highlighted by Monier et al. (2014) there is a consensus that EPR schemes (and PRO fees) should cover the costs of collecting and treating waste that is pre-sorted by households minus the revenues from recovered materials. However, there is debate concerning the allocation of other costs: costs for the collection and treatment of waste that is not pre-sorted by households (e.g. collected by municipalities via the residual mixed waste bag); costs for awareness raising campaigns; costs for clean-up of litter; costs for enforcement and monitoring of the EPR scheme (including auditing and measures against free-riding); and finally R&D investment in DfE.

An argument to restrict the costs allocated to producers is the distribution of control in the waste management chain. Producers are hesitant to compensate all costs made by municipalities because it would take away incentives for cost efficiency. Also, considering that enforcement is a public authority, producers feel powerless for such tasks. Conversely, an argument in favour of allocating full costs of waste management to producers relates to the aim of EPR to shift away waste management costs from the municipalities. In addition, considering that the magnitude of the PRO fee determines the incentives for DfE, full cost internalisation is key. From a DfE perspective, cost-sharing with municipalities and other policy levels should be avoided. E.g. in France producers of packaging material only have to pay for 80% of the waste management costs, the remainder is financed by municipalities (OECD, 2014). In Japan, the national government has to finance educational and awareness-raising campaigns to stimulate collection of compact rechargeable batteries (see Annex G). Although these cost-sharing arrangements may foster collaboration between stakeholders and avoid financial liability without financial control by producers, incomplete internalisation leads to weakened incentives for DfE. By assigning the operational or financial responsibility of awareness campaigns, clean-up of littering and administrative monitoring to producers, rather than to governments, additional costs would be internalised and DfE incentives strengthened.

PROs are increasingly active in R&D consortia. For example the European association of PROs for consumer electronics, WEEE Forum, actively participates in European research projects. Whereas their role is now only focused on supplying expertise, Didier and Sittler (2014) are pleading to extend their role to a supplier of research funds. As illustrated by the numerical model of Brouillat and Oltra (2012), additional PRO research funds would enhance DfE, in a rather evident way, by fostering technological breakthroughs. In addition, PRO participation can stimulate collaboration between all actors of the value chain, which will enable better alignment of the different processes in all lifecycle phases. Furthermore, increasing the PRO costs raises PRO fees, whichmakes DfE incentives of CPR more effective as discussed earlier. Some PRO expenditure for R&D can therefore improve EPR efficiency if the external costs of products are not fully internalised yet.

Another cost internalisation issue is related to the take-back targets that are used in most EPR schemes (Acuff and O'Reilly, 2013). As shown by Palmer and Walls (1997) and Dubois (2012), these collection and recycling targets only lead to partial internalisation. Indeed, if a take-back target induces recycling of 35% of plastics packaging put on the market, 65% of waste will be disposed of via the municipal mixed waste collection or via littering. On the one hand, increasing the take-back target any further may be inefficient because the environmental gains may not compensate the additional expenditure. More specifically, imposing a 100% take-back target would not only be close to impossible to achieve but also excessively expensive. On the other hand, keeping the take-back target at the current level does not lead to full cost internalisation or incentives for further innovation with respect to recycling. This stalemate can be circumvented by taxing producers (or PROs) for non-collected waste fractions. Such a combination of EPR take-back targets and taxes on non-collected waste fractions would combine the operational advantages of EPR with full cost internalisation. Until recently, Slovakia combined take-back targets with a product tax (or Advance Disposal Fee – ADF) for several products (Grgulovà, 2014). Although such a tax can remediate incomplete cost internalisation of take-back targets (Dubois and Eyckmans, 2014), Slovakia has decided to abolish the ADF due to its alleged redundancy.

5.4.4. *Scope*

To enhance cost internalisation by producers, the scope of EPR should be widened to include more products (Van Rossem et al., 2006; Mayers and Butler, 2013). For example, considering that printed press amounts to 34% of graphic paper put on the market in France, the exemption of EPR obligations for printed press significantly hampers EPR effectiveness (Didier and Sittler, 2014). Moreover, a perverse incentive is at play for products that are hard to recycle. Since most governments impose take-back targets, non- recyclable goods are rarely targeted by EPR regulation. If, however, thanks to new models or DfE, recycling becomes feasible, producers suddenly risk imposition of EPR. In such a situation, rational producers would avoid investing in DfE. Actually, Korea is one of the few countries that use ADF to internalise waste management costs of products that are hard to recycle such as chewing gum, disposable diapers, cigarettes, non-packaging plastics or kitchenware (see Annex J). Such ADF for non-recyclable products can restore incentives to design products that can be recycled in the future. ADF can also internalise costs for small waste streams for which setting up a take-back scheme would induce excessive administrative costs.

EPR schemes for durable goods typically have low collection rates (Nash and Bosso, 2013). Excluding Scandinavian countries, collected volumes of e-waste in Western Europe are below 10 kg per inhabitant per year (Eurostat, 2013 – data 2010). The recent Australian scheme for television and computer recycling collects about 2 kg while the states in the US collect between 0.3 and 4 kg e-waste per inhabitant per year (see Annex K and Department of the Environment 2014). Considering that yearly sales of consumer electronics per inhabitant and per year are approximately 25 kg in Western Europe, close to 30 kg in Australia and above 30 kg in the US, the need for additional collection and registration is high (Eurostat, 2014 – data 2010, *http://step-initiative.org/*). In a similar way, collection rates for waste compact batteries in Japan are below 30% (see Annex G). There are multiple reasons for the gap between collected volumes and volumes put on the market: in a growing market, sales of durable goods are higher than waste production; small consumer electronics and batteries are hoarded in-house rather than disposed of by the available collection schemes; there is no registration for vehicles and consumer electronics that are exported as second-hand goods; small recycling companies collect metal containing waste without registration. This is mainly an issue for waste streams with positive market value, such as car batteries or some consumer electronics; small e-waste and batteries are disposed via mixed household waste; and finally, waste can be disposed of illegally (Hotta et al., 2014). As discussed above, full internalisation is needed to strengthen DfE. Consequently, a better registration of e-waste and more ambitious EPR aims are needed. In this perspective, it was long due that the European e-waste directive (2012/19/EU) finally substituted the meagre target of 4 kg per inhabitant. The new targets for 2019 aim for collection of 65% of consumer electronics put on the market or 85% of available e-waste.

Fundamentally, there is no difference between incentives from a voluntary or a legally imposed EPR scheme since both internalise costs. However, as highlighted by a historical review of voluntary EPR programs in the US for consumer electronics, rechargeable batteries, mercury thermostats and auto switches, the scope of voluntary schemes tends to be limited (Nash and Bosso, 2013). For example, although consumer electronics are made up of hundreds of products, many voluntary schemes only focus on specific products with high visibility, such as TVs or computers. Voluntary programs also tend to achieve low collection rates (Hickle, 2013). Evidently, even a voluntary program with low collected volumes is better than no program. Nonetheless, due to the typical narrow scope, it seems that, at least in some cases, voluntary programs are strategic behaviour to pre-empt more rigorous legal action (Maxwell et al., 2000).

5.4.5. *Global market vs local EPR schemes*

Consumer products such as mobile phones are designed for the global market. Conversely, EPR schemes are organised by states, countries or provinces. Considering that incentives from small local markets will not be taken into account for the design of global consumer products, the influence of individual EPR schemes is limited (Didier and Sittler, 2014; Vanderstricht, 2014; Séguin, 2014). Moreover, since international producers typically use local retailers for commercialisation, DfE incentives are only indirectly felt by international producers.

Due to differences in international consumption patterns, many products are already adapted to local preferences. For example, different food consumption patterns across countries, have led to nationally differentiated food packaging. Also, assembly of cars allows significant flexibility in order to fulfil national consumer demands, such as the side of the steering wheel or the tax-related optimal size of an engine. Clearly, the impact of local EPR schemes on design will be more important for products that are adapted in function of local consumption patterns. In this perspective, EPR schemes may have significant spill-over effects. Learning lessons from an environmentally designed product in one market can easily lead to innovations in other markets.

International harmonisation can enhance the impact of modulated fees for global consumer products. Such harmonisation may be facilitated by an international categorisation of products according to their environmental performance, such as ease of disassembly. Actually, RoHS has induced design changes worldwide thanks to its transparent focus on six hazardous materials. Defining similar clear DfE priorities which enable to categorise products in an international context will help local policy makers and PROs to modulate their fees based on both this product categorisation and local boundary conditions, such as labour cost. This can result in more harmonised and more effective DfE incentives.

5.4.6. *Technological innovation*

The development of Brand/model recognition technologies such as Radio Frequency Identification (RFID) opens possibilities for better cost internalisation. RFID uses radio waves to remotely read tags embedded in products. A cost-effective way of product recognition helps recyclers to detect products that are easy to disassemble or recycle. In addition, product information such as material composition and dismantling instructions can be easily recalled either directly from the RFID or through a database. Most importantly, recognition allows allocation of the waste management costs between producers (O'Connel et al., 2013). Unfortunately, although markets exist for these technologies, deployment of this technology for waste management is still in an early stage (Dempsey et al., 2010).

5.5. Conclusions

Assessing the extent to which EPR fosters DfE is complicated by three factors. First, due to the diversity and interaction of environmental aspects, comparing the DfE level of two products is not straightforward. For example, comparing easy to recycle vehicles with more energy-saving models requires a detailed analysis and subjective weights for distinct environmental impacts. Secondly, DfE can be driven by many other elements than EPR, such as material or transport costs. Disentangling the contribution of each driver is challenging. Thirdly, EPR can contribute to DfE by financial incentives but also by less tangible effects such as improved communication between producers and recyclers. Although anecdotal evidence highlights the importance of an improved information exchange, it is hard to measure the impact. In order to overcome the barriers for quantitative assessment, qualitative research methods are used to investigate the impact of EPR on DfE. The results suggest that EPR contributes to DfE but is seldom the triggering factor.

With respect to the financial incentives for DfE, EPR schemes can be grouped into three types: Individual Producer Responsibility (IPR), variable-fee Collective Producer Responsibility (CPR) and fixed-fee CPR. Under IPR, each producer is responsible for the waste originating from its own products. Although literature highlights that IPR provides efficient DfE incentives, few IPR schemes have been implemented due to the economies of scale and convenience of industry-wide CPR schemes. The weight-based cost allocation of variable-fee CPR is typically implemented for fast-moving mono-material goods, such as packaging and graphic paper. Whereas these variable-fee CPR schemes can give effective incentives to make products lighter or to opt for different materials, they rarely stimulate other DfE aspects such as the use of secondary resources. Fixed-fee CPR schemes are typically adopted for durable or multi-material goods, such as cars or consumer electronics. Considering that all models of the same category face an identical unit cost, fixed-fee CPR schemes provide only weak incentives for DfE.

In addition to the EPR type, six features significantly affect the financial DfE incentives of CPR schemes. First, fixed fees that are often used in CPR schemes for multi-material or durable goods, weaken DfE incentives. To address this challenge, modulated fees, differentiated in proportion to the environmental impact of a product, could be introduced. For example, modulation may lead to lower fixed fees for products that contain less toxic materials. However, to limit additional administrative costs, modulated fees should only focus on design aspects with significant environmental or financial gains. Second, the magnitude of the financial incentive is decisive. Low fees, such as applicable for many small consumer electronics, will hardly give incentives to make design changes. Third, the importance of the magnitude of the financial incentive stresses the relevance of the ongoing debate on full cost internalisation. The principle of full cost internalisation means that all costs related to waste management of products should be internalised via EPR. There is a consensus that EPR schemes should cover the costs of collecting and treating waste that is pre-sorted by households minus the revenues from recovered materials. However, there is debate concerning the allocation of other costs: costs for the collection and treatment of waste that is not pre-sorted by households (e.g. collected by municipalities via the residual mixed waste bag); costs for awareness raising campaigns; costs for clean-up of litter; costs for enforcement and monitoring of the EPR scheme (including auditing and measures against free-riding); R&D investment in DfE. From the perspective of DfE, more costs allocated to producers means higher financial incentives for DfE. Fourth, in order to improve DfE incentives, the ambition with respect to product range and take-back targets should be further improved for many EPR schemes. For example, the worldwide collection rates of consumer electronics via EPR are low. Fifth, EPR is implemented by countries, regions or provinces. For some products, such as food products, design and packaging are already adapted to local preferences. In such a context, incentives for DfE from a local EPR scheme will be taken into account by producers. In addition, environmental designs in one country can easily spill over to other markets. Conversely, due to the international disparity of policies, the impact of local EPR schemes is weak for consumer products that are designed for the global market. For example, financial incentives from one local EPR scheme will not be taken into account for the design of mobile phones. In order to strengthen DfE incentives for global consumer products, more international harmonisation of EPR incentives is required. Finally, since cost allocation between competitive producers is key for efficient DfE incentives, emerging technologies such as Radio Frequency Identification (RFID) can, in the future, induce design changes and may even lead to cost-efficient IPR implementation.

Note

1. RoHS aims to phase out six hazardous materials in consumer electronics.

References

Acuff, K. and D.T. Kaffine (2013), "Greenhouse gas emissions, waste and recycling policy", *Journal of Environmental Economics and Management*, Vol. 65, pp. 74-86.

Agrawal, V.V. et al. (2012), "Is leasing greener than selling?", *Management Science*, Vol. 58/3, pp. 523-533.

ARN (2012), *Waste Disposal Fee*, *www.arn.nl/english/ARN-Auto-Recycling/Waste disposal fee*, retrieved 6/10/2012.

Department of the Environment (2014), *The Australian National Television and Computer Recycling Scheme: Case Study* prepared for the OECD, *www.oecd.org/env/waste/gfenv-extendedproducerresponsibility-june2014.htm*.

Bio IS (2012) *Use of Economic Instruments and Waste Management Performances*, Report for the European Commission DG Environment, Bio Intelligence Service, Paris.

Bouteligier, S., R. Bracke and K. Dierick (2014), *Extended Producer Responsibility: The Case of Used Tyres in Flanders (Belgium)*, Case Study prepared for the OECD, *www.oecd.org/env/waste/gfenv-extended producerresponsibility-june2014.htm*.

Brezet, H. and C. Van Hemel (1997), *Eco-design. A Promising Approach to Sustainable Production and Consumption, United Nations Environmental Programme* (UNEP), France.

Brouillat, E. and V. Oltra (2012), "Extended producer responsibility instruments and innovation in eco-design: An exploration through a simulation model", *Ecological Economics*, 83, 236-245.

Bury, D. (2013), "Canadian Extended Producer Responsibility", *Journal of Cleaner Production*, 17, 2: 167-169.

Calcott, P. and M. Walls (2000), "Can downstream waste disposal policies encourage upstream 'Design for environment'?", *American Economic Review*, 90, 233-237.

Calcott, P. and M. Walls (2005), "Waste, recycling and 'Design for environment': Roles for markets and policy instruments", *Resource and Energy Economics*, 27, 287-305.

Damgaard, A., A.W. Larsen and T.H. Christensen (2009), "Recycling of metals: Accounting of greenhouse gases and global warming conditions", *Waste Management & Research*, 27, 773-780.

Dempsey, M. et al. (2010), *Individual Producer Responsibility: A Review of Practical Approaches to Implementing Individual Producer Responsibility for the WEEE Directive*, INSEAD, Fontainebleau.

Didier, E. and E. Sittler (2014), *Mieux concevoir les produits pour mieux valoriser les déchets*, Rapport d'information fait au nom de la Commission du développement durable, des infrastructures, de l'équipement et de l'aménagement du territoire.

Dubois, M. (2012), "Extended Producer Responsibility for consumer waste: The gap between economic theory and implementation", *Waste Management & Research*, Vol. 30(9) Suppl. 36-42.

Dubois, M. (2013), "Towards a coherent European approach for taxation of combustible waste", *Waste Management and Research*, Vol. 8, 1776-1783.

Dubois, M. and J. Eyckmans (2014), *Efficient Waste Policies and Strategic Behaviour with Open Borders*, Mimeo, KU Leuven, Leuven.

Duflou, J.R. et al. (2009), "Environmental impact analysis of composite use in car manufacturing, CIRP Annals", *Manufacturing Technology*, 58, 1, 9-12.

Fleckinger, P. and M. Glachant (2010), "The organisation of extended producer responsibility in waste policy with product differentiation", *Journal of Environmental Economics and Management*, 59: 57-66.

Fost Plus (2014), *Tarifs 2014, www.fostplus.be/fr/entreprises/votre-declaration/tarifs*.

Gottberg, A. et al. (2006), "Producer responsibility, waste minimization and the WEEE Directive: Case studies in eco-design from the European lightning sector", *Science of the Total Environment*, 359: 38-56.

Gui, L. et al. (2013), "Implementing extended producer responsibility legislation", *Journal of Industrial Ecology*, 1-15.

Grgulova, A. (2014) *Extended Producer Responsibility – Packaging and Packaging Waste in Slovakia*, Case study prepared for the OECD, available at *www.oecd.org/env/waste/gfenv-extendedproducerresponsibility-june2014.htm*.

Hickle, G.T. (2013), "Comparative analysis of extended producer responsibility policy in the United States and Canada", *Journal of Industrial Ecology*, 17, 2, 249-261.

Hosoda, E. (2004), *Evaluation of EPR Programmes in Japan, in Economic Aspects of Extended Producer Responsibility*, OECD, Paris.

Isaksson, O., T.C. Larsson and A. Ohrwall Ronnback (2009), "Development of product-service systems: Challenges and opportunities for the manufacturing firm", *Journal of Engineering Design*, 20, 4: 329-348.

Khetriwal, D.S., P. Kraeuchi and R. Widmer (2009), "Producer responsibility for e-waste management: Key issues for consideration – learning from the Swiss experience", *Journal of Environmental Management*, 90, 1, 153-165.

Lifset, R. and T. Lindhqvist (2000) "Does leasing improve end of product life management?", *Journal of Industrial Ecology*, 3,4, 10-13.

Lindhqvist, T. and R. Lifset (2003), "Extended Producer Responsibility, can we take the concept of individual EPR from theory to practice?", *Journal of Industrial Ecology*, 7, 3-6.

Matsueda, N. and Y. Nagase (2012), "An economic analysis of the packaging waste recovery system in the United Kingdom", *Resource and Energy Economics*, Vol. 34, pp. 669-679.

Maxwell, J.W., T.P. Lyon and S.C. Hackett (2000), "Self-regulation and social welfare: The political economy of corporate environmentalism", *Journal of Law and Economics*, 43, 2: 583-618.

Mayers, K. (2007), "Strategic, financial and design implications of extended producer responsibility in Europe", *Journal of Industrial Ecology*, 11, 3, 113-131.

Mayers, K. et al. (2012), "Implementing individual producer responsibility for waste electrical and electronic equipment through improved financing", *Journal of Industrial Ecology*, 17, 2, 186-198.

Mayers, K. and S. Butler (2013), "Producer responsibility organizations development and operations", *Journal of Industrial Ecology*, 17, 2, 277-289.

McKerlie, K., N. Knight and B. Thorpe (2006), "Advancing extended producer responsibility in Canada", *Journal of Cleaner Production*, 14, 616-628.

Monier, V. et al. (2014), *Development of Guidance on Extended Producer Responsibility*, Report for the European Commission, DG Environment.

Nash, J. and C. Bosso (2013), "Extended producer responsibility in the United States", *Journal of Industrial Ecology*, 17, 2, 175-185.

Neto, J.Q.F. and L.N. Van Wassenhover (2013), "Original equipment manufacturers' participation in take-back initiatives in Brazil", *Journal of Industrial Ecology*, 17, 2, 238-248.

O'Connel, M. et al. (2013), "Feasibility of using radio frequency identification to facilitate individual producer responsibility for waste electrical and electronic equipment", *Journal of Industrial Ecology*, 17, 2, 213-169.

OECD (2014), "The state of play on Extended Producer Responsibility (EPR): Opportunities and challenges", available at *www.oecd.org/environment/waste/Global%20Forum%20Tokyo%20Issues%20 Paper%2030-5-2014.pdf.*

OECD (2013), *What have we learned about Extended Producer Responsibility in the last decade? A survey of the recent economic EPR literature*, ENV/EPOC/WPRPW(2013)7/FINAL, OECD, *http://spot.colorado.edu/ ~daka9342/OECD_EPR_KO.pdf.*

OECD (2005), *Analytical Framework for Assessing the Costs and Benefits of Extended Producer Responsibility Programmes*, OECD, *http://dx.doi.org/10.1787/oecd_papers-v5-art13-en.*

OECD (2004), *Economic aspects of extended producer responsibility*, OECD Publishing, *http://dx.doi.org/ 10.1787/9789264105270-en.*

OECD (2001), *Extended Producer Responsibility – A Guidance Manual for Governments*, OECD Publishing, *http://dx.doi.org/10.1787/9789264189867-en.*

Palmer, K. and M. Walls (1997), "Optimal policies for solid waste disposal: Taxes, subsidies and standards", *Journal of Public Economics*, 65, 193-205.

Pangburn, M.S. and E. Stavrulaki (2014), "Take back costs and product durability", *European Journal of Operational Research*, 238, 175-184.

Peeters, J.R. et al. (2014), "Design for disassembly for consumer products", *Working Paper*, KU Leuven, Leuven.

Plambeck, E. and Q. Wang (2009), "Effects of e-waste regulation on new product introduction", *Management Science*, 55, 3, 333-347.

Recupel (2013), *List of Appliances* (translated from: Apparatenlijst), Brussels.

Roine, K. and C.-Y. Lee (2006), "With a little help from EPR? Technological change and innovation in the Norwegian plastic packaging and electronics sectors", *Journal of Industrial Ecology*, 10, 1-2: 217-237.

Rotter, V.S. (2011), "Waste management and producer responsibility: A score behind – A new ahead", *Waste Management & Research*, 29, 889.

Sachs, N. (2006), "Planning the funeral at the birth: Extended Producer Responsibility in the European Union and the United States", *Harvard Environmental Law Review*, Vol. 30, No. 51.

Séguin (2014), *Promoting Sustainable Materials Management Through Extended Producer Responsibility: Canadian Waste Electrical and Electronic Equipment*, Case study prepared for the OECD, available at *www.oecd.org/env/waste/gfenv-extendedproducerresponsibility-june2014.htm.*

Short, M. (2004), "Taking back the trash: Comparing European extended producer responsibility and take-back liability to US environmental policy and attitudes", *Vanderbilt Journal of Transnational Law*, 37, 1217-1254.

Subramanian, R., S. Gupta and B. Talbot (2009), "Product design and supply chain co-ordination under extended producer responsibility", *Production and Operations Management*, 18, 3, 259-277.

Tojo, N. (2004), "Extended Producer Responsibility as a driver for design change – utopia or reality?", IIIEE Dissertations 2004: 2, Lund University, Lund.

Tong, X and L. Yan (2013), "From legal transplants to sustainable transition", *Journal of Industrial Ecology*, 17, 2, 199-212.

Van den Abeele, P. (2006), "Pay as you go recyclage", *Industrie – Technisch & Management*, No. 32, Agoria, Brussels.

Van Hemel, C. and J. Cramer (2002), "Barriers and stimuli for eco-design in SMEs", *Journal of Cleaner Production*, 10, 439-453.

Van Rossem, C., N. Tojo and T. Lindhqvist (2006), *Extended Producer Responsibility*, Report commissioned by Greenpeace International, Friends of the Earth and the European Environmental Bureau (EEB).

Vanderstricth, C. (2014), *What Have We Learned about Extended Producer Responsibility in the Past Decade? Case study Chile*, Case study prepared for the OECD, available at *www.oecd.org/environment/waste/20140526%20Case%20Study%20-%20EPR%20Chileamended.pdfOECD Intranet.*

Walls, M. (2006), *EPR Policies and Product Design: Economic Theory and Selected Case Studies*, document prepared for OECD, *http://papers.ssrn.com/sol3/papers.cfm?abstract_id=901661.*

Yeo, J.H. (2005), *Evaluation of Extended Producer Responsibility Performance: Korean Case Study*, Report prepared for OECD, Korea Environment Institute.

Websites

http://app.leg.wa.gov/RCW/default.aspx?cite=70.95N&full=true#.

http://designguiden.grontpunkt.no/.

http://global.epson.com/SR/csr_topics/201408_01.html.

http://group.electrolux.com/en/category/sustainability/products-and-sustainability.

http://proeurope4prevention.org/packagings-trends.

http://reduction.ecoemballages.fr/catalogue/.

http://step-initiative.org/index.php/Overview_USA.html.

http://worldwide.hyundai.com/wcm/idc/groups/sggeneralcontent/@hmc/documents/sitecontent/mdaw/mdm2/~edisp/hw033151.pdf.

www.apple.com/benl/environment.

www.bestbuy.ca/en-CA/for-kids-environment.aspx#tab-1.

www.conai.org/hpm00.asp?idcanale=99.

www.dell.com/learn/us/en/uscorp1/dell-environment-greener-products?c=us&l=en&s=corp&cs=uscorp1.

www.ecoemballages.fr/.

www.ecofolio.fr/adherents/le-bareme.

www.eco-systemes.fr/documents/Bareme.pdf.

www.epbp.org/.

www.fostplus.be/SiteCollectionDocuments/Leden/GP%20tarieven/Groene%20Punt%20tarieven%202014.pdf.

www.fujitsu.com/global/about/environment/index.html.

www.lighting.philips.com/main/projects/rau.wpd.

www.nedvang.nl/#!/bedrijven/preventievoorbeelden/overzicht.

www.oneteamoneplanet.com/americas/tires4ward.

www.preventandsave.ie/CaseStudyexamples.html.

www.preventpack.be/exampleswww.recylum.com/enjeux/eco-conception-des-equipements/.

www.samsung.com/us/aboutsamsung/sustainability/environment/takebackrecycling.

www.sony.net/SonyInfo/csr_report/environment/recycle/policy.

www.unesda.org/.

www8.hp.com/us/en/hp-information/environment/product-recycling.html#.U_MU5GM6Lm4.

PART II

Chapter 6

Extended producer responsibility and the informal sector

This chapter examines the role that the informal sector plays in extended producer responsibility (EPR) systems in middle-income countries. It is intended to supplement the 2001 OECD guidance manual on EPR which had focused on EPRs in OECD countries and which did not examine the role of the informal sector in any depth. The main findings of the chapter are that while there are serious concerns about downstream informal dismantling and recycling which can generate negative economic and environmental impacts, the potentially positive contribution of informal waste collection and sorting activities is increasingly recognised. As a result, the policy objective has shifted from "rescuing" to integrating informal workers into formal waste management systems. Recent experience also shows that failure to doing so can seriously undermine EPR systems.

6.1. Introduction

Many middle-income countries are considering the development or expansion of EPR systems for different waste streams in the context of the rapid modernisation of their waste management systems. A common challenge faced by emerging and developing countries in applying EPR policies is the absence of well-established waste management systems. These systems as well as important stakeholders such as manufacturers, PROs, municipalities and recyclers that are presumably available in developed countries and have potentially significant roles in EPRs are not always available (Akenji et al., 2011). Another challenge they often face is how to deal with the large informal sector that relies on these waste streams for their livelihoods. Several sets of policy makers are interested to learn from experience gained and identify best practices: policy makers in middle-income countries where most of the estimated 20 million informal waste workers are located; donors supporting middle-income countries to develop their solid waste management sector (Lerpiniere et al., 2014) and some higher income OECD countries where anecdotal evidence suggests that the global financial crisis has fostered the development of a large and active informal recycling and re-use sector.

While the informal sector is usually relatively small in OECD countries and is frequently associated with interfering negatively with well-functioning formal waste management systems (e.g. by illegally removing high-value products and materials from the waste stream), it often performs useful functions that are not provided by the formal sector in middle-income and developing countries with limited waste management systems (such as collecting valuables and recovering the material from them). When an EPR system is introduced in these countries, it interferes with the livelihoods of informal waste pickers who will compete for valuable materials. This report focuses on the latter case and the opportunities and challenges that it presents.

The main message of the chapter is that there are significant potential benefit from including the informal sector in EPR systems in terms of effective waste management operations, achieving recovery targets, and facilitating affordable and sustainable financing in middle-income countries where waste management systems are limited. Inclusion also provides opportunities for providing informal workers with sustainable livelihoods as well as improved health and social protection.

The first section of this chapter introduces the key concepts of public service and private value chains in solid waste and recycling, and outlines the historical development of waste management in OECD countries. Section 2 focuses on recycling and waste management, including the role of the informal sector, in middle-income countries. The example of Botswana is used to illustrate some of the features of recycling and waste management in these countries. Section 3 examines the various forms of informal recycling, and summarises what is known about the workers in this sector. Alternative approaches for integrating informal workers in waste management systems are presented. The experience of an inclusive approach to informal workers in Colombia is contrasted with experience from Bulgaria where the informal sector was excluded from the establishment and operation of an EPR system. Section 4 concludes by summarising the main lessons learned from the involvement of informal workers in EPR systems and makes some recommendations for how this relationship can be made beneficial both for the informal workers and waste management systems.

6.2. Solid waste management and recycling

6.2.1. *Service and value chains*[1]

Modern, integrated, sustainable waste management consists of two quite different sectors:

- The service chain: This sector is primarily managed by the public sector and includes a range of services including urban cleaning, street sweeping, public space management, waste collection, transport and disposal.
- The value chain: This sector is primarily managed by private sector entities, including micro, medium, large, and multi-national commodities businesses. The businesses in the recycling or industrial value chain are involved in the extraction and commercialisation (valorisation) of valuable materials from the waste stream. Valorisation includes the extraction or separation of valuable materials from waste, cleaning and processing, and trading them on markets. A useful distinction can be drawn between the *industrial value chain*, handling recyclables, the *agricultural value chain*, through which kitchen, garden, and food processing wastes are valorised as animal feed or soil nutrients, and *the re-use or second-hand goods system*, a sector emerging as important in upper-middle-income countries. This chapter focuses on the industrial value chain, which is of primary relevance to EPR systems.[2]

The two chains and their relationship to each other are shown in Figure 6.1 a highly conceptualised process flow diagram.

Figure 6.1. **Service chain and value chain**

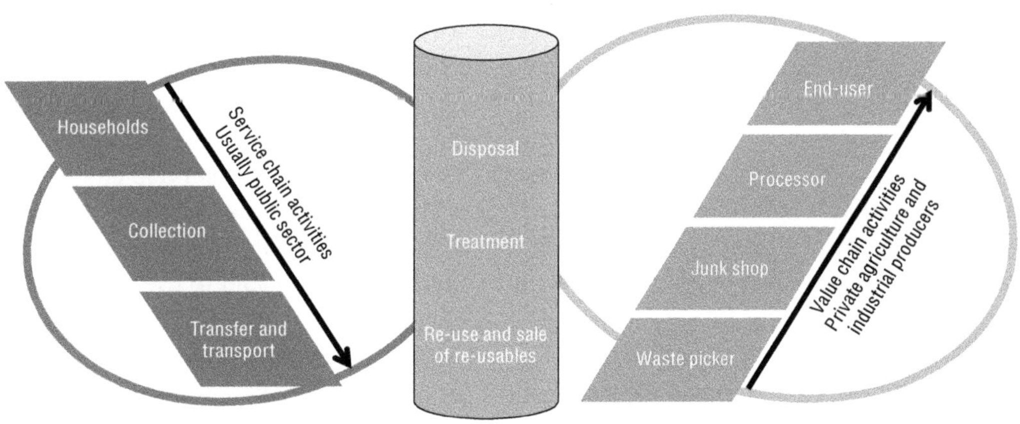

Source: Adapted from WASTE (2010), "Training Materials in Integrated Sustainable Waste Management", Gouda, the Netherlands.

6.2.2. *Historical perspective: policy drivers and modernisation*

Wilson (2007) distinguishes three policy drivers for waste management: public health, environmental protection, and resource management. Protecting public health was the main driver of the 19th century modernisation of sanitation; concerns about environmental protection was the driver for the wave of environmental research and legislation starting in the 1970s; and more efficient management of natural resources, including waste, is the main driver for current policies to promote resource productivity and sustainable materials management in OECD countries and beyond.

Figure 6.3 describes value chain recycling. This was the norm in urban areas in OECD countries until the emergence of Integrated Waste Management in the 1980s. This is the context in which middle-income countries have been developing their waste management systems. It is also the situation in which most low-income countries find themselves before they establish a basic solid waste law or an environmental ministry (De Swaan, 1988; Scheinberg and van de Klundert, 2005; Scheinberg, 2011; Strasser, 1999; Poulussen, 1987; Velis et al., 2009; Melosi, 1981; Wilson, 2007; Gille, 2007).

Figure 6.2. **Value Chain Recycling Framework**

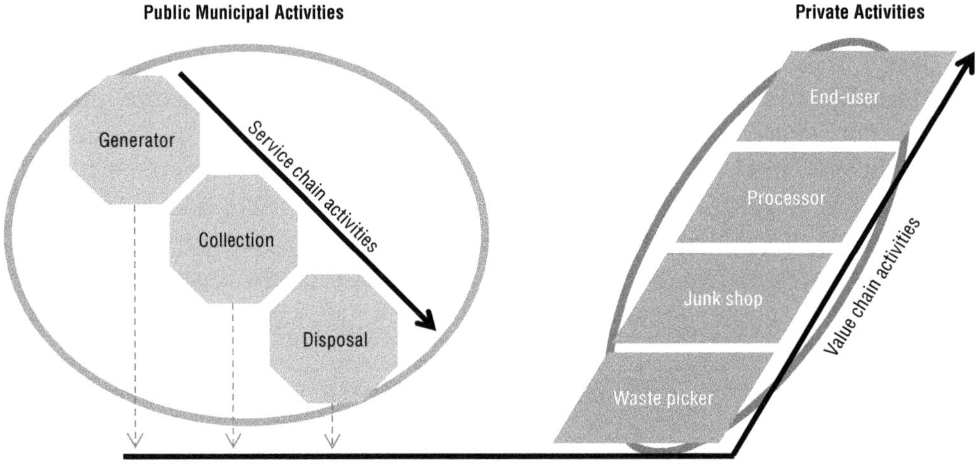

Source: Adapted from WASTE (2010), "Training Materials in Integrated Sustainable Waste Management", Gouda, the Netherlands.

In the value chain recycling system, public waste management and private sector recycling belong to two separate universes, which barely interact or acknowledge each other. On the right of the figure, private enterprises collect materials and valorise them in the value chain. The only movement of recyclables from the public service chain to the private value chains is through informal activities. The public sector solid waste system on the left, keeps the city clean, and ignores the small amounts of material that are extracted, either by waste pickers or by sanitation workers themselves, and sold to the value chain.

Since the early 1980s, most OECD countries have introduced some variant of municipal recycling. This was in response to high disposal costs associated with regionalised sanitary landfilling and/or mass incineration, as well as public demand and lobbying from environmental NGOs on minimizing landfill and incineration. Municipal recycling was motivated by the need to reduce disposal and the related costs that were financed by the service chain. For this reason, high recycling rates are often found where population densities are high, land is scarce, and disposal is expensive. The resource conservation benefits of recycling or composting are often mentioned as strong environmental motivators. However, in practice, the key driver for cities to find alternative value chain destinations for municipal waste has been when the price of disposal rose above about USD 40 per tonne (Scheinberg, 2011; Scheinberg et al., 2010b; UNEP, 2015). At this point, and as illustrated in Figure 6.4, the municipal service chain expands and may ultimately eliminate and replace the lower levels of the private value chain.

Figure 6.3. **Classic Municipal Recycling as Developed in the 1980s in the United States and Canada**

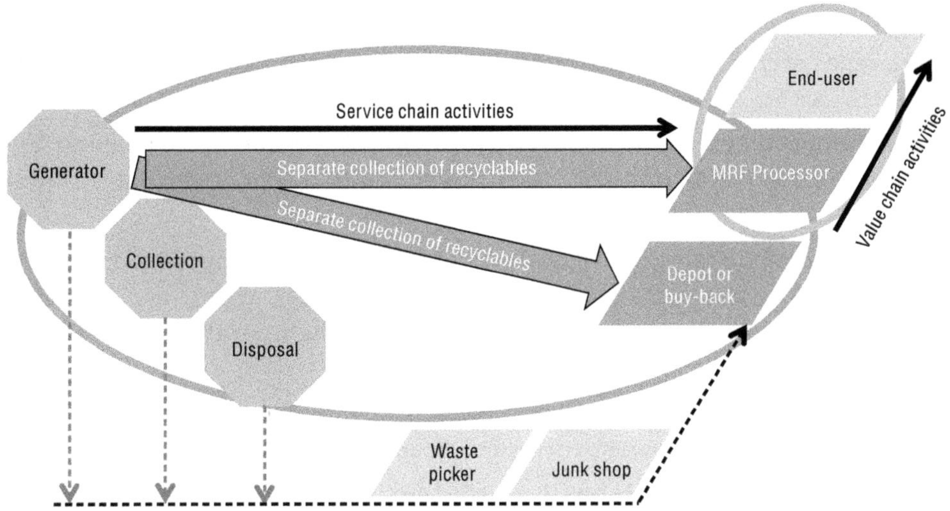

Several aspects of Figure 6.3, which draws on the experience in Australia, Canada and the US, deserve special attention.

First, the municipal sector has replaced the lower levels of the value chain with two new public institutions and facilities: the depot or buy-back centre, and the Materials Recovery Facility (MRF). The depot or buy-back centre replaces small junk shops. The MRF introduces a new function, the post-collection separation of comingled recyclables, which is at the same level as a mid-level value chain dealer or intermediate processing facility (IPF). These two institutions, taken in combination with a third innovation, public-sector separate collection of recyclables, made it possible in the 1980s for local authorities to start to extract clean recyclables from household waste. These developments depended on changes in the behaviour of households, the users of the system. However, they also benefited from a period of high levels of innovation in communications systems and household storage technology (Scheinberg, 2011).[3]

The second consequence was that the sphere of influence of the private value chain shown in Figure 6.2 is greatly reduced in comparison with that shown in Figure 6.3. The private sector junk shops and the network of waste pickers that supply them are marginalised and find themselves outside of the system. Their numbers reduced greatly in large cities in OECD countries in the 1990s.

The third result was increased pressure on the end-user industries in the existing value chain, to absorb the continually growing volume of recyclables, and to eliminate dangerous or difficult to recycle materials altogether. This can be seen as the origin of the hierarchy of waste management concept, and the motivation for a focus on waste prevention on the one hand, and on cradle-to-cradle (C2C) or design for recyclability, on the other. This development also fostered the emergence of producer responsibility systems, or product stewardship schemes, the voluntary equivalent. Experience gained by the first generation of EPR systems provided the basis for the 2001 OECD EPR Guidance Document (OECD, 2001).

The development of municipal recycling also led to the elimination and/or integration of informal recycling in the 1980s and 1990s in many OECD countries. When scavenging became illegal, many individual waste pickers were absorbed into the service chain. In the US many found work in MRFs, and in European countries like the Netherlands, many became part of the second-hand goods sector.

These series of changes are now beginning to play out in middle-income countries, but the circumstances are somewhat different than those in OECD countries in the 1980s:

- GDP and average household income levels are lower in middle-income countries today than they were in OECD countries in the 1980s, and, in many, income inequality is greater.
- The service chains are more unevenly developed, and the user-pays principle is less well established. Public waste management systems have less money and less access to public financing, and many are not able to cover the operational costs of waste infrastructure.
- There is less public pressure to protect groundwater and other environmental resources from inadequate waste management and to close dumpsites.
- Nearly 30 years of development has gone into the current generation of high-technology solid waste facilities and equipment, making them far more expensive in both absolute and relative terms than they were in the 1980s and 1990s.
- There are more diverse products and packaging on the market, made of more complex and less recyclable materials, than 40 years ago. These complex materials create large volumes of difficult-to-handle wastes.
- The category of household (and commercial) appliances and electronics has expanded dramatically, due to the larger number of devices and applications available today at low cost and higher rates of replacement.
- The number of informal recyclers and their impacts on the system is large and growing in middle-income countries. It is estimated that globally about 20 million people work in the informal recycling sector – plus their family members (ISWA, 2014).

These differences form the backdrop to the discussion of EPR and the informal sector in this report.

6.3. Waste management and recycling in middle-income countries

This section examines some of the key features of service and value chains in middle-income countries, including the role of the informal sector.

6.3.1. *Value chain recycling and the informal sector*

In middle-income countries, recycling is quite limited to valuable fractions of waste, while non-valuable fractions of waste is discarded, and it is largely a private sector rather than a municipal activity. Its complexity and intensity is illustrated in Figure 6.4. Recycling markets, as well as private sector activities in the value chain, vary significantly depending both on the proximity of large value chain buyers and the intrinsic economic value of the waste stream. This intrinsic value is generally a factor of GDP, purchasing power, and socio-cultural factors which define the content of these wastes (Gille, 2007; Bauman, 2002).

Figure 6.4. **Schematic of the value chain**

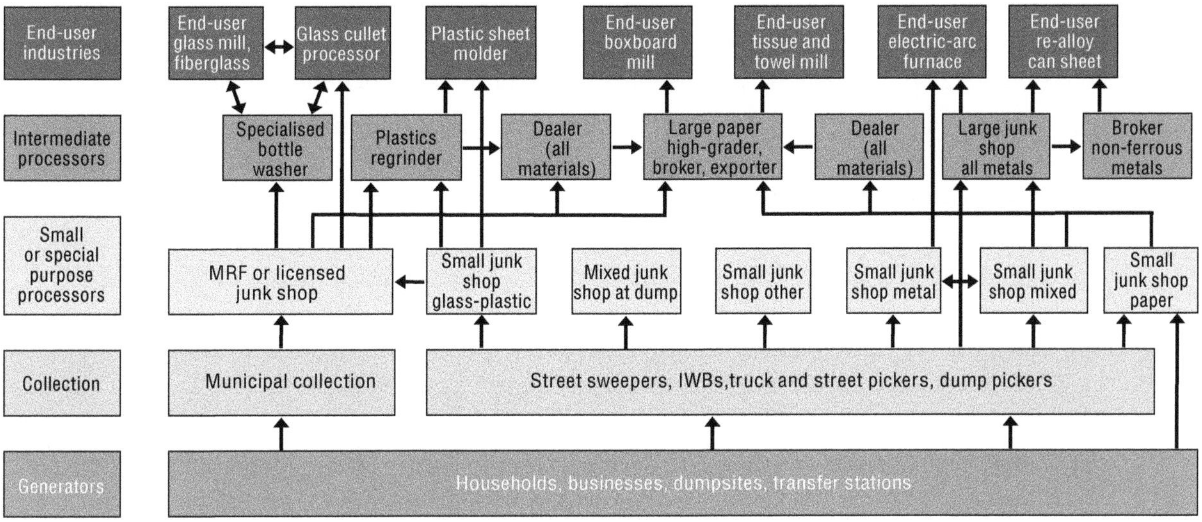

Source: Adapted from WASTE (2010), "Training Materials in Integrated Sustainable Waste Management", Gouda, the Netherlands

Large capital cities attract thousands of informal recyclers, whereas the situation in smaller cities is more variable (Gerdes and Gunsilius, 2010; Lardinois and Furedy, 1999). In Asia and Latin America, there is a basic level of value chain recycling, which usually does not exceed 15% of domestic waste, in many medium-sized cities[4].

In addition to earning livelihoods for themselves and their families and thereby participating in the economy, informal recyclers provide a number of other economic, environmental, and social benefits: they help to reduce the amount of waste going to landfill; they provide an alternative, free collection service, which is a direct economic benefit to businesses and households, and helps to reduce collection and disposal costs for municipalities; they help to reduce demand for the extraction of primary resources, and their activity contributes to lowering the emission of greenhouse gases (Chaturvedi et al., 2009).

Thus, informal private recycling in middle-income countries creates positive economic and environmental externalities. In a sense, informal recyclers are subsidising the municipal waste system by providing alternative sinks at no cost to the city (DEAT, 1998; Scheinberg et al., 2010a, 2010b; Wilson et al., 2006). While informal recyclers receive direct economic benefit, municipalities are benefiting indirectly. In recognition of these indirect benefits, municipalities such as in Bogota, Colombia, pay recognised informal recyclers for the recyclables they collect (Information from ARB and CEMPRE Colombia).

6.3.2. *Materials and market failures*

Value chain recycling is a viable business model when materials found in the waste stream have *intrinsic economic* value and when there is a market or practical economic demand for them. Metals have a consistently high value and can be profitably extracted and marketed in most parts of the world. High-grade paper is also in demand for a good price but it is perishable and the markets are not always close enough to make the collection and marketing activities profitable. Plastics have both a global and a local market value and therefore marketability depends on the type of plastics, the process method, and the quality of the separation. Bottle glass is a reliable low-value material for local recycling, but has a shrinking share of the packaging market.

The economic viability of value chain recycling is often constrained by market failures that result in little or no local economic demand even for materials that are technically recyclable. Three common types of market failures can be distinguished. Demand for producer responsibility schemes is generally in response to one or more of such market failures.

- The bottleneck: In this type of market failure, there are abundant supplies of materials and also a robust economic demand, but there is a lack of collection and processing capacity to connect them. Investment, micro-financing, and EPR-facilitated agreements can eliminate this failure and bring demand and supply together.
- The round peg in a square hole: This market failure represents a failure of the seller to meet the buyer's specifications and is often created by a lack of local knowledge about the needs of the end-user industry. Capacity development and some forms of co-operative marketing can address this type of issue, often without an EPR or PS.
- Flooding the market: This type of market failure occurs when there is a rapid increase in the supply that exceeds the demand. Markets can be flooded locally or globally and the situation may arise from a local authority or national government enacting a pro-active recycling law, or from a socio-economic crisis that drives many people to engage in recycling as a means of survival. This happened to the United States paper market in 1990, when large numbers of maturing municipal recycling programmes started to produce large volumes of high-grade (high-value) paper. In general this type of market failure solves itself over time, as the value chains find ways to use the supply to increase their production (Scheinberg, 2011; WASTE, 2010).

Table 6.1 presents a matrix showing the linkages among types of materials, their economic value, typical market failures and preferred policy responses.

6.3.3. *Exclusive recycling – competition for recyclables*

National governments and local authorities become interested in recycling when it promises to generate additional revenues to finance their solid waste systems or when legal obligations come into place. Their objective is to reduce the costs of disposal. However, any expectation of net positive revenue is unrealistic. (Scheinberg et al., 2012, 2010a; Chikarmane and Narayan, 2007).

Evidence from the analysis of 20 cities in the 2010 UN-Habitat publication Solid Waste Management in the World's Cities[5] suggests that public sector recycling without priced disposal can create a downward spiral of decreasing recycling rates and increasing volumes of waste subject to disposal (Scheinberg, 2011; Scheinberg et al., 2010b). The total amount of recycled material tends to decrease because cities are inexperienced in commercialising recyclables. This reduces revenues and increases environmental risks due to increased disposal. At the same time, the value chain loses its access to well-prepared recyclables and therefore waste pickers and their families lose livelihoods.

This approach, which can be labelled as exclusive recycling is illustrated in Figure 6.5. It shows the expansion of the municipal sphere of influence without the corresponding investment in recycling which is shown above, in Figure 6.3. Recycling in the exclusive framework is weak, and exists in two variants symbolic recycling and cash cow recycling (see Box 6.1). More money goes to recycle fewer materials through formal channels, informal recyclers work longer hours under worsening conditions to assure their livelihoods, and conflicts between formal and informal systems grow. As the level of conflict grows, cash cow recycling may lead to police harassment, property damage or

Table 6.1. **Understanding values of different types of materials in recycling**

Type	Examples of materials	Economic value	Market failures	Policy aspects
Type 1: High intrinsic value, globally traded commodities	High grades of waste paper, aluminium UBC, ferrous & non-ferrous metals, In recent years, clean PET to China. Is ± 10 to 20% of household waste.	Price paid for the materials covers or exceeds the cost of labour and equipment involved in extracting, collecting, transporting and commercialising them.	Market failures are usually of the inundation type, and are often solved with time.	These materials are almost always recycled at high levels, so interventions risk perverse effects. When there is a low rate of recovery, some form of intervention may help by increasing economy of scale or financing export.
Type 2: Moderate intrinsic commodity value, locally traded commodities	Glass, tin, steel cans, middle and low-grades of paper, rubber, PET and non-PET polyolefines (PP, LDPE, HDPE), some WEEE, textiles, low-grade paper. Total is about 10-15% of household waste.	Have some value but not consistently enough to cover the cost of extraction, processing and marketing. Recycling is not "profitable", or even able to cover costs on its own.	Market failures of the wineglass type, can be solved by investment in local processing capacity.	Requires cross-subsidy from avoided cost of disposal through "municipal recycling". Typical products for EPR/PS systems with price supports or other forms of facilitation.
Type 3: Poor recyclability, lack of clear demand or product definition, in both the agricultural and industrial value chains, no clear price points, can be valorised with subsidies, avoided cost of disposal, and/or subject to 'market development'.	3A: Non-separated C&D waste re-usables, durable goods and components with local options for 'beneficial product reuse', down-cycling, second-hand use, organised repair, etc. Examples: refrigerators, parts of vehicles and automobiles, architectural elements. Currently plastic "pillow" water packaging is in this category in some countries.	Products can be repaired, re-used, sold at greatly reduced prices. Costs of dismantling and pre-processing are usually more than the intrinsic commodity value, but with some use value and some environmental value.	All three types of market failure, but mostly lack of any viable market. Wineglass model interventions can help in some cases.	Auto parts and some parts of e-waste can be recycled when producers pay part or all of the cost of dismantling. Some deposit/return systems work for tyres, take-back for oils, advance disposal fees for electro-domestic appliances. Design for Environment (DfE)/ Waste prevention for packages
	3B: Kitchen, garden, restaurant, green space management and small livestock waste for composting (± 20-70% of household waste).	Valorisation heavily dependent on hygienic rules, for example, whether small livestock are allowed in urban areas. Also high tendency of no EPR experience but could involve suppliers.	Flooding the market or simply lack of a market is typical for compost, which is usually not a recognised product with a clear price.	Government actions to create demand and 'purchase' compost for public uses: cemeteries, parks, sports fields, mine reclamation, erosion control, landfill cover and highways, Donor- or government-subsidised programmes for biogas or other forms of recovery
Type 4: Special/contaminated/hazardous materials associated with negative environmental externalities. Use of valorisation as a "safe sink" depends on transfer payments from the service chain.	4A. Bio-sanitary waste (used sanitary protection, nappies); laminates (thin multi-layer packaging for chips, processed foods), healthcare waste.	Some residual value added, but not enough to cover cost of safe management with or without recovery.	Wineglass model may operate because of the high cost or health risks of extraction of marketable fractions.	Main candidates for waste prevention, minimisation, re-design or design for Environment/Design for Recycling. EPR indicated.
	4B. Hazardous wastes, packages contaminated with oil, contaminated soils, chemicals, fluorescent light bulbs, used engine oil, end of life E-waste (WEEE), refrigerators with CFCs. accumulators, batteries, non-CFL white-/brown goods, approximately 5% of household waste.	Negative value, can be monetised through an advanced disposal fee or through pro-active labelling or deposit-return systems.	In relation to WEEE, some fractions were wineglass failures and EPR take-backs have helped.	Regulation, licensing and certification of users, and labelling and use requirements set the boundary conditions for sale, handling, and also for safe end of life management.
Type 5: Residual MSW fractions: contaminated/not recoverable	pet litter, shoe soles, light bulbs, styrofoam, fridges	negative economic value, not useful for any type of re-processing or re-use	no market, these materials are not (yet) valorisable	waste prevention, product taxes, DfE,

Source: Adapted and updated from: Scheinberg et al., *Closing the Circle, Bringing Integrated Sustainable Waste Management Home. Association of Dutch Municipalities*, The Hague, the Netherlands, 2008.

Figure 6.5. **Exclusive recycling – public-sector recycling when disposal is not priced**

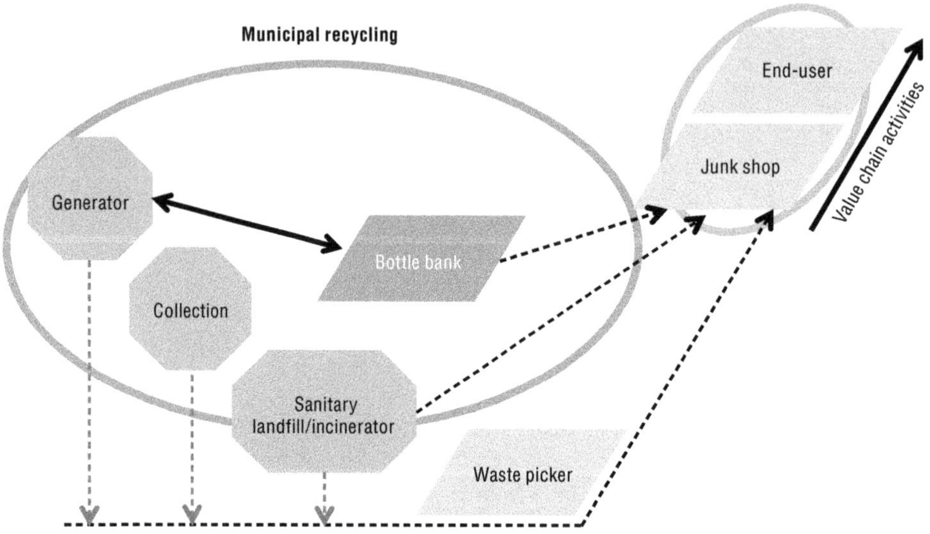

Box 6.1. **Increasing recycling target rates**

Symbolic recycling typically results from donor or NGO pressure. The main investment is in awareness raising, rather than a change of the collection system. As a result, most households do not participate. Recycling remains weak and becomes marginalised in planning and investment. Globally designed symbolic bottle banks and depots may attract small quantities of materials at very high cost per tonne. Formal recycling is uneconomic. Informal recycling may be inhibited by new landfill fences, gate controls and regulations. Some form of value chain recycling may continue at the margins of the solid waste system, but under constant threat of elimination.

Cash cow recycling aims to generate revenues to finance disposal and collection costs. In search of revenues, municipalities claim a monopoly on the capture of recyclable materials. Informal recycling is criminalised, value chain businesses are black-listed. Local authorities build a parallel formal recycling system through bottle banks, buy-back, reverse vending, or separate collection, supported by awareness campaigns. They often look to producers' organisations for financial support. Waste pickers, junk shops, and itinerant buyers have reduced legal access to materials. Sometimes, they resort to vandalising locked containers or ignoring landfill fences. Formal recovery decreases, waste going to disposal increases, and costs rise, with increasing resources going to legal enforcement of exclusivity. Pressure on municipalities to raise revenues and enforce against waste pickers increases.

violence. As explained below, this is currently the situation in some Balkan countries (ISWA/EXPRA/RDN, 2014; Scheinberg et al., 2010a; Gunsilius et al., 2010; Chaturvedi and Scheinberg 2010; Chaturvedi, 2009; Bhaskar and Chikarmane, 2012; Scheinberg, 2011).

6.3.4. *Context for EPR in middle-income countries*

As mentioned earlier in the introduction, this chapter mainly focuses on middle-income countries with limited waste management systems where EPR systems are being introduced along with a pre-existing and active informal sector. Countries with well-established EPR systems based on proper waste management systems present a different set of challenges and are not discussed in this report.

The common situation in middle-income countries that do not have well established waste management systems can be summarised as follows:

- Fast growing economies generate:
  - Growing quantities and increasing complexity of waste, associated both with growing consumption and with changing distribution patterns of wholesalers, which produces lots of visible waste with few proper disposal routes for them.
  - Strong demand from citizens, especially those breaking into the middle class, for a better life, which generally includes more consumer goods, better household infrastructure, and a cleaner living environment; in some cases this translates to strong political pressure for improvement, as well as resistance to new placement of disposal facilities.
  - Generally high household willingness to pay for solid waste services, especially removal, as long as the price is below 2% of household income.[6]
- A service chain in transition to a modern system, with:
  - well-organised waste collection, approaching 100% coverage in inner cities
  - a mixed disposal capacity with sanitary landfills in a few places, some open dumps, some controlled disposal facilities, many leaks in the system, and un-priced or under-priced disposal
  - unstable institutions with limited disposal capacities, financing from user fees and public sources that often does not cover operational costs, and weak ministerial oversight.
- At the municipal level, latent or open conflicts for access to recyclable materials between the waste utility or public works department, informal recyclers, and small junk shops.
- Uncontrolled waste dumps outside urban centres; significant waste in water; masses of plastic in the environment.
- A well-functioning private value recycling chain with the informal sector extracting up to 90% of everything that reaches the supply chain, with recovery levels differing widely depending on proximity to the global centres of the recycling industry, and limited involvement of the service chain (Scheinberg et al., 2010a).
- Mistrust, mutual suspicion, and lack of knowledge between stakeholders in the value chains, waste pickers, service chain, NGOs, and public officials.

This is the backdrop for EPR in middle-income countries such as Botswana, Colombia, India, Malaysia, Brazil, and Caribbean and Pacific SIDS (small island developing states). In some ways this resembles the situation in North-western Europe and North America in the early 1980s, when the current wave of modernisation was gaining momentum, but with important differences in income levels and number of informal recyclers.

6.3.5. *Service chains and value chains – the case of Botswana*[7]

The solid waste system in Gaborone, Botswana has many features common to middle-income countries, and can help to illustrate how waste management, recycling, institutional development, and financial flows interact in a middle-income country. Botswana also provides an example of a voluntary product stewardship scheme which makes provision for the involvement of the informal sector.

Figure 6.6 represents the Botswana solid waste system with service chains and value chains largely distinct. The waste pickers at the landfill connect the two systems through a well-designed and managed system of authorised waste picking. This involves junk shops obtaining a license from district councils to send authorised waste pickers to a landfill under their supervision with proper equipment, protective clothing, and access to medical care.

Figure 6.6. **Product Stewardship Plus Inclusive Recycling, Gaborone, Botswana, 2012**

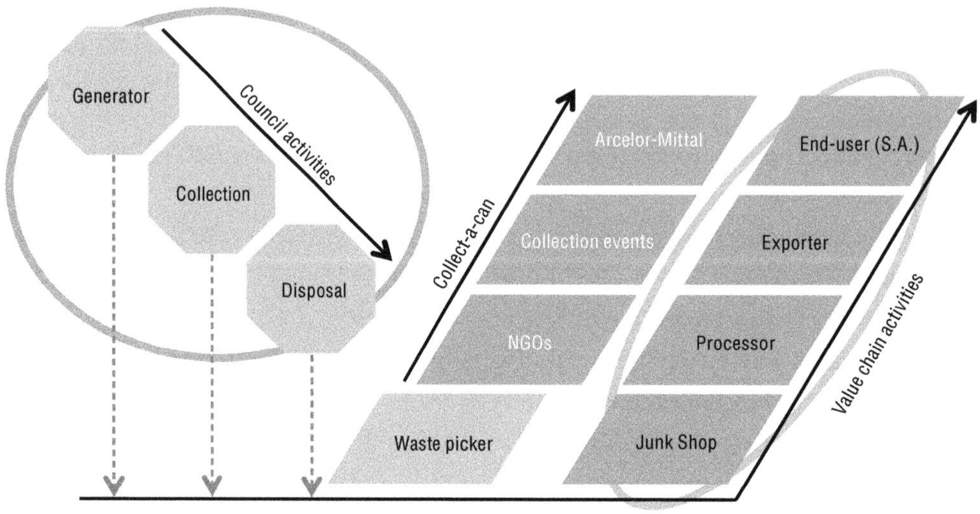

The *service chain* delivers limited municipal waste management services. The Gaborone City Council, in co-operation with private contractors, is responsible for waste collection from households, commerce and institutions. However service is irregular and subject to disruptions resulting in accumulations of uncollected waste. The collection service in rural and tribal areas is very limited. There is confusion at household level about the roles of the competent authority and implementation agency, as well as about collection schedules and payment obligations. Disposal facilities are being upgraded to become sanitary landfills with controlled gate access, a weighbridge, a geotextile liner and separate areas for specific materials, but without effective boundary control or leachate collection.[8] There is on-site incineration of medical waste and collection of crankcase (automotive) oil. Although disposal is priced and there is a schedule of differentiated tipping fees for different materials, the disposal price does not cover the full costs of operation.

Value chain recycling in Botswana is functioning quite well with markets for almost all materials. The healthy Botswana private recycling sector purchases materials, processes them, and sells them to the large South African recycling industry. However, Botswana national and district officials mistrust the operations of the value chain, in part because it is seen as creating wealth in South Africa rather than domestically. The lack of trust between the public and private sectors is also reflected in incomplete information management and poor reporting. The Department of Solid Waste Management and Pollution Control registered 821 tonnes of waste recovered by authorised waste pickers at landfills equivalent to 1.9% of total waste in 2012. Business-to-business recycling transactions go unreported and unrecognised. (Scheinberg, 2011; Scheinberg et al., 2010b, 2012).

Botswana lacks municipal recycling but there is a Product Stewardship (PS) system for steel used beverage containers (UBC) called Collect-a-Can which promotes inclusive recycling as shown in Figure 6.6 and also profiled below in Box 6.2.

Box 6.2. **The collect-a-can system in Botswana**

Botswana participates in a regional, voluntary product stewardship (PS) system for beverage cans which benefits NGOs and the informal sector.

Collect-a-Can aims to stimulate the collection of steel and increasingly other metals including aluminium used in beverage containers (UBC), for beer, water, juices, and soft drinks. This PS system is a joint venture of the regional beverage industry and the South African steel company Arcelor Mittal. The participation of the steel giant Arcelor-Mittal suggests that this is a value-chain driven system. However, it seems that the steel collected does not go back into steel production. Rather, the collected cans are delivered to the Botswana and Zambia copper mining industry for use in copper smelting, although little information is available on how the materials are actually used. According to global definitions, this is considered down-cycling or incineration, rather than recycling, but it does qualify as end-of-life management.

The main instrument used in this system is price support, whereby a fixed price is guaranteed to informal recyclers, NGOs, or local authorities for the steel cans they collect. The price offered exceeds the global price for thin steel can sheet. Price supports create a kind of implicit take-back commitment that is designed to flow through existing value chain channels, with junk shops and processors handling large volumes of cans; charitable and NGO events also stimulate direct purchase. An unusual feature of this system is that virtually no one collecting or handling the cans has any idea what happens to them,

Sources: www.arcelormittalsa.com/portals/0/collect-a-can.pdf; and *www.collectacan.co.za/index.php/our-partners/beverage-industry*.

6.4. The informal recycling sector

In the late 1990s, informal recyclers were generally considered as a weak and vulnerable group and the policy concern was to provide them with social support and they were encouraged to move to other, more traditional forms of employment, often against their will. While this discourse still dominates the discussion of informal recycling in some parts of the world, particularly in Europe, attitudes have changed in other regions, notably in Asia and Latin America. As a result, the policy discussion has broadened to include four aspects of informal recycling.

- occupations and labour characteristics, occupational recognition
- social position, working conditions, and health issues and interventions
- operational, economic and environmental attributes of informal recycling
- integration, formalisation and interventions in the service and value chains.

6.4.1. *Occupations and labour characteristics*

Occupations

The United Kingdom charity, Women in the Informal Economy, Globalising, Organising (WIEGO) has helped improve understanding of the basic occupations and the related characteristics of workers in informal value chains. The most common occupations in the informal recycling (valorisation) sector are listed in Table 6.2.

Table 6.2. **Breakdown of informal recycling occupations – in selected middle-income countries**

Informal occupation	Corresponding formal occupation, if any	Function/earning model
Itinerant waste buyers (IWBs) who move along a route, and pay households (or businesses) for separated recyclables.	In value chain recycling: agents of the recycling businesses who collect materials from households.	Sale of recyclable materials to dealers or end-users.
Itinerant Waste Collectors (IWCs) who collect recyclables, re-usables, clothing, furniture, and metals with no money exchanged. Mainly observed in, but not limited to, South-eastern Eastern European cities.	Second-hand shop and charitable collection of re-usables, e-waste, clothing, shoes for a good cause or to raise funds for charity. Rural localities in New England (United States) have a take it or leave it area for exchange.	Re-sale of items, recycling of materials.
Itinerant collectors of swill (food waste) for feeding pigs and other animals. Mainly observed in, but not limited to, Latin America and Asia.	Municipal kitchen / garden organics waste collection in high-income countries., collection of frying fat.	Processed for animal feed or industrial uses.
Street pickers collect recyclables and bulky waste set out for formal collection.	Municipal crews for separate collection of recyclables or management of waste facilities.	Pick up valuable materials from streets, gutters, public spaces.
Container pickers remove materials from community or commercial containers or secondary collection sites.		
Truck Pickers are formal waste crews (or their friends) who skim valuables during the formal waste collection.	Collection crews, bulky waste collection crews, second-hand shop pick-ups, workers at private or public second-hand shops or recycling centres.	
Re-usables pickers, also sometimes public employees, extract WEEE, furniture, toys, and clothes from municipal recycling centres or depots.		
Dump and landfill pickers work and often also live on the landfill or dumpsite, meeting trucks, and sort through the waste as it is disposed there.	Landfill supervisor, labourers, equipment operators.	Extract valuable materials.
Small and large junk shop owners and workers that buy material from the IWBs and different types of pickers or even employ waste pickers, often paid by the piece/kg.	Workers at MRFs (public sector) or IPCs (private sector) with sorting lines and processing.	First and second level buyers.
Mobile Re-usables Collectors. Individuals, families, or small groups of entrepreneurs who collect re-usables, generally electronic or electric appliances, furniture, or clothing, from households or from waste set-outs.	Second-hand shop employees, antique collectors, workers at charities such as the Salvation Army or Humanitas, in the United States, Take it or leave it workers.	Capture of re-usables from households or the street.
Second-hand Traders. Individuals, families, or small groups of entrepreneurs who buy and sell re-usables, often having stands or booths at open markets like the Porta Portese market in Rome, 'Russian markets' in the Balkans, 'European clothing' markets or street vending in Africa, or 'flea markets' in North America.	Flea market entrepreneurs, professional E-bay traders, second-hand shop owners, used textile brokers, antique furniture shops.	Marketing of re-usables for their product value.
Second-hand Transporters and Processors: semi-formal companies that sort, grade, and trade used clothing, WEEE collectors and exporters who sell second-hand computers in Africa.	Formal traders, transporters, and processors.	Movement of second-hand goods from richer countries to poorer ones.
Non-professional street person collectors This category of casual or occasional street pickers was for the first time quantified in Bogotá, Colombia, where they account for 10% of the recycling attributed to the informal sector.	Not applicable as it is not a profession or occupation.	Subsistence activity for food or shelter.

Source: Elaborated by the author with some reference to:

Scheinberg et al. (2010a), *Economic Aspects of the Informal Sector in Solid Waste*, Final Report and Annexes, GIZ (German International Co-operation), the CWG (Collaborative Working Group on Solid Waste Management in Low- and Middle-income Countries), and the German Ministry of Foreign Affairs, Eschborn, Germany. *www.GIZ.de*.

Scheinberg, A. et al. (2007b), *Economic Aspects of the Informal Sector in Solid Waste*, unpublished draft research report, German Technical Co-operation, Eschborn, Germany; available at *www.waste.nl* and *www.giz.de* (NB: rewritten and published under the same authors in 2010).

Scheinberg, A. and Nesic, J. (2014), "Engaging informal recyclers in Europe: Status and report of a consultation", paper delivered at the 2014 ISWA Global Congress, 7-9 September 2014, São Paolo, Brazil, *www.ISWA.org*.

Information on reuse categories from:

Occhio del Riciclone (2012), *Rapporto Nazionale sul Riutilizzo 2012*, Ministero dell'Ambiente, 2012 (National Report on Reuse 2012, Ministry of the Environment).

Occhio del Riciclone (2009), *La Seconda Vita delle Cose*, ("The second life of things", Environment Issues) Edizioni Ambiente.

Occhio del Riciclone (2008), *Impatti Occupazioni di un riuso sistemico nella cittá di Roma*, (Occupational impacts of a reuse system in the city of Rome, the City of Rome in 2008), Comune di Roma.

Social status

The social status of informal recyclers includes legal identity, housing, medical care, and access to education for children. While there has been a great deal of activity in relation to the social aspect of waste picking since the late 1990s, few facts and little analysis is available.

Labour characteristics and risks[9]

The informal sector is not uniform, and involves different occupations and statuses. For instance, waste pickers, and their status, can be distinguished depending on whether they are a professional who has been working in recycling for many years, whether they are first, second or third generation on the landfill, an amateur or a new entrant, an occasional worker, a family member who contributes without considering themselves a waste picker, or an indigent or street person who picks waste with no other means of income.

Outside Europe, recycling is the main source of income for most professional waste pickers. In contrast, in the Balkans, a significant proportion of the primarily Roma informal sector recyclers switch to agricultural labour during planting and harvesting seasons. Professional waste pickers can generally earn the equivalent to, or significantly more than, the minimum wage.

It is estimated that about a quarter of the informal recyclers are women. Women outnumber men in the service chain only in Africa where there is a strong and growing tradition of women-owned micro-enterprises promoted by the International Labour Organisation. The value of unpaid family labour is high in all cases. Of the six cities reported in a GIZ study, child labour was reported to be highest in the mega-cities of Quezon and Lima, lowest in Cairo and Pune (which have the longest history of social activism and the organisation of waste pickers), and moderate in the two other cities. (Scheinberg et al., 2010a, 2007b; Gunsilius et al., 2010).

Waste picking is hard and dirty work which can be associated with serious health risks, social isolation, illegal activities and organised crime. The following is a summary of labour conditions experienced by informal workers and micro enterprises in the above-mentioned occupations. Some of these are specific to informal recyclers, whereas others, such as the lack of identity cards or social services, are common in other informal sectors such as home-based workers, agricultural labourers or street vendors:

- higher levels of income than formal occupations with the same level of qualification
- long working hours, no holidays or vacations
- long-distance biking, pushing, pulling or carrying: itinerant pickers without transportation walk up to 20 km per day, those with non-motorised transport bike up to 35 km per day
- muscle and other physical strain from lifting and carrying
- accidents, abrasions, infections
- exposure to all kinds of weather
- contact with medical wastes, body parts, dead animals, human and animal faeces
- exposure to toxic chemicals in the waste or as products of reclamation processes, such as burning of polyvinyl chloride (PVC) coatings to extract copper and other metals from insulated wire and cable
- social isolation, ridicule, vandalism, other forms of harassment

- no recognised occupation
- lack of access to medical facilities, education, housing
- income loss when sick, pregnant, or too old to work; no pension or social security
- child labour (usually accompanied by parents)
- police and sexual harassment, blackmail, violence
- no access to credit and often no access to banking or financial services
- barriers to formalising enterprises or acquiring personal identity documents.

In spite of these labour characteristics, waste picking is frequently passed on as a reliable income source from parents to children and even to grandchildren. It is not uncommon to find three generations of a family working on a dumpsite or in a junk shop, but it is unusual to find four.[10] In Asia and Latin America, there is evidence that the second generation of organised waste pickers are achieving educational levels that allow them to function as book-keepers, supervisors, and project managers in the co-operatives where their parents work.

Working capacity and competence

There are mixed views on the operational capacity of the informal sector and its economic and environmental impacts. In particular, the informal sector may produce positive results for collecting and sorting of waste with economic value while there are serious concerns over informal dismantling and recycling processes which can create negative economic and environmental impacts.

Informal dismantling and recycling often fail to prevent emissions or to capture hazardous substances, resulting in occupational and environmental risks (Romero, 2014). For instance, de-soldering and wet chemical leaching of printed circuit boards lead to contamination by heavy metals such as mercury, lead, cadmium and zinc (Kojima et al., 2009). In particular, hydrometallurgical processes to extract gold and other valuable metals from printed circuit boards use acids and cyanides which are sometimes dumped into local water ways, causing severe pollution. Similarly, open burning of coated wires to extract copper and other metals emit dioxins, furans and other toxic chemicals which can lead to serious health issues such as skin infections, breathing problems, and cancer (Akenji et al., 2011; Williams et al., 2013). Furthermore, residuals with non-economic value are improperly managed in general and can be discarded or dumped leading to negative environmental impacts (Akenji et al., 2011).

Moreover, informal recycling processes are usually less effective at recovering material from waste. In an evaluation from India, it was reported that informal hydrometallurgical processes with integrated metal refineries could only recover up to 25% of gold imbedded in printed circuit boards while modern integrated smelters could recover over 95% of them (Keller, 2006; Rochat et al., 2007). While some informal recycling is more effective (as reported for China by Williams et al., 2013), the general level of informal sector material recovery is limited, capturing only 20-50 % of the valuable fractions (Romero, 2014).

On the other hand, the informal sector is seen to contribute to effective collection, sorting and sometimes dismantling of waste with positive economic value in developing countries that do not have well developed waste management systems (Williams et al., 2013), despite some trade-offs against high levels of working capital and low levels of traceability.

Informal sector operations typically involve high levels of working capital, high transaction costs and low to moderate levels of traceability. High transaction costs are related to the quality of recyclables and market standards. For this reason the unit price for large lots of material is higher than that for small quantities. Small recycling shops must pay cash to waste pickers, and after selling loads of up to 10-30 tonnes, often only receive payment up to 90 days after delivery. These micro-enterprises need space to accumulate loads. All this requires high levels of working capital. The low to moderate levels of traceability are partly due to the mix of legal and semi-legal sourcing channels. The largely unaccountable inventories of the entire recycling industry leave all levels of private recyclers vulnerable to workers stealing materials. It is these characteristics of the value chain which make it difficult for local authorities or NGOs to "eliminate the middle man". Underestimation of the importance of storage, working capital, and value chain knowledge cause many symbolic, cash cow, and donor-stimulated recycling efforts to fail dramatically and dismally.

At the same time, the informal waste collection and sorting sector in middle-income countries can achieve effective collection, sorting and dismantling of materials with high economic efficiency, low energy footprint, and high tolerance to risks.

High economic efficiency and low costs per tonne are partly due to long working hours, poor labour conditions, and the use of unpaid family labour in processing. The low energy footprint is associated with the widespread use of manual labour for carrying materials (pushing or pulling hand carts, riding bicycles or tricycles with or without trailers,) and the use of animal traction especially horses in Europe and donkeys in Africa. Motorised transport used by the informal sector involves low-consumption motorised vehicles such as small tractors or auto rickshaws in contrast to the high fuel consumption of compactor trucks and other vehicles used by the formal sector. However, these vehicles may also generate relatively high levels of pollution. The informal sector also has a high tolerance for operational and commercial risks, particularly when, for example, the risk of injury or a fall in prices is offset by the potential for high rewards through the sale of materials.

Several global studies have compiled convincing evidence that informal systems in middle-income countries collect more materials than formal recycling systems in circumstances where both exist side by side. The GIZ informal sector study based on Scheinberg et al. (2010a, 2007b) analysed recycling activities in six cities and revealed that no formal system collected more than 13% of the total waste generated while the informal sector collected up to 30%. The scale and volume of informal activities in these six cities are described in Table 6.3. The UN-Habitat study confirmed that, with the exception of one country, the recovery rate of formal systems in low and middle-income countries was low. (Scheinberg et al., 2010a, 2010b). This is emerging as a major issue in South-eastern Europe, the Balkans, and Turkey, where producer responsibility systems set up in conformity with the EU waste directives generally capture less than 10% of materials, with the informal sector capturing the rest.

Recognition – workers or enterprises?

Informal recyclers usually identify themselves as individuals or family enterprises in the Americas and South-eastern Europe, or as self-employed workers in South and East Asia. Identification as a kind of primary producer similar to a farmer might be the best descriptor for African waste pickers. These are important distinctions because they offer

Table 6.3. **Basic quantitative information on informal and formal sectors**

City	Number of informal sector workers	Annual tonnes collected (per worker)	Average earnings of an informal sector worker (Euros per day)	Ratio of informal sector workers to formal sector workers	Number of city inhabitants per informal sector worker	Number of city households per informal sector worker	Number of informal sector workers per km^2
Cairo	40 000	54	4.3	6 : 1	432	88	6
Cluj	3 226	4.6	6.28	10 : 1	120	36	18
Lima	11 183	48	5.4	1.28 : 1	694	162	4
Lusaka	390	13.9	6.52	0.66 : 1	1 238 227	498.4	1.23
Pune	9 509	12.4	2.8	3 : 1	315	69	69
Quezon	10 102	14	6.26	2 : 1	246	54	63

Source: Gunsilius, E. et al. (2011), *Recovering Resources, Creating Opportunities. Integrating the Informal Sector into Solid Waste Management*, Eschborn, Deutsche Gesellschaft für Internationale Zusammenarbeit (German International Co-operation) (GIZ).
Scheinberg, A. and A.P.J. Mol (2010), "Multiple modernities; transitional Bulgaria and the ecological modernisation of solid waste management", *Environment and Planning C: Government and Policy*, Vol. 28, No., 1, pp. 18-36.
Scheinberg, A. and Anschütz, J. (2007), "Slim pickin's: Supporting waste pickers in the ecological modernisation of urban waste management systems", *International Journal of Technology Management and Sustainable Development*, Vol. 5, No. 3, pp. 257-270.

some guidance on how to conceptualise and design interventions to integrate informal workers into the waste management system.

Interventions in Asia appear to have started in India in the 1980s, with the Self Employed Women's Association (SEWA) in Gujarat and Kolkata. The first waste picker union, KKPKP, was established in Pune, India in the 1990s (Hill, 2001; Lardinois and Furedy, 1999; Chikarmane and Narayan, 2009). Outside India and South Africa, labour unions for informal recyclers are rare.

Other forms of worker organisations have proven to be an effective approach to support both the daily activities and broader objectives of waste pickers and recyclers in Europe, Africa and Latin America. These efforts have taken a variety of forms such as movements, co-operatives, associations and labour unions. Brazilian waste pickers Catadores have been organised since the 1980s as a political movement both through the National Movement of Waste Pickers *(Movimento Nacional de Catadores de Reciclaje)* and represented within the Waste and Citizenship Movement *(Fore de Lixo e Ciudania)*. In Colombia and other Latin American countries, waste picker organisations in the 1990s resulted in co-operatives and associations, forms of organisation more consistent with the identity of micro-entrepreneurs (Price et al., 1998).

Outside Southern Africa and the Middle East and North Africa (MENA) regions, the numbers of informal recyclers in Africa are small, in part because the formal value chains are weak, with only a few buyers and low prices. There is limited documentation on the organisation of waste pickers in East or West Sub-Saharan Africa. Ghana, for example, apparently has substantial numbers of waste pickers whereas only 200 are estimated to be working in Tanzania.

The numbers of informal recyclers and re-use entrepreneurs in Southern Europe, the Balkans, Turkey, Central Asia, and the MENA region are large and growing. There are 80,000 informal recyclers reported in Morocco, concentrated in the industrial zone between Rabat and Casablanca, and perhaps as many as a million in Egypt. In these regions, informal recyclers overwhelmingly consider themselves as small enterprises. Within the borders of the European Union, there is an association of re-use entrepreneurs in Hungary, a strong NGO initiative in Italy, and business or cooperative-based organisations and projects in a number of former Yugoslav republics and Albania. A very small union of waste pickers, the

Syndicaat, has organised waste pickers in Serbia's second city, Nis, under the guidance of a high-profile Roma politician, but it is not clear whether this is a labour organisation or a political platform.

The WIEGO (Women in the Informal Economy, Globalising, Organising) became active in the organisation of waste pickers in the period between 2006 and 2008 (WIEGO, 2015). WIEGO's organising strategy supports both local waste picker movements and the formation and strengthening of the Global Alliance of Waste Pickers (Global Rec) and the Latin American Waste Picker Network (RED Lacre). The InteRa system also provides support for the organisation of waste pickers (Velis et al., 2012).

The modernisation of waste management systems in middle-income countries since 2000 also led to increased attention to informal recycling within development co-operation organisations. The 2004 review of the ILO initiative to eliminate child labour in scavenging was a turning point in the global discussion of informal recycling (ILO/IPEC, 2004). Desk research indicated that neither welfare rights nor other development-co-operation-based approaches actually improved the socio-economic position of informal recyclers. The problem was that social support programmes, while in some sense necessary, were operating from the point of view that waste pickers were social victims who needed rescuing. The result was that they were invited – in many cases pushed – to exit waste picking, which was assumed to be an unacceptable occupation. However, this approach failed to recognise that waste picking was often a relatively lucrative and stable source of income. The employment opportunities available when exiting the sector generally did not come close to replacing the income of a waste picker. As a result most families (and their children) returned to picking waste as soon as the project was over.

6.4.2. Interventions in informal recycling

As mentioned earlier in 6.1.4, although informal recycling processes can create negative impacts to the environment and may not be as efficient as the formal processes in recovering materials, the informal sector can be extremely effective in collection, sorting and possibly dismantling waste. Whereas more effective enforcement policies are needed to upend sub-standard informal recycling, the challenge is to find ways to make use of efficient informal waste pickers and sorters, i.e. to secure the economic and social benefits of informal operators while mitigating their environmental impacts (Williams et al., 2013).

A variety of approaches have been pursued to support informal waste pickers depending on whether the dominant form of recycling is value chain recycling, municipal recycling, or a pre-existing EPR/PS system, and whether the motivation for working with the informal sector is primarily social, economic, political, operational or environmental.

It is possible to classify efforts to integrate informal waste pickers into the following categories, the first two of which dominated until about 2004.[11] These different interventions are not exclusive, but complementary.

1. *Welfare-based interventions*, also sometimes referred to as social integration. These interventions focus on the individuals and families involved in informal recycling, and are based on the premise that waste pickers are socio-economically disadvantaged people requiring social support.
2. *Rights-based interventions*, including labour organisation. Rights-based interventions support groups of waste pickers and their families to claim labour or citizenship rights and to build associations that strengthen this claim. Such interventions generally consider

waste pickers as workers. Formation of unions and political lobbying for occupational recognition are typically key goals of rights-based interventions.

3. *Informal Sector Integration*, sometimes also referred to as inclusive recycling. Integration usually involves a two-way process between the service chain and informal recyclers, including some form of accommodation and recognition of informal waste pickers. Formal contracts or agreements, identity cards, or licensing are ways that integration has been operationalised. Integration stimulates the formation of co-operatives and business associations in which informal recyclers are considered as entrepreneurs or enterprises. Integration initiatives often result in the establishment of semi-formal recycling operations in co-operation with the formal waste authorities. Integration may also include supportive measures such as the creation of umbrella contracts for legalising business relationships, and finding options for registering informal enterprises in the tax system. Integration is increasingly proposed in support of inclusive municipal recycling. This involves informal recyclers going door-to-door to collect source-separated recyclables which they then have the right to sell. The source of funds for organising and capitalising the separate collection is a combination of avoided disposal costs, external financing, the recyclers' own resources, and, in Latin America, contributions from Product Stewardship systems.
4. *Formalisation* involves waste pickers accepting to comply with tax laws and business norms, to register as enterprises, and to agree to follow laws and rules. It does not imply any reciprocal measures by the public authorities. Formalisation is useful as a way of integrating pure value chain operations into a wider formal solid waste landscape. The two main instruments for promoting formalisation are formation of business associations, and capacity building and training. The International Finance Corporation's private sector recycling initiative in the Balkans between 2005 and 2008 – IFC Recycling Linkages – focused on training informal recyclers. Informal recyclers gained status and recognition, along with knowledge and skills (IFC, 2008). Particularly in the re-use/second-hand trade part of the informal sector, creation of associations of traders has become an important way of gaining recognition and a legal space for economic activities.
5. *Professionalisation and access to financing*. This approach aims to support informal recyclers as autonomous micro or family enterprises which, because of their ethnic, social, or immigration status, are excluded from access to financial and business services. It is primarily demand-driven. It is particularly relevant in situations where there is value chain recycling and some interest in municipal recycling, and where there is no active confrontation between local authorities and informal recyclers. Its focus is on increasing knowledge, capacity, business skills, and access to materials and financing. It can also enhance occupational recognition which in turn can help to improve access to micro-financing and business lines of credit. In some countries like Tunisia, occupational recognition is also a path to social inclusion: to register for a health card, the applicant must fill in a recognised occupation.
6. *B2B (business to business) value chain activities*. This approach involves value chain support in the form of pre-financing inventory and/or providing infrastructure and equipment. It operates best in situations where the value chain is strong. Globally, larger junk shops and end-users form a focal point for their suppliers, who often have a strong vertical bond with the purchaser, even when they lack horizontal linkages with other informal

recyclers. Using a junk shop as a focal point for introducing a light form of organisation may provide a way around the difficulties of organising in Europe and the MENA regions, where waste pickers are strongly individualistic.

6.4.3. *EPRs and informal workers in Colombia and Bulgaria*

The following two sections briefly describe contrasting approaches to working with the informal sector. The first examines the experience in Colombia where an inclusive approach had been pursued. While there is no detailed analysis of the environmental and socio-economic impacts of this approach, it is clear that the status, remuneration and working conditions of informal workers has improved and that they play a constructive role in the Colombian product stewardship scheme. In contrast, informal workers have been largely excluded from the Bulgarian scheme. This has led to overt and covert conflict between the formal and informal sectors. The outcome has been both inefficient and ineffective: 90% of recyclable materials are captured by the informal sector while the bulk of financial resources are concentrated in the formal sector (ISWA/EXPRA/RDN, 2014). While the Colombian and Bulgarian approaches are quite different, with Colombia operating a voluntary product stewardship systems and Bulgaria a mandatory take-back scheme, these two cases provide some valuable insights into the merits of inclusiveness.

Inclusive recycling in Colombia

CEMPRE (Compromiso EMpresiarial Para el REciclaje) – the Entrepreneurial Commitment to Recycling – promotes inclusive recycling in Colombia. It is a membership organisation registered as an NGO, with approximately 20 private sector members ranging from Coca Cola to several of Colombia's industrial value chain stakeholders. The membership fee covers CEMPRE's operating costs, and supports national and regional initiatives on inclusive recycling.[12] CEMPRE defines inclusive recycling as systems of recovery and re-utilisation of waste materials as well as those organisations, institutions and individuals that strengthen the participation of informal recyclers in public service chains and private value chains.

The activities of CEMPRE are outlined in Table 6.4. They cover activities at national and municipal governmental level, and include promoting public policies, inclusive value chains and an inclusive culture of recycling.

Table 6.4. **The three building blocks of CEMPRE**

	Public Policies	Inclusive Value Chains	Inclusive Culture of Recycling (strengthening political will)
National System of Inclusive Recycling	Inclusion of the considerations of CEMPRE in solid waste law.	National Observatory for Inclusive Recycling.	Support national government initiatives.
	Regulation of shared responsibility.	Sectorial Agreements for Inclusive Recycling.	
Municipal approaches to inclusive recycling	Adoption of inclusive recycling models in the municipal and regional strategic plans.	Development of inclusive municipal recycling systems that supply materials to the value chain with effective participation of organised informal recyclers.	Support municipal government initiatives.

Source: Moreno, F. (2014), "Colombian alliance for inclusive recycling", PowerPoint presentation, CEMPRE Colombia.

CEMPRE promotes a shared responsibility approach and works closely with the main ministries, private sector, waste picker associations and other stakeholders. As in many

Latin American countries, sectoral accords are widely used, in this case to promote product stewardship, including by:

- dissemination, operationalisation and institutional anchoring of the provisions of decisions by the constitutional court to secure the rights of informal recyclers
- advocacy for recycling and better waste management
- support to waste pickers as political and operational stakeholders in inclusive recycling, including training in lobbying, business plan development, association management, and micro-credit
- facilitation of agreements between waste picker organisations and national institutions to ensure the practical implementation of inclusive recycling
- improvement in implementation through enhanced participation of stakeholders
- facilitation of communication and relationships between donor agencies, industry, and associations of recyclers, and mobilisation of funds.

These activities align most closely with the professionalisation agenda described above, with a strong additional emphasis on multi-dimensional integration of waste pickers in policies and institutions, and the integration of waste pickers' perspectives in municipal recycling plans.

CEMPRE members work to ensure moderate levels of recycling and end of life management for their packaging waste streams, primarily by strengthening existing value chains, and supporting organised informal recyclers. Informal recyclers have an incentive to join recognised associations; if they do not, they remain outside the inclusive recycling system and risk marginalisation. The most visible instruments used are price support and the authorisation of junk shops to pay recyclers through their co-operatives. The burden of paying for these instruments largely falls on the municipalities. The largely private waste collection companies objected strongly, but were unable to prevent the introduction of this system.

The Bogotá public service company, UAESP pays a diversion credit of USD 50 per tonne to waste pickers selling to authorised junk shops. Authorised junk shops register the quantities of all materials purchased from individual recyclers, and forward this information to the city administration. The city administration then deposits the diversion credit directly in the waste picker's bank account based on the quantities collected in the previous two months. Thus waste pickers must be part of the formal economy in order to receive their credit.

Independent junk shops, with at least one competing system of authorisation, operate in parallel. The ECORED network for PET collection is a collaboration of two CEMPRE members, Coca Cola and ENKA, a PET recycler. ECORED operates a small number of authorised, subsidised junk shops specialising in the purchase and processing of PET beverage packages. Operations include buy-back with price support to associated waste pickers, collection, processing, and marketing to the value chains.

The Colombian product stewardship system is inclusive but incomplete. Not all producers are members, not all packaging materials are covered, not all waste pickers are members of co-operatives, and many co-operatives and junk shops remain independent and unauthorised. This means that they cannot claim price support from ECORED nor administer the UAESP diversion credit.

Effectively there are at least three parallel channels involved in the purchase of recyclables. In this context, instruments such as price support or diversion credits provide a competitive advantage to the junk shops that can offer them. The effectiveness of this approach in diverting packaging waste from disposal to recycling is not yet clear. An Observatory for Inclusive Recycling which includes a benchmarking component, aims to shed light on this and the broader questions of whether the voluntary, facilitative approach based on sectoral accords and shared responsibility is successful in increasing recovery rates, stimulating innovation, and promoting sustainable materials management. It should also analyse the impact on informal workers and assess their contribution to improved waste management.

EPR system for packaging waste in Bulgaria

The Bulgarian waste management system went through extensive reform following the transition from a centrally-planned to market-oriented economy.[13] Amongst other things, this involved privatisation and the break-up of the waste management state monopoly. A new EPR system for packaging waste was established in 2004. At that time, neither well-functioning public sector service nor private sector value chains had been established. The introduced EPR system achieved to a certain extent growing recovery targets with well-established recycling practices of the large industrial and commercial sectors while the separate collection of recyclables from households was facing significant challenges. In parallel to the EPR system, a considerable number of informal waste pickers were engaged in collecting recyclable materials and selling them to buy-back centres, and were alternatively achieving significant levels of recycling rates (Doychinov and Whiteman, 2013).

A report by the Ministry of Environment and Water, Republic of Bulgaria (2003) illustrates the situation before they introduced the new EPR system, and estimated that around 10 000 informal waste pickers pre-existed and relied on their waste collection activities to sustain their livelihoods.[14] However, these informal waste pickers were not consulted when the new EPR system was established. As a result, the targets that were set for recycling packaging materials failed to take account of the amount of recycling that was already underway. Ironically, the targets that were set were *lower* than the actual rates of recycling in the first years of implementation (Doychinov and Whiteman, 2013).

The primary instruments in the new EPR system were a packaging tax and collective, industry-financed physical compliance scheme with 100% producer responsibility for end of life management of packaging waste.[15] Individual producers, including distributors and importers, had a mandated, individual take-back obligation unless they became members of the collective compliance scheme. The packaging industry opposed the scheme and argued for a shared responsibility arrangement with the municipalities. The mandated take-back scheme was introduced, setting the stage for 12 years of conflict over collecting recyclables from households, particularly between the formal and informal sectors, and undermining of the effectiveness of the system (Doychinov and Whiteman, 2013).

The primary means of physically collecting recyclable material within the EPR system was plastic, colour-coded 2 m^3 drop-off containers placed in the street. Recyclables could also enter the system through occasional door-to-door separate collection initiatives. Traditional recovery channels were ignored, including several types of buy-back centres, pensioners collecting recyclables from the street to supplement their pension income, and the recovery activities of professional street, container and dump pickers. The formerly state-owned recycling industry was not involved.

As a result, the new EPR system failed to benefit from the activities and networks of informal waste pickers for household collection. In addition, the designers of the new system failed to anticipate the resistance of households and small businesses to abandoning their preferred recycling channels. Consequently, the buy-back recycling centres continued to operate in parallel to the newly established separate collection containers (Doychinov and Whiteman, 2013, p. 21). Moreover, the existing informal recyclers found the containers to be a convenient and free source of materials that they could extract from the containers and sell to the buy-back centres. It is estimated that 90% of materials has continued to flow through informal channels (ISWA/EXPRA/RDN, 2014).

The formal and informal sectors became locked in an overt and covert conflict. The formal sector argued that the buy-back centres should be closed and the informal workers arrested for theft of their materials. Informal workers whose livelihoods were threatened retaliated by vandalising infrastructure and doing whatever they could to undermine the operations of the formal sector.

The disfunctionality of the EPR system has also come at a high financial cost. The system has been well-capitalised by EU and national funds, particularly for waste infrastructure. However, because it captures so few – 10% of – materials, and because the investments in infrastructure are so expensive, it has very high operating costs and disappointing revenue streams. This creates a vicious circle in which there is a continued need for infusions of funds to keep the infrastructure operating. On the other hand, the 90% of materials collected in the informal sector does not benefit from EPR funding.

A regional workshop in Bucharest, Romania, in October 2014, Challenges to separate collection systems for different waste streams – barriers and opportunities established that Bulgaria's experience is widely shared across the region (ISWA/EXPRA/RDN, 2014). Representatives of EPR schemes in 10 Balkan and Mediterranean countries including Greece, Turkey, Malta, Tunisia, Romania, Bulgaria, and Macedonia presented their challenges, which were in large measure about the difficulties of competing with the informal sector recycling. As in Bulgaria, these efforts were largely unsuccessful. Clearly new approaches are needed which treat the informal sector as partners rather than competitors.

6.5. Towards inclusive EPR systems in middle-income countries

6.5.1. *Lessons learned*

Previous sections of this chapter presented some findings from case studies in Botswana, Bulgaria and Colombia. In preparing this chapter, case studies from several other countries were also reviewed: for China (e-waste), Costa Rica (e-waste), Tunisia (packaging) and a Latin American regional initiative (pesticide packaging).[16] The findings of all seven case studies are summarised in Table 6.5. Several conclusions can be drawn from these case studies.

It is possible to have well-functioning EPR and PS systems in middle-income countries without full scale legislation or national government leadership that replaces informal waste collection activities. That said, the support, oversight, monitoring, and reporting functions of national and local government are often important elements in well-designed EPR/PS systems. Clear goals, broad socio-political consensus, and transparency help to establish effective systems by strengthening good environmental governance. Consultation and co-operation are essential prerequisites for identifying innovative approaches for sharing costs and responsibilities.

Table 6.5. **Examples of the extended producer responsibility and product stewardship cases**

Country	Tunisia	Bulgaria	LAC[18]	Colombia	China	Costa Rica	Botswana
Name/description	EcoLef/ANGED EPR System for Packaging	EPR packaging system in EU Accession Process	Triple-wash Product Stewardship for Pesticide Packaging	CEMPRE, Product Stewardship for Inclusive Recycling	Ordinance EPR National E-Waste Recycling System 2012	EPR National E-waste based on a multi-stake-holder process	Collect-a-Can
Intervener	National government ANGED	EU via the Ministry of environment	Producers/distributors of pesticides and agricultural chemicals	CEMPRE and its members	National government represented by multiple ministries	Technical Committee as part of Co-operation Project Netherlands	Arcelor-Mittal, Breweries, Coca Cola
Domain	Value chain	Service chain	Service chain (pollution avoidance), value chain	Value chain	Value chain and service chain	Value chain	Value Chain
Type of system	EPR	EPR	PS	PS	EPR/Environmental Regulation	EPR	PS
Collection mechanism	Existing informal system, junk shop is first point of intervention	Street containers, door to door collection, buy-back	Industry reverse logistics to farmers	Existing informal system, junk shop is first point of intervention	Was Old for New, now unclear	Existing recycling centres, point of sale collections	Buyback with price supports via value chain businesses and charitable events
Aspects of the System	EcoLef points, patented traders, price supports at junk shop level	Packaging tax, licensing, compliance schemes	On-going educational and promotional campaign, price supports for recycling?	Price supports for PET/ packaging, Constitutional Court decision	Operating subsidies for licensed authorised recycling centres	Reinforcement of existing institutions all along the chain, stake-holder platform	Organises a subsidised market for steel UBC
Main instrument	Tax, price supports	Ecotax, Take-back requirement	Take-back system (potentially with price supports)	Sector agreement	Invisible Advanced Recycling Fee	Invisible advanced disposal/recycling fee, sectorial accords	Price supports, implicit take-back
Other instruments	Issuance of patents	Infrastructure and operations subsidies	Industry financing of infrastructure, operations	Price supports, diversion credits	Regulation, certification, and inspection	ASEGIRE a producers' and stakeholders platform	Sector agreements
Financing	Creation of packaging tax,	Creation of packaging licensing fees and penalty tax, street collection systems	Internal funds mobilisation by producers	Internal funds mobilisation by producers, service fee per tonne	National government, with some funds recovery from invisible product charges	Producers, retailers, importers, large generators, revenues from recycling	Members of the PS system
Owner-ship	ANGED/National Government	National Environmental Ministry, compliance schemes	Producers, Foundation CropLife America	Platform of producers at national and global level, Municipalities	National government	Committee/ASEGIRE, B2B construction with only light regulation,	Collect-a-Can consortium
Politico-institutional Level	National	National	Supra-national, continental	National, supra-national	National, provincial	National	National/Regional
Product and packaging groups	Packaging with focus on PET and HDPE	Packaging	Pesticide packaging	All recyclables, packaging	WEEE; emphasis on domestic electric appliances	WEEE	Steel UBC packaging

Table 6.5. **Examples of the extended producer responsibility and product stewardship cases** *(cont.)*

Country	Tunisia	Bulgaria	LAC[18]	Colombia	China	Costa Rica	Botswana
Results and Effects	Creates parallel channel, competes with value chain	Competition, conflict, probable reduction of levels of recovery	65% of pesticides sold are covered, no other reported results	improvement of policy and institutions for inclusive recycling	improved recycling operates, modest implementation results	Functioning WEEE system recycling/EoL (end-of-life) paid by producers	Functioning system for steel cans
Materials	Types 2	Type 1-2	Type 4	Types 1-2	Types 3-4	Types 3-4	Type 3
Legislated?	Legislated	Legislated and required by the EU	Non-legislated	Legislation in process, court decision exists	Fund is legislated, compliance unclear	Created non-legislated, legislation followed	Non-legislated
Targets	No	Yes	Yes	No	No	No	Yes
Reporting licensing/	Unclear	Yes	Yes	In process of being designed and negotiated	Yes	In process	Yes to the members only
Sources of funds	Producers, revenues from sales	Producers, licensing fees, European Union (punitive product tax)	Producers, sales of materials	Producers, municipal waste budget	Producers, licensing fees, revenues from recycling	Sector agreement	Producers, revenues
Uses of funds	Control buying and selling prices, processing infrastructure	Infrastructure, compliance payments, other?	Infrastructure, reverse logistics, no further info	Facilitation, technical support, policy and capacity development, observatory	Recycling operations. safe disposal	Reverse logistics, safe disposal through Basel Convention	Buy-back and price supports
The good	Supports waste pickers through existing channels	Is being re-designed	Prevents pollution, includes user training and monitoring	Works on system at macro level, regional co-ordination	Addresses working conditions and environmental results	Operates a system in a small country, taking Basel Convention into account	Creates a functional demand for steel UBC and keeps it out of the landfills
The bad	Appears to distort the value chain markets; EcoLef points are inefficient	System failed in part because it ignored existing stakeholders	Not transparent, recovery and capture varies widely per country	Weak relationship to real operations	Low recovery and few categories of covered items	Limited items covered, small scale	Not transparent, presented as recycling but is something else
The special	National ministry deeply involved in value chain	Multiple compliance systems	Exists since 2001, highly effective, good communication	Producers fully support and co-operate with informal recyclers	Addresses conditions of work in recycling factories	Came into existence through bottom-up platform	Significant financial support to informal recyclers

Source: Elaborated by the author.

The role of private value chain enterprises is much more important in middle-income countries than it was in OECD countries in the 1980s and 1990s. This is partly because the EPR systems in middle-income countries are heavily dependent on revenues in order to function. The best-functioning systems are those which embrace an open strategy that includes both informal collectors and the existing value chain enterprises in the system. This is the case even when new EPR or PS institutions are established in parallel to existing institutions as was done in Colombia, Costa Rica and Tunisia. In contrast, the exclusivity of the Bulgarian system seems to have been one of problems in the system design.

Consistent with the importance of private value chains, EPR and PS systems in middle-income countries make more extensive use of economic instruments such as price support and diversion credits than has been the case in OECD countries. In addition, interviewed waste pickers have suggested that they prefer payments which offer a freedom to choose their own protective clothing, for example, rather than programmes that donate gloves and shoes. Advanced disposal fees were used in the Costa Rica, China, Tunisia and Bulgaria case studies. However they are always invisible. The experience in China involving subsidised take-back at the level of consumers suggests that price support and diversion credits might be a potentially useful instrument in middle-income countries.

As a corollary to the higher importance of economic instruments, regulatory requirements such as take back and recycling targets are less important in these countries than in OECD countries. For example, the role of take-back requirements appears to be negligible in middle-income countries, except as a means to reinforce buy-back channels for low-value or non-recyclable materials. This makes sense as a take-back requirement is primarily an instrument to relieve the service chain of responsibilities for expensive end of life management, but it is of little interest to the value chain enterprises where there is no benefit or payment.

In designing EPR/PS schemes in middle income countries with limited waste management systems, it is important to explore policy alternatives to tap into the potential economic and social benefits from the informal sector while managing environmental impacts. It should be recognised that the introduction of EPR systems are likely to take away the livelihoods and the associated benefits of the pre-existing informal operators, and their possible integration into formal systems should be considered. Informal waste pickers often play a critical role in EPR systems in middle-income countries.[17] Waste picking can provide a stable source of income for the workers involved, better than alternatives. Informal waste pickers can also generate positive economic and environmental impacts including by: reducing the amount of waste going to landfill, providing an alternative, free collection service, and helping to reduce collection and disposal costs for municipalities. A few research studies establish a hypothesis that informal waste picking can provide benefits in reducing the CO_2 footprint of the waste management system (Chaturvedi 2009, Gunsilius et al., 2011). There is some evidence that informal systems in middle-income countries collect more materials than formal recycling systems when they work in parallel. However, informal recycling processes such as informal foundries and open burning often generate negative impacts for the environment and public health and therefore should be upended. In this respect, enforcement, institutional monitoring and oversight from national administration is necessary.

In order to integrate informal waste pickers into EPR systems where producers and importers bear physical or financial responsibility of the schemes, registration, gradual integration, formalisation and professionalization of waste pickers will be a key approach for them to fulfil necessary compliance measures. As seen in the case of Bulgaria, rapid introduction of EPR schemes can create severe conflict and competition between informal and formal sectors, therefore a gradual approach is likely to be more effective (Akenji et al., 2011; Williams et al., 2013).

Failure to effectively include the informal sector in EPR systems can undermine the efficiency and effectiveness of EPR systems as is the case in Bulgaria and neighbouring countries. Thus, further efforts are needed to identify ways in which informal workers can contribute to waste management systems to the benefit of both.

6.5.2. Next steps – 10 principles

The following principles could contribute to the inclusive design of EPR initiatives.[19]

1. **Recognition:** The role that informal recyclers play should be recognised: in middle and low-income countries, they are responsible for most of the materials that are captured, processed and sold in the recycling value chain. Recognition of their role can be enhanced by compiling and documenting information on what informal recyclers actually do and the contribution they make to achieving waste management policy objectives.
2. **Competence:** Cities in middle-income countries should consider how they could best draw on the knowledge of waste pickers and junk shops. They are often the only stakeholders with practical experience, the knowledge to maximise recycling under local market conditions, and the incentive to adapt quickly to new value chains and market opportunities.
3. **Participation:** Informal recyclers should be invited to contribute their experience and expertise in all relevant public decision making processes. They should be engaged in the design, monitoring and evaluation of recycling and valorisation systems, as well as the definition of quality standards.
4. **Source separation**: Producers, city authorities and informal recyclers should work together (experiment) to strengthen, or introduce, upstream separation of recyclables, organics and residuals at the level of businesses and households. Producers and local authorities in middle-income countries often consider that such separation is not feasible. However, upstream separation provides important support for EPR systems. There are cases (e.g. door-to-door collection in Pune) which suggest that well-designed source separation is feasible as part of inclusive recycling. However it will be organised differently than municipal recycling and needs to take account of local circumstances.
5. **Find and document what is working:** Public authorities and private producers should work with informal recyclers to collect data on waste generation and recycling rates. The assumption that no recycling is taking place should be avoided. Look for, analyse, and build on existing collection, processing, and marketing activities, as well as successful small and medium-scale experiments rather than assuming that everything has to be built from scratch, or that existing informal systems should be replaced by (parallel) formal systems.
6. **Safe and dignified working conditions:** Integrate the local insights and ambitions of informal recyclers with global ideas of good practice to achieve adequate environmental, health and safety working conditions for informal recyclers. Promote dignified working conditions, occupational recognition, and appropriate and fair business models.

7. **Eliminate child labour:** Work with informal recyclers and their children to ensure that children go to school and that recycling activities by children under the legal age of adulthood in the country are constrained, supervised, reduced or eliminated.
8. **If it's working, don't fix it:** EPR systems should avoid intervening in the recycling of materials where the private value chain is likely to be functioning well. Poor environmental or occupational performance can be addressed independently, without disturbing market relationships in this part of the value chain. EPR systems provide more opportunities for stakeholders, including informal recyclers, when they address market failures, including: dangerous waste streams, low-value materials, recyclables difficult to dismantle, or recycling in areas where there are few value chain buyers within a reasonable transport distance such as sub-Saharan Africa, Caribbean or Pacific SIDs. Give priority to improving marketability and improving working conditions.
9. **B2B before PPP:** Give priority to developing business partnerships with informal and micro and small recycling enterprises over a PPP approach which *a priori* involves government more than the host community.
10. **Maintain Inclusivity:** In developing EPRs, engage local authorities, municipal associations, national governments, regional economic communities, bilateral and multi-lateral global institutions, and all related stakeholders (e.g. producers, importers, collectors, sorters, and processors); evaluate, disseminate, and transfer sound practices of partnership with informal recyclers into public policy and legislation; and use these partnerships and activities to promote recognition of the informal recycling sector.

Notes

1. The information in this chapter is developed and derived from data and reports of development projects between 2004 and the present, enriched by the author's direct or indirect engagement in research and projects in Integrated Sustainable Waste Management (ISWM) and inclusive recycling in a number of middle-income countries. These countries include: Costa Rica, Botswana, Zambia, the Maldives, India, Brazil, Bulgaria, Serbia, Montenegro, Tunisia, South Africa, Mauritius, Péru, Romania, Egypt, Hungary, as well as Italy and Greece. The chapter builds on discussion at the Thematic Forum on Product Stewardship and Inclusive Recycling financed by German International Co-operation (GIZ), Berlin 2012.
2. EPR systems will be used in this chapter to refer to both publicly supported systems as well as product stewardship schemes, the voluntary equivalent usually led by the private sector
3. Source separation and separate collection of garden waste in North America, and of kitchen and garden waste in North-western Europe, were also developed in the same period and in response to the same drivers. These innovations produced the institution of *municipal composting* which is in all senses but one completely parallel to municipal recycling. The main difference is that source separated recyclables collected separately and processed at a MRF are recognised industrial feedstock products which can meet pre-existing industrial market standards and therefore have a price. In contrast, municipally produced compost is a useful substance but does not correspond to any recognised input in the agricultural value chain. Therefore, markets for compost have had to be developed.
4. Figure based on author's estimation.
5. The 20 cities analysed in this report are as follows: Adelaide (Australia), Bamako (Mali), Belo Horizonte (Brazil), Bengaluru (India), Canete (Peru), Curepipe (Republic of Mauritius), Delhi (India), Dhaka (Bangladesh), Ghorahi (Nepal), Kunming (China), Lusaka (Zambia), Managua (Nicaragua), Moshi (Tanzania), Nairobi (Kenya), Quezon City (Philippines), Rotterdam (Netherlands), San Francisco (United States), Sousse (Tunisia), Tompkins County (United States), Varna (Bulgaria
6. Figures are based on author's estimation.

7. Botswana information courtesy of WASTE, advisers on urban environment and development, Gouda, the Netherlands, UNDP Botswana, and the Botswana Department of Environmental Protection, and gathered in 2012 in the UNDP Botswana financed project The Botswana Recycling Guidelines.
8. Botswana is a dry country, so there is little leachate to monitor, although most landfills have a few monitoring wells.
9. Compiled by the author based on observations/discussions/research/participation in the global informal sector integration discourse and the global recyclers movement, the TransWaste project, Occhio del Riciclone, and other sources. Information not attributable to any single source, and also not directly related to the GTZ study.
10. Based on anecdotal information and a range of discussions with waste pickers worldwide.
11. In 2012, some scholars worked together to create the InteRa system for classifying interventions to integrate informal recyclers into solid waste systems. In this article, social integration, technical integration into the service chain, and economic integration into the value chain were considered distinct interventions, while organising was an overall pre-condition to any of the other forms of integration (Velis et al., 2012).
12. See: *www.cempre.co.*
13. The Directive 94/62/EC on packaging and packaging waste was adopted in 1994 and revised ten years later (Directive 2004/12/EC) (Doychinov and Whiteman, 2013). The new system, re-designed in 2012, is not profiled here.
14. Although it can be argued that informal waste pickers in Bulgaria increased their numbers from 10 000 as of 2003 to 15 000 as of 2013 and some of them emerged after the introduction of the EPR system, this case shows that there were considerable pre-existing activities carried out by the informal waste pickers before the introduction of the EPR system (Ministry of Environment and Water, Republic of Bulgaria, 2003; Doychinov and Whiteman, 2013).
15. Ecopack Bulgaria was the first EPR scheme established in 2004. Later, when the compliance market fragmented, Ecopak became only one of a number of compliance schemes competing with EkoBulPak, and among others.
16. Information on the case studies have been drawn from the following sources: China (Annex D), Costa Rica (Vega, V.R., 2007), Tunisia (Abdeljaoued, 2014; Soos et al., 2014) and Latin American regional initiative (Stewardship Community, n.d.).
17. To secure the benefits provided by the informal waste pickers, the negative effects that could possibly be related to their activities need to be controlled. For instance, illegal collection attributed to stealing from other properties, scavenging from illegal dumpsites or smuggling from abroad need to be mitigated through better monitoring and enforcement measures.
18. Latin America and Caribbean.
19. Principles adapted from the output of the GIZ-sponsored 2013 workshop: Extended Producer Responsibility and the Informal Sector, Berlin, November 2013.

References

Abdeljaoued, I. (2014), "The extended producer responsibility: The Tunisian experience", Paper presented at SWEEP-NET's Fourth Regional Forum Amman, Jordan, 13-15 May 2014.

ACEPESA (2004), *Estrategia Nacional para el Manejo Integrado y Sostenible de Desechos de Artefactos Eléctricos y Electrónicos* (National Strategy for Integrated Management and Sustainable Waste of Electrical and Electronic Appliances), *www.acepesa.org/documentos.*

ACR+ (2009), "Municipal waste in Europe", *Collection Environment*, Victoires Editions, Association of Cities and Regions for Recycling and Sustainable Resource Management, Paris.

Akenji, L. et al. (2011) "EPR policies for electronics in developing Asia: An adapted phase-in approach", *Waste Management & Research*, Vol. 29, No. 9, pp. 919-930.

Anschütz, J. (ed.) (2005), "UWEP City Case Studies, City Case Studies of Bamako, La Ceiba, Bangalore and Batangas Bay, the Four PPS Cities of the UWEP Plus Programme (2001-2004)", WASTE, Gouda, *www.waste.nl.*

Anschütz, J. et al. (2004), "Putting integrated sustainable waste management into practice: Using the ISWM assessment methodology as applied in the UWEP Plus Programme (2001-2003)", WASTE, Gouda.

Aziz, H. (2004), "Improving the livelihood of child waste pickers: experiences with the Zabbaleen in Cairo, Egypt, an evaluative field study", WASTE, Gouda.

Ball, J. (1998), *Minimum Standards for Disposal by Landfill*, Second Edition, Department of Water Affairs and Forestry, Private Bag X313, Pretoria.

Bauman, Z. (2004), "Wasted lives. Modernity and its outcasts", Polity Press, Cambridge.

Bhaskar, A. and P. Chikarmane (2012), "The story of waste and its reclaimers: Organising waste collectors for better lives and livelihoods", *The Indian Journal of Labour Economics*, Vol. 55, No. 4, *www.swachcoop.com/kkppandswachpublications.html.*

Chaturvedi, B. (2009), "Cooling agents: the impact on the informal recycling sector on carbon emissions", *Chintan-Environmental*, Delhi, India, *www.chintan-india.org/documents/research_and_reports/chintan_report_cooling_agents.pdf.*

Chaturvedi, B. (2007), "Privatization of solid waste collection and transportation in Delhi: The impact on the informal recycling sector", Paper prepared as partial fulfilment of course on *Urban Issues in Developing Countries*, School for Advanced International Studies, John Hopkins University. Washington, DC.

Chaturvedi, B. (2006), "60 Kilos, Film", *Chintan-Environmental*, New Delhi, India, *www.chintan-india.org//publications_films.php?p=1.*

Chaturvedi, B. (2003), "Space for waste, planning for the informal recycling sector", Brochure, *Chintan-Environmental*, New Delhi, India, *www.chintan-india.org/publications_research_reports.htm.*

Chaturvedi, B. et al. (2009), "City Presentation Document for Delhi/New Delhi, India", Unpublished report used in the production of Scheinberg, Wilson and Rodic (2010), *Chintan-Environmental*, Delhi, India.

Chen, M.A. (2012), "The informal economy: Definitions, theories and policies", *WIEGO Working Paper* No. 1, August 2012, Report V.

Chikarmane, P. and L. Narayan (2009), *Rising from the Waste – Organising Wastepickers in India, Thailand and the Philippines*, Committee on Asian Women (CAW), Bangkok, Thailand, *www.swachcoop.com/kkppandswachpublications.html.*

Chikarmane, P. and L. Narayan (2007), "City Report for the City of Pune", India, Unpublished report used in the production of Scheinberg, Simpson and Gupt (2010), KKPKP, Pune, India, *www.giz.de.*

Chintan-Environmental (2009), "Wastepickers and climate change: Greenhouse gas mitigation by the informal recycling sector in Delhi", Unpublished study, *Chintan-Environmental*, New Delhi, India.

Cohen, P. et al. (2013), *Preparing Inclusion Plans for Informal Recyclers at Final Disposal Sites: An Operational Guide*. Inter-American Development Bank (IDB), Washington, DC.

Cointreau, S. (1994), *Private Sector Participation in Municipal Solid Waste Services in Developing Countries*, The Private Sector, World Bank, Urban Management Programme, Washington, DC, Volume 1.

Cointreau, S. (1989), "Provision of solid waste services in developing countries", *Resource Paper* presented at the International Seminar on the Provision of Municipal Public Services in Developing Countries, August 21-29, 1989 Assenovgrad, Bulgaria.

D'Alisa, G. et al. (2010) "Conflict in Campania: Waste emergency or crisis of democracy", *Ecological Economics*, Vol. 70, pp. 239-249.

De Swaan, A. (1988), "In care of the State, health care, education and welfare in Europe and the United States in the modern era", Polity Press, Cambridge, United Kingdom.

DEAT (1998), "Working With Waste, Guidelines on Recycling of Solid Waste, Department of Environmental Affairs and Tourism of South Africa, Tswane, South Africa.

Dias, S.M. (2010), "Overview of the legal framework for social inclusion in solid waste management in BrazilWIEGO, Cambridge, Massachusetts, United States, *www.inclusivecities.org.*

Dias, S.M. (2006), "Waste and citizenship forums – Achievements and limitations", *Paper* No. 11, CWG-WASH Workshop, Kolkata, India, 1-5 February 2006, *http://wiego.org/related/publications/3545/25/1615?page=5* (accessed 25 April 2012).

Dias, S.M. (2000), "Integrating waste pickers for sustainable recycling", Paper delivered at the Manila Meeting of the Collaborative Working Group (CWG) on Planning for Sustainable and Integrated Solid Waste Management, Manila, Philippines, *www.cwgnet.net.*

Dimino, R. (2010), "Beyond waste, a sustainable materials management strategy for New York State", New York State Department of Environmental Conservation, Albany, N.Y. United States.

Doychinov, N. and A.Whiteman (2013), *Economic Instruments in Solid Waste Management: Case Study Bulgaria*, Produced by RWA for the GIZ Operator Models study, 2012-2013, *www.giz.de/giz2013-en-economic-instruments-bulgaria*.

Doychinov, N. (2008), "The Bulgarian EkoPak Experience", Presentation given at the IFC Recycling Linkages Summit, Belgrade, November 2008. IFC PEP SE, IFC, Ulica Makedonija 9-11/III, 1000 Skopje, Macedonia, *www.ifc.org/pepse*.

European Environmental Agency (2011), *Earnings, Jobs, and Innovation: The Role of Recycling in a Green Economy*, EEA, Copenhagen.

Fernandez, L. (2011), "Towards a global waste pickers network: WIEGO and its collective efforts organizing waste pickers", ISWA Beacon Conference on Waste Prevention and Recycling, Buenos Aires, Argentina, ARS.

Folz, D. (1998), "The many important ways recycling has changed", *Resource Recycling*, September 1998, pp. 26-30.

FAO and WHO (2008), *International Code of Conduct on the Distribution and Use of Pesticides. Guidelines on Management Options for Empty Pesticide Containers, Food and Agriculture Organisation (FAO) and World Health Organisation*, May 2008.

Gerdes, P. and E. Gunsilius (2010), *The Waste Experts: Enabling Conditions for Informal Sector Integration in Solid Waste Management. Lessons Learned from Brazil, Egypt and India*, Eschborn, GTZ.

Giddens, A. (1994), "Beyond left and right, the future of radical politics", Polity Press, London.

Gille, Z. (2007), "From the cult of waste to the trash heap of history: The politics of waste in socialist and postsocialist Hungary", Indiana University Press, Bloomington, Indiana, US.

Gunsilius, E. et al. (2010), *The Economics of the Informal Sector in Solid Waste Management*, Deutsche Gesellschaft für Internationale Zusammenarbeit (German International Co-operation) (GIZ), Eschborn, Germany.

Gunsilius, E. et al. (2011), *Recovering Resources, Creating Opportunities. Integrating the Informal Sector Into Solid Waste Management*, Eschborn, Deutsche Gesellschaft für Internationale Zusammenarbeit (German International Co-operation) (GIZ).

Hill, E. (2001), "Women in the Indian informal economy: Collective strategies for work life improvement and development", *Work, Employment and Society*, September 2001; Vol. 15, No. 3, pp. 443-464.

IFC (2008), "The IFC Recycling Linkages Programme", Presentation given at the CWG Workshop in Cluj-Napoca, Romania, February 20-22, 2008, International Finance Corporation, *www.greenpartners.ro*.

Ijgosse, J. (2012), "Paying waste pickers for environmental services: A critical examination of options proposed in Brazil", *WIEGO Technical Brief* (Urban Policies), No. 6, November 2012.

ILO (2013), "Sustainable development, decent work and green jobs", 102nd Session of the International Labour Conference, *Report V*, Geneva, June 2013, International Labour Office.

ILO/IPEC (2004), "Addressing the exploitation of children in scavenging: A thematic evaluation of action on child labour", A Global Synthesis Report for the ILO prepared by WASTE, Gouda, the Netherlands, ILO, Geneva, Switzerland.

Ishengoma, A. and T. Lyimo (2002), "The Dar es Salaam experience to reduce poverty – promoting employment through urban services", Paper presented at the World Urban Forum; Nairobi Kenya, 29 April-4 May 2002.

Ishengoma, A. and K. Toole (2003), "Jobs and services that work for the poor; Promoting decent work in municipal service enterprises in East Africa; the Dar es Salaam project and the informal economy", Paper presented at the Knowledge-sharing Workshop organized by INTEGRATION, ITC Turin Italy; 28 October-1 November 2003.

Iskandar, L. et al. (2007), *City Report for Cairo*, Background document prepared by CID, Cairo Egypt, summarised in Scheinberg, Simpson and Gupt (2010), *www.giz.de*.

ISWA (2014), "Globalisation and waste", Series of presentations made at the World Congress of the International Solid Waste Association (ISWA), São Paolo, Brazil, September 2014, *www.iswa.org*.

ISWA/EXPRA/RDN (2014), "Challenges to separate collection systems for different waste streams - barriers and opportunities", Workshop organised on 9 October by the International Solid Waste Association (ISWA), the Extended Producer Responsibility Association (EXPRA) and the Regional Development Network (RDN), *www.iswa.org*.

Keita, M.M. (2003), "Diagnostique de la filière de récuperation de déchets dans la Commune IV du District de Bamako, apport final" (Diagnosis of the waste recovery pathway in Commune IV of Bamako district, final report), COPIDUC and Commune IV of Bamako, WASTE, Gouda, the Netherlands.

Keller, M. (2006), *Assessment of Gold Recovery Processes in Bangalore, India and Evaluation of an Alternative Recycling Path for Printed Wiring Boards, A Case Study*, Institute for Spatial and Landscape Planning, Regional Resource Management at the ETH Zurich, Zurich, Switzerland.

Kojima, M. et al. (2009), "Difficulties in applying extended producer responsibility policies in developing countries: case studies in e-waste recycling in China and Thailand", *Journal of Material Cycles and Waste Management*, Vol. 11, No. 3, pp. 263-269.

Lardinois, I. and C. Furedy (1999), "Separation at source", WASTE, Gouda.

Lerpiniere, D. (2014), "A Review of development co-operation in solid waste management", presentation given at the 2014 ISWA Global Congress in São Paolo, Brazil, 10 September 2014.

Lifuka, R. (2007), *City Report for Lusaka*, resource document for Scheinberg, Simpson and Gupt (2010), Riverine Associates, Lusaka, Zambia.

Medina, M. (2009), "Global supply chains in Chinese and Indian industrialization: impact on waste scavenging in developing countries", *UNU/WIDER Working Paper*, The United Nations University/ Institute of Advanced Studies, Tokyo, Japan.

Medina, M. (1997), "Informal recycling and collection of solid wastes in developing countries: Issues and opportunities", *UNU/IAS Working Paper*, No. 24, The United Nations University/Institute of Advanced Studies, Tokyo, Japan.

Melosi, M. (1981), "Garbage in the cities, refuse, reform and environment, 1880-1980", Texas A&M Press, College Station, Texas (United States).

Ministry of Environment and Water, Republic of Bulgaria (2003), *Program for Implementation of Directive 94/62/EC on Packaging and Packaging Waste, http://old.europe.bg/upload/docs/CONF_BG_02_03_ad07.pdf.*

Moreno, F. (2014), "Colombian alliance for inclusive recycling", PowerPoint presentation, CEMPRE Colombia.

Morris, J. (2000), "What works best to increase waste diversion?", *Resource Recycling*, January 2000, pp. 37-40.

Motsumi, S. (2007), *Botswana Waste Management Strategy*, revised from 2004, Department of Environmental Affairs, Botswana.

Occhio del Riciclone (2012), *Rapporto Nazionale sul Riutilizzo 2012*, Ministero dell'Ambiente, 2012 (National Report on Reuse 2012, Ministry of the Environment).

Occhio del Riciclone (2009), *La Seconda Vita delle Cose*, ("The second life of things", Environment Issues), Edizioni Ambiente.

Occhio del Riciclone (2008), *Impatti Occupazioni di un riuso sistemico nella cittá di Roma*", (Occupational impacts of a reuse system in the city of Rome, the City of Rome in 2008), Comune di Roma 2008.

OECD (2014a), "What have we learned about Extended Producer Responsibility in the past decade?", Documents from an Expert Meeting, *www.oecd.org/officialdocuments/publicdisplaydocumentpdf/ ?cote=ENV/EPOC/WPRPW(2013)7/FINAL&docLanguage=Fr.*

OECD (2014b), *What Have We Learned about Extended Producer Responsibility in the Past Decade? Case study Chile*, document from an Expert Meeting, OECD, available at *https://www.oecd.org/environment/ waste/20140526%20Case%20Study%20-%20EPR%20Chileamended.pdfOECD Intranet.*

OECD (2014c), "How does the Chinese EPR System Work? Case Study – China", Document prepared for the OECD, *www.oecd.org/environment/waste/China%20case%20study%20final.pdf.*

OECD (2013), *What Have We Learned about Extended Producer Responsibility in the Past Decade? A Survey of the Recent EPR Economic Literature*, OECD Paper ENV/EPOC/WPRPW(2013)7/FINAL available at *http:// spot.colorado.edu/~daka9342/OECD_EPR_KO.pdf.*

OECD (2001), *Extended Producer Responsibility: A Guidance Manual for Governments*, OECD Publishing, Paris, *http://dx.doi.org/10.1787/9789264189867-en.*

Oelofse, C. et al. (2006): "Shifts within ecological modernization in South Africa: Deliberation, innovation and institutional opportunities", *Local Environment*, Vol. 11, No. 1, pp. 61-78.

Poulussen, P. (1987), *Van Burenlast tot Milieuhinder: het stedelijk leefmilieu 1500-1800* (Neighbours burden to environmental pollution: The urban environment 1500-1800), DNB/Uitgeverij Pelckmans, Kapellen, Belgium.

Price, J. et al. (eds.) (1998), "Micro and small enterprises, the case of Latin America", WASTE, Gouda.

Republic of Botswana (1998), *Botswana's Strategy for Waste Management*, First Edition, 1998.

Rochat, D. et al. (2007), "Optimal recycling for Printed Wiring Boards (PWBs) in India", in proceedings of the 2007 Conference on Recovery of Materials and Energy for Resource Efficiency, Davos, Switzerland, 3-5 September 2007.

Romero, V. (2014), "Steps towards EPR implementation in developing countries", (MSc thesis), Imperial College London.

Samson, M. (ed.) (2009), *Refusing to be Cast Aside: Waste Pickers Organising Around the World*, Women in Informal Employment: Globalising and Organising (WIEGO), Cambridge, Massachusetts, United States, *www.wiego.org*.

Scheinberg, A. (2011), "Value added: Modes of sustainable recycling in the modernisation of WASTE management systems", PhD Dissertation, Wageningen University, published by WASTE, Gouda, the Netherlands.

Scheinberg, A. (2001a), *Financial and economic issues in integrated sustainable waste management; Integrated sustainable waste management, set of five tools for decision-makers*, published by WASTE, Gouda, the Netherlands.

Scheinberg, A. (2001b), *Micro and small enterprises in integrated sustainable waste management: Integrated sustainable waste management, set of five tools for decision-makers*, published by WASTE, Gouda, the Netherlands.

Scheinberg, A. and J. Anschütz (2007), "Slim pickin's: Supporting waste pickers in the ecological modernisation of urban waste management systems", *International Journal of Technology Management and Sustainable Development*, Vol. 5, No. 3, pp. 257-270.

Scheinberg, A. and J. Ijgosse (2004), *Waste Management in the Netherlands*, Report prepared for UNITRABALHO, Recife, Brazil, published by WASTE, Gouda, The Netherlands.

Scheinberg, A. and A.P.J. Mol (2010), "Multiple modernities; transitional Bulgaria and the ecological modernisation of solid waste management", *Environment and Planning C: Government and Policy*, Vol. 28, No. 1, pp. 18-36.

Scheinberg, A. and A. van de Klundert (2005), *ISWM Case Study: Developing the Dar es Salaam – UNIDO Recycling Processing Centre*, UN Industrial Development Organisation (UNIDO), Vienna, *Austria*, *www.waste.nl*.

Scheinberg, A. and J. Nesic (2014), "Engaging informal recyclers in Europe: Status and report of a consultation", Paper delivered at the 2014 ISWA Global Congress, 7-9 September 2014, São Paolo, Brazil, *www.ISWA.org*.

Scheinberg, A. et al. (2012), *The Botswana Recycling Guidelines. Advice on Valorisation for Middle-Income Countries.* Volume 1: *Inception Report and Recycling Frameworks*; Volume 2: *Recycling Guidelines*, UNDP Botswana, Gaborone and New York.

Scheinberg, A. et. al. (2010a), *Economic Aspects of the Informal Sector in Solid Waste*, Final Report and Annexes, GIZ (German International Co-operation), the CWG (Collaborative Working Group on Solid Waste Management in Low- and Middle-income Countries), and the German Ministry of Foreign Affairs, Eschborn, Germany, *www.GIZ.de*.

Scheinberg, A. et al. (2010b), *Solid Waste Management in the World's Cities, UN-Habitat's Third Global Report on the State of Water and Sanitation in the World's Cities*, Earthscan Publications, Newcastle-on-Tyne, United Kingdom.

Scheinberg, A et al. (2008) *Closing the Circle, Bringing Integrated Sustainable Waste Management Home*, VNG International, The Hague, the Netherlands.

Scheinberg, A. et al. (2007a), *Needs of Roma Collectors and Other Stakeholders in the PEP SE Region for Training, Technical Assistance, and Financial Services and Recommendations for Programmatic Response*, an assessment report prepared for the recycling linkages Private Enterprise Programme South East Europe (PEP SE) of the International Finance Corporation, Skopje, Macedonia.

Scheinberg, A. et al. (2007b), *Economic Aspects of the Informal Sector in Solid Waste Management*, unpublished draft research report, German Technical Co-operation, Eschborn, Germany; available at *www.waste.nl* and *www.giz.de* (NB: rewritten and published under the same authors in 2010).

Schmied, E. et al. (2011), "Formalisation options for informal sector activities and their legal requirements in Central Europe", paper delivered at the Thirteenth International Waste

Management and Landfill Symposium, Sardinia 2011. BOKU, Institute of Waste Management, University of Natural Resources and Life, Sciences, Vienna, Austria; *www.transwaste.eu.*

Shove, E. (2003), *Comfort, Cleanliness and Convenience: The Social Organization of Normality (New Technologies/New Cultures)*, Berg Publishers, United Kingdom.

Simpson-Hebert, M. et al. (2005): *A Paper Life*, WEDC, Loughborough University, Loughborough, United Kingdom.

Simpson, M. (1993), "Lapaks and Bandars Convert MSW in Indonesia", *Biocycle*, June 1993, pp. 78-80.

Smink, C.K. (2002), "Modernisation of environmental regulations. End-of-life vehicle regulations in the Netherlands and Denmark", PhD Dissertation, Aalborg University, Aalborg, Denmark.

Soos, R. and C. Popovici (2008), "Planning in the real world, challenges of sustainable and affordable modernisation in the solid waste sector", Report delivered at the CWG International Workshop, Cluj-Napoca, Romania, 22 and 23 February 2008, Green Partners, str. Fintinele 18, 400294 Cluj-Napoca, Romania; *www.greenpartners.ro.*

Soos, R. et al. (2014), *Structural Integration of the Informal Sector into the Municipal Solid Waste Management Sector in Tunisia*, Inception Report, Project; unpublished document, German International Co-operation, Eschborn, Germany.

Spaargaren, G. (1997), "The ecological modernization of production and consumption", Essays in Environmental Sociology, PhD thesis, Wageningen University, Wageningen.

Spaargaren, G. and B. van Vliet (2000) "Lifestyles, consumption and the environment. Ecological Modernisation of Domestic Consumption", *Environmental Politics*, Vol. 9, No. 1, pp. 50-77.

Spaargaren, G. et al. (2005), "Mixed modernities: Towards viable urban environmental infrastructure development in East Africa", *Position Paper*, Environmental Policy Group, Wageningen University and Research Centre, Wageningen, The Netherlands.

Strasser, S. (1999), *Waste and Want. A Social History of Trash*, Henry Holt and Company, New York.

Stewardship *Community (n.d.) Best Spraying Practices, Triple-rinse System, www.stewardshipcommunity.com/en/best-spraying-practices.html.*

Tasheva, E. (2012), "Presentation on informal recycling in Bulgaria", given at the TransWaste Closing Meeting in Budapest, September 2012.

Tompkins County New York (2012), *Case Study for the USEPA Zero Waste Initiative.*

Tompkins County, New York (1995), *Tompkins County Solid Waste Management Plan.*

UNDP (2002), *Country Profile of Botswana*, prepared for the Johannesburg Local Agenda 21 Summit, 2002.

UNEP (2015), *Global Waste Management Outlook*, United Nations Environment Programme, International Environmental Technology Centre (UNEP-IETC), Osaka, Japan, *www.unep.org/ietc/InformationResources/Events/GlobalWasteManagementOutlookGWMO/tabid/106373/Default.aspx.*

Van de Klundert, A. and I. Lardinois (1994), "Informal resource recovery: The pros and cons", WASTE, Gouda.

Van Eerd, M. (1996): "The occupational health aspects of waste collection and recycling: A survey of the literature", *UWEP Working Document*, No. 4, Part I, WASTE, Gouda.

Van Vliet, B. (2002) "Greening the grid. The ecological modernisation of network-bound systems", PhD Dissertation, Wageningen University, the Netherlands.

Van Vliet, B. et al. (2005), *Infrastructures of Consumption*, Earthscan Publications Ltd, Newcastle-on-Tyne, United Kingdom.

Vega, V.R. (2007), "Gestión de residuos electrónicos en Costa Rica: sistematización de la experiencia Proyecto Bilateral Costa Rica-Holanda Fase I y II, 2003-2007" (E-waste management in Costa Rica: Systematization of experience in Costa Rica-Holland bilateral project Phases I and II). ACEPESA, San José, Costa Rica, and WASTE, Netherlands.

Velis, C.A. et al. (2012), "An analytical framework and tool ('InteRa') for integrating the informal recycling sector in waste and resource management systems in developing countries", *Waste Management Research*, Vol. 30, pp. 43-66; *http://wmr.sagepub.com/content/30/9_suppl/43.*

Velis, C.A. et al. (2009), "19th century London dust-yards: a case study in closed-loop resource efficiency", *Waste Management*, Volume 29, pp. 1282-1290.

WASTE (2010), "Training Materials in Integrated Sustainable Waste Management", Gouda, the Netherlands.

WIEGO (2015) Women in the Informal Economy Globalising, Organising (WIEGO), *www.wiego.org*.

Williams, E. et al. (2013), "Linking informal and formal electronics recycling via an interface organisation", *Challenges*, Vol. 4, No. 2, pp. 136-153.

Wilson, D. (2007), "Development drivers for waste management", *Waste Management and Research*, Vol. 25, pp. 198-207.

Wilson, D. et al. (2010), *Comparative Analysis of Solid Waste Management in Cities around the World*, United Kingdom Solid Waste Association, November 2010.

Wilson, D. et al. (2006), "Role of informal sector recycling in waste management in developing countries", *Habitat International*, Vol. 30, pp. 797-808.

Extended Producer Responsibility
Updated Guidance for Efficient Waste Management

ANNEX A

Television and computer recycling in Australia[1]

by
Bruce Edwards *and* Declan O'Connor-Cox[2]

SUMMARY BOX	
Cost allocation	No data on local government's expenditure
Cost coverage	Market driven competitive scheme, Australian Government not involved in contracting or fee setting.
Role of government	• Accreditation of co-regulatory arrangements; • Administration of the scheme including ensuring compliance by liable parties, and approving and monitoring the outcomes of co-regulatory arrangements including collection and material recovery targets; • Authority to impose civil and financial sanctions for non-compliance; • Monitoring and consultation with stakeholders in view of possible reviews.
Environmental performance	In 2012-2013, 40 813 tonnes of e-waste recycled, equivalent to 98.8% of the target (i.e. 41 327 tonnes) and nearly double the volume of TVs and computer recycling per annum prior to the implementation of the scheme. In 2013-14, 52 736 tonnes of e-waste was recycled, exceeding the target of 43 430 tonnes by over 9 300 tonnes.
DfE incentive	90% material recovery target as from 1 July 2014 (almost met already)
Cost efficiency	• Net benefit estimated between USD 517 million and USD 742 million over the period 2009 to 2030. • Limited data is available on e-waste recycling occurring outside the Scheme. Pre-implementation data indicated an initial recycling rate of 17 %, which in some cases may have been absorbed into Scheme activities.

1. Description of EPR set-up

a) Legal context

The *National Waste Policy: Less waste, more resources* was endorsed by the Australian government, and the state, territory and local governments in November 2009. It identified 16 key areas for collaborative efforts, including the establishment of a national product stewardship legislation to formalize the number of voluntary schemes existing at that time. This led to the *Product Stewardship Act 2011* (the Act) in August 2011, which provides a national framework to support voluntary, co-regulatory and mandatory product stewardship schemes. The *National Television and Computer Recycling Scheme* (the scheme) started in 2012 as the first co-regulatory scheme under the Act. Specific requirements and outcomes of the scheme are specified in a subordinate legislation, the *Product Stewardship (Televisions and Computers) Regulations 2011*, which can more easily be amended than the overarching Act.

a) Allocation of responsibilities (distribution of roles, financial flows)

The scheme requires manufacturers and importers (liable parties) of televisions, computers and computer peripherals to jointly fund the collection and recycling of a portion of products disposed of in Australia every year (a gradually increasing percentage figure which is a proportion of waste arising, 35% in 2014-15). Companies importing or manufacturing volumes under a certain threshold (liability is triggered by import of 5 000+ units of televisions and/or computers, and 15 000+ units of computer parts/peripherals) are not required to participate. Recycling services under the scheme are provided free of charge to households and small businesses that drop-off end-of-life products at industry-provided collection services. Liable parties are required to join and fund co-regulatory arrangements that are responsible for organising and delivering collection services and to provide recycling services on behalf of the liable parties. They are also in charge of raising awareness to the public about the availability of collection services and the products being accepted. Co-regulatory arrangements must apply to the government for accreditation. Three organizations were approved in 2011-12, providing 635 collection sites services. A further two arrangements were approved in 2012-13. Co-regulatory arrangements may choose to contract out collection, logistics or recycling services to third party providers, which would then sell the recovered materials to national and international markets. E-waste not covered by the Scheme remains the constitutional responsibility of state and territory governments, and through them local governments. In 2013-14, these entities are responsible for managing 67% of TV and computer e-waste. These figures will progressively decrease as industry-funded collection targets increase. Over 200 local governments have partnered to date with co-regulatory arrangements to provide services under the Scheme. They are also in charge of approval and licencing conditions for waste collection and recycling infrastructure, in order to safeguard consumer and environmental protection.

b) Governance system

The scheme is a co-regulatory scheme, as opposed to mandatory and voluntary approaches that are also envisaged under the Act. Under the co-regulatory approach, product stewardship operations are delivered by the industry and regulated by the Australian Government. The Australian Government administers the scheme by ensuring compliance by liable parties, and by approving and monitoring the outcomes of co-regulatory arrangements including the collection and material recovery targets. Co-regulatory arrangements report directly to the Australian Government. Liable parties that do not meet their obligation to join a co-regulatory arrangement can be issued civil penalties, injunctions, or financial sanctions up to the total costs that they would have disbursed by joining an arrangement administered by the Australian Government. The Act also provides for sanctions when a co-regulatory arrangement fails to meet one of the required outcomes of the scheme; these sanctions include an improvement notice or, as a last resort, the cancelation of the arrangement's approval.

2. Environmental effectiveness

a) Collection and recycling rates

A review of the scheme in 2013 introduced a single recycling target for all products covered in order to provide greater flexibility to co-regulatory arrangements to manage risk and to ensure better alignment between collection activities and e-waste disposed by households and small businesses.

The industry e-waste recycling target was set at 30% in 2012-13. In the first year of the scheme, 40 813 tonnes of TV and computer waste was recycled, equivalent to 98.8% of the scheme target (i.e. 41 327 tonnes) and nearly double the volume of TV and computer recycling per annum prior to the implementation of the scheme. The rate of recycling increased throughout the financial year as co-regulatory arrangements established collection services and entered into contracts with recycling service providers. Annual recycling targets are set to evolve in proportion to the total waste estimated to be generated in a given year, according to a formula detailed in the Regulation.[3] Targets are expected to increase slowly to 80% in 2021-22. Figure A.1 shows the proportion of computer and TV waste covered by the targets in relation to overall waste expected to be generated per year.

Figure A.1. **Pre-implementation projections of waste arising from 2011-12 to 2023-24, showing the proportions covered and not covered by the scheme's annual recycling targets**

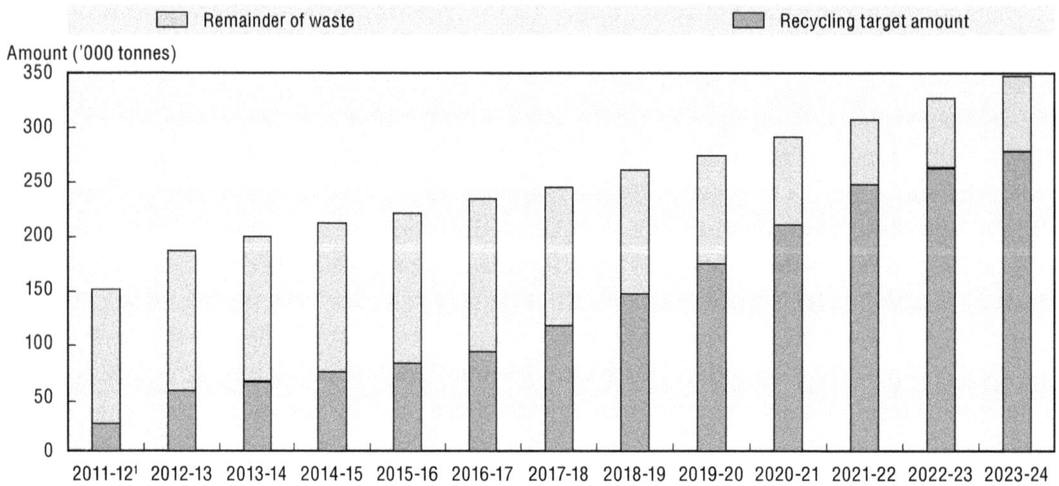

Note: Projections are being revised as part of the current operational review.
Source: Bruce Edwards and Declan O'Connor-Cox (2014), The Australian National Television and Computer Recycling Scheme, Case study prepared for the OECD, *www.oecd.org/env/waste/gfenv-extendedproducerresponsibility-june2014.htm.*

b) Design-for-Environment (DfE)

As from 1 July 2014, a material recovery target of 90% was set. This target refers to the proportion of components and materials recuperated from TVs and computers after recycling to be sent for processing into useable products and materials. It is designed to achieve a reduction in the post-recycling materials being sent to landfills and to increase the quality of recycling. The current performance of recycling providers generally exceeds 90% but some improvement is needed by individual e-waste recycling providers in order to ensure this standard is consistently met from July 2014.

Following the commencement of the scheme, Standards Australia[4] developed the Australia-New Zealand Standard "*Collection, storage, transport and treatment of end-of-life electrical and electronic equipment*" *(Standard 5377)* in 2013, which specifies requirements for the safe and environmentally sound collection, storage, transport and treatment of end-of-life electrical equipment. Application of this standard is expected to maximise re-use and material recovery in the recycling process, reduce or eliminate the amount of waste going to landfill, safeguard the health of industry workers and minimise potential harm to the environment. Incentives for companies are financial – encourages sourcing of market for recovered materials, to offset recycling costs. However, mandatory accreditation to the standard is being considered as part of the current operational review.

3. Economic efficiency (including competition aspects)

a) Cost efficiency

A cost-benefit analysis was undertaken in 2009 by the Australian Government under the Regulatory Impact Statement (RIS), which estimated that the scheme for TVs and computers would provide a net benefit of between USD 517 million and USD 742 million over the period 2008-09 to 2030-31.

The financial liability of individual importers and manufacturers is determined by a specific product code assigned to imported products. Every code has an associated conversion factor, which is an estimated average weight of the product type, and is updated over time as technological advances occur.[5] Co-regulatory arrangements set fees for liable parties in a competitive and open market, and no set proportion of costs must be covered.

b) Leakages and free riders

Monitoring and data received by the Australian Government since the introduction of the scheme has revealed a thorough implementation of the scheme's requirements and it was observed that a majority of liable importers and manufacturers have met their obligation to join a co-regulatory arrangement.

c) Trade and competition

The scheme has stimulated competition in the e-waste recycling industry. As the industry grows stakeholders including co-regulatory arrangements and recyclers have formed partnerships to take advantage of the expanding industry and increased recycling activity. In addition to developing and growing local recycling capability, some parts of the industry are specialising in particular areas of resource recovery and recycling.

4. Key issues and possible reforms

Television and computer manufacturing industries continuously innovate, improve technologies and make new products available to consumers. This can cause variations in both the applicable product codes and the conversion factors which assign an estimated weight to each product type. Scheme product codes and conversion factors will be subject to ongoing assessment and consideration, including an industry survey of products and weights every 2-3 years to inform further amendments.

Notes

1. Full source available at: Bruce Edwards and Declan O'Connor-Cox (2014), The Australian National Television and Computer Recycling Scheme, Case study prepared for the OECD, *www.oecd.org/env/waste/gfenv-extendedproducerresponsibility-june2014.htm*.
2. Department of Environment, Government of Australia.
3. Based on the average weight of imports of the last three years and multiplied by a factor of 0.9, which assumes that the majority of imported TVs and computers are purchased to replace another product that enters the waste stream.
4. A non-government organisation that works with government and industry to comply with Australia's internationally aligned standards and that publishes specifications and procedures designed to ensure products, services and systems are safe, consistent and reliable.
5. The purpose of this conversion factor is to approximately align the data collected by the Australian Customs Service, which is calculated in number of units imported, with the measurement metrics used by the recycling industry, which are calculated in weight of items recycled. Converting product data from numerical imports to weighted products helps provide more consistent and meaningful information to inform recycling targets.

Extended Producer Responsibility
Updated Guidance for Efficient Waste Management

ANNEX B

EPR for used tyres in Flanders (Belgium)

by

Ovam (the Public Waste Agency of Flanders)[1]

SUMMARY BOX	
Cost allocation	Waste management operations are entirely financed through environmental fees paid by consumers and transferred to Recytyre. Municipalities are not financially liable for used tyres as these are not categorized as household waste.
Cost coverage	Environmental fees paid by consumers cover 100% of collection and treatment operations for used tyres.
Role of government (OVAM)	• Observer in Recytyre's Board of Directors and General Assembly. • Advisor and evaluator of Recytyre's management plan, financial plan and communication plan. • Monitoring of stakeholders' activities (producers, intermediaries, vendors, Recytyre) and evaluation of the achievement of targets. • Conducts inspections on a regular basis and in case of non-compliance. • A division of OVAM can fix financial sanctions for non-compliance.
Environmental performance	• Amount of waste collected in Flanders: 51 375 tonnes. • Of amount collected: ❖ amount of material recovery (= reuse, retread and material recycling); in Belgium: 84.96% ❖ amount of energy recovery; in Belgium: 15.04% ❖ amount disposed of in landfills: 0.
DfE incentive	• communication about tyre pressure • no production of tyres in Belgium, which makes it difficult to promote DfE.
Cost efficiency	• Economic efficiency: net benefit of EUR 3 976 971 in 2012.

1. Description of EPR set-up

a) Legal context

In Flanders, the Acceptance Duty was initiated as an EPR instrument in waste policy in 1994, ensuring that those who produce and market products bear responsibility for the collection and processing of the resulting waste. In order to implement the duty of acceptance, a first environmental policy agreement (MBO) was established between the government and the producers of tyres on 22 May 2000, which was replaced by a new one in 2003. In 2010, a new agreement – covering not only tyres from replacement, but also tyres from original equipment with the exception of bicycle tyres – entered into force. On 1 July 2012, the Flemish regulation on the sustainable management of materials and waste (the Materials Decree or VLAREMA) further developed the legal framework for the Acceptance Duty and EPR principles.

b) Governance and Enforcement

OVAM is the Public Waste Agency of Flanders and is responsible for enforcing waste policy. OVAM is an observer in the Board of Directors and the General Assembly of Recytyre,[2] and gives advice and approval on the management plan, action plan, financial plan, and communication plan of Recytyre on a yearly basis. The producers, intermediaries, vendors, and waste management organization hand over all necessary data (including the total amount of used tyres and tyres for reuse that have been collected and the way they have been handled) to OVAM. OVAM then evaluates the implementation of the Acceptance Duty and whether targets are met; and reports back to the Flemish Government and the Flemish Parliament. Furthermore, OVAM conducts inspections on a regular basis in order to ensure that the Acceptance Duty is implemented (two inspectors controlled the transport sector in 2011). If a company is deemed non-compliant, it receives recommendations to comply with the relevant legislation. An external company audits the financial situation. Finally, OVAM is responsible to report to the Flemish Government and the Flemish Parliament on the EPR scheme for used tyres.

c) Allocation of responsibilities (distribution of roles, financial flows)

Consumers pay an environmental fee as they buy a tyre, which will be used for the collection and treatment of used tyres. The Acceptance Duty implies that the vendor is obliged to take back for free the product returned by the consumer. The intermediaries are obliged to take back for free the returned product from the vendor in proportion to the amount of products they had delivered to vendors. The responsibilities of intermediaries and vendors of tyres also include charging the environment fee to consumers and contacting transport operators acknowledged by Recytyre.

The producers are obliged to take back for free the waste collected by the vendors and intermediaries and have to make sure that collection and recovery of used tyres is undertaken for an amount of products proportional to the amount initially delivered to the market. In 1998, the major tyre producers operating on the Belgian market founded the non-profit organization Recytyre (no accreditation is required under the legislation) in order to organize waste management operations (collection, sorting and processing of used tyres) as well as reporting, prevention and awareness raising activities on their behalf. In 2012, Recytyre represented 663 active members, 51 partners (collectors), 5 357 active collection points and 234 treatment facilities. Recytyre has an agreement with the collectors to collect and process used tyres and remunerates them for the collection and recycling activities. Private recyclers chosen at the discretion of collectors are in charge of the reuse and retreatment operations of used tyres.

Municipalities do not have an obligation to organize the collection of used tyres as these are not categorized as household waste. However, most municipalities have signed an agreement with Recytyre for the free collection of used tyres.

2. Environmental effectiveness

a) Collection and recycling rates

All used tyres that are offered for take-back are collected with a rate of maximum 100% of the amount of new tyres placed on the market in the same year. The percentage of reuse, retread and recycling of the collected used tyres should be at least 55% and the remaining part would be incinerated with energy recovery at a maximum rate of 45%. The existence of a tax on incineration and the legislation enabling the use of rubber granulate as infill material in synthetic sport fields further encourages reuse and recycling. Landfilling of used tyres is not allowed.

The EPR for used tyres has significantly contributed to decreasing the amount of used tyres that are dumped to almost zero. As the consumer pays an environmental fee when buying the tyres and does not have to pay an extra fee to have them collected, almost no used tyres are now dumped in Flanders. The 2011 evaluation report[3] states that an average of 88% of all waste tyres is collected, meaning that the target for collection is met. When considering only the replacement market, a collection rate of 99% was achieved. This indicates that tyres of original equipment are less widely collected, which can be explained because these vehicles are typically exported and are not processed in Belgium as end-of-life vehicles (except from trucks, tractors and construction machines). Regarding recycling, targets are usually met. Used tyres are often recycled abroad however as Flanders has little recycling treatment capacity and because there is a high demand for Belgian used tyres that are considered to be of high quality, availability and quantity, which is reflected in the increasing amount of used tyres being reused and re-treaded. As illustrated in Figure B.1, when Recytyre initially started to finance and collect more types of tyres including agriculture, civil engineering, industry and moto tyre in 2006, used tyres where mostly sent to energy recovery, and material recycling progressively increased over the years and now is now the most used option.

Figure B.1. **Belgian percentage of material recycling and energy recovery of used tyres 2006-11**

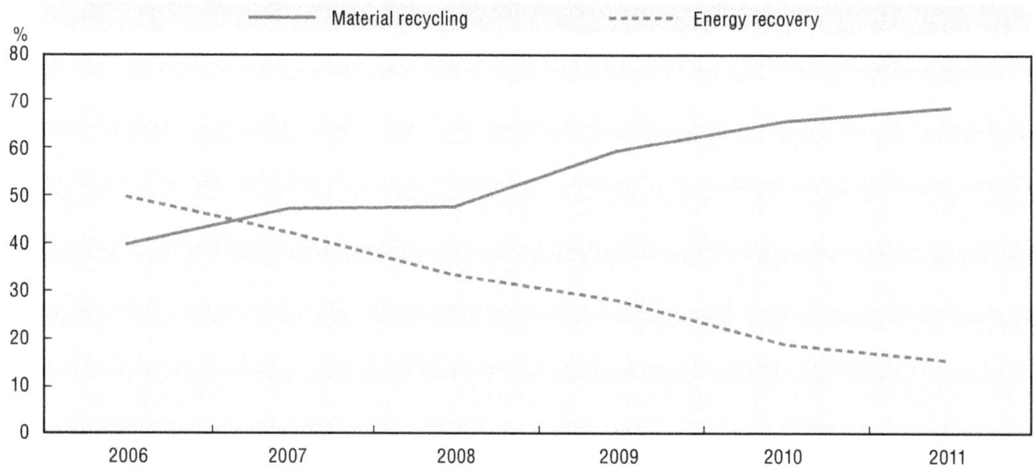

Source: OVAM (2014), Extended producer responsibility. The case of used tyres in Flanders (Belgium), Case study prepared for the OECD, *www.oecd.org/env/waste/gfenv-extendedproducerresponsibility-june2014.htm.*

b) Design for Environment (DfE)

The strategic aim of the EPR for used tyres is reducing the environmental impact of tyres in the waste stage as much as possible by Eco design and waste prevention and by separate collection and processing, prioritizing reuse, retreating and recycling wherever possible. Recytyre communicates about the importance of tyre pressure to extend the lifecycle of the tyre. It is difficult to weigh on eco-design itself because there is no production of tyres situated in Belgium.

3. Economic efficiency

a) Cost efficiency

When buying tyres, consumers pay a fee that is transferred by producers to Recytyre and that fully covers collection and treatment operations for used tyres. The fee depends on the type of tyre and ranges from EUR 1.32 including 21% VAT (e.g. motorcycle tyres) to EUR 794 37 including VAT (e.g. tyres from some vehicles for public works and construction services).[4] The fee is mentioned separately on the consumer's receipt.

In 2012, Recytyre had a net benefit of EUR 3 976 971, which is rather stable compared to the profit and loss account in the last years (Table B.1). Both revenues and expenses have decreased by about 15% between 2011 and 2012. The decrease in revenues can be largely attributed to the decrease in tyres sold on the replacement market due to the economic crisis, which at the same time decreased the costs for collection and treatment as fewer

Table B.1. Financial situation of Recytyre 2010-12

		2010	2011	2012
BALANCE SHEET	Fixed assets	357 638	464 463	466 989
	Accounts receivable	5 476 725	4 695 648	4 870 402
	Cash and cash equivalents	5 492 845	11 622 078	15 063 611
	Total assets	**11 327 208**	**16 782 190**	**20 401 002**
	Equity	284 055	4 913 089	8 890 060
	Provisions	0	6 888 000	7 499 000
	Liabilities	11 043 153	4 981 101	4 011 941
	Total liabilities and equity	**11 237 208**	**16 782 190**	**20 401 002**
PROFIT AND LOSS ACCOUNT	**Revenues**	**24 247 611**	27 058 201	22 627 668
	Operational costs	19 823 214	14 837 540	17 098 614
	Provisions	0	6 888 000	611 000
	Other company costs	514 234	734 292	1 060 671
	Total expenses	**20 337 448**	**22 459 832**	**18 770 285**
	Financial result	-60 719	30 665	119 588
	Profit	**3 849 444**	**4 629 033**	**3 976 971**

Source: OVAM (2014), Extended producer responsibility. The case of used tyres in Flanders (Belgium), Case study prepared for the OECD, *www.oecd.org/env/waste/gfenv-extendedproducerresponsibility-june2014.htm*.

products were collected. Company costs increased slightly however due to the increase in personnel as Recytyre is investing in tighter monitoring of the operators' activities. Regarding the balance sheet, a remarkable increase was achieved in the last years as the equity had become negative following considerable losses in the years 2006, 2008 and 2009. The turnabout was achieved thanks to a number of measures including an increase in the environmental fee and some cost savings. Recytyre now has a solid financial buffer that can ensure the continuation of its activities in the coming years. The organization is currently undergoing an operational and financial restructuring, which implies that current numbers as outlined in the table may not be a reference for the financial results for the coming years.

b) Leakages and free riders

Free-riders (producers who do not comply with the acceptance duty for used tyres) receive an inspection visit from OVAM, and receive a written reminder listing their legal obligations. If the producer is still not in compliance, the inspectors make a written statement about the non-compliance. The division Environment Preservation, Environmental Damage and Crisis Management (Afdeling Milieuhandhaving, Milieuschade en Crisisbeheer or AMMC) then fixes a penalty that is case-specific, the lowest being around EUR 200 and up to about EUR 5 000.

c) Trade and competition

Recytyre is the only coordinating waste management organization for used tyres in Flanders[5] so that it does not face competition. The collection and treatment markets for used tyres are subject to the free market. Every collection operator can participate in the system and will be financed by Recytyre if it works according to the Recytyre guidelines. Used tyres collected in Flanders are often recycled abroad as Flanders has little recycling capacity for this waste stream. This has not been a problem so far however as all collected used tyres end up being treated especially as there is a high demand for used tyres from Belgium due to the availability and quality of the product.

4. Key issues and possible reforms

Further improvement is still possible when it comes to setting the right fee for truck tyres. However, due to the fees applicable in surrounding countries, Recytyre is reluctant to increase these fees as this might stimulate the purchase of tyres in countries where the fees are lower or where there are no fees for truck tyres. Also, so far, there is no market for re-treaded tyres for passenger cars.

The revision of the MBO takes place every five years and allows for further improvements in line with technological advances.

Notes

1. Full source available at: OVAM (2014), *Extended producer responsibility. The case of used tyres in Flanders (Belgium)*, Case study prepared for the OECD, *www.oecd.org/env/waste/gfenv-extendedproducer responsibility-june2014.htm*.
2. Non-profit waste management organization for used tyres that fulfils the tasks for the representative organizations in the Flemish region, as well as in the Brussels Capital and the Walloon regions.
3. See "Evaluatierapport 2011" at the line "banden" under *www.ovam.be/stand-van-zaken-aanvaardingsplicht*.
4. A detailed list of the most recent fees for 2009 per tyre type and size can be found on the website of Recytyre, *www.recytyre.be/fr*.
5. With the exception of the EPR system for used tyres of end-of-life vehicles where Febelauto is the waste management organization.

Extended Producer Responsibility
Updated Guidance for Efficient Waste Management

ANNEX C

EPR for waste of electric and electronic equipment in Canada

by
Jacinthe Séguin, Waste Reduction and Management Division, Environment Canada[1]

SUMMARY BOX	
Governance	• Provinces (environment ministries or mandated organizations) lead the establishment of waste electrical and electronic equipment (WEEE) extended producer responsibility (EPR) regulations.
Roles of provincial government	• to develop and administer regulations for the management of WEEE; • to set performance targets for the designated materials • to review and approve industry stewardship plans • to monitor and oversee the programs' operations, and • to provide the compliance and enforcement measures.
Cost allocation	• Fees paid by obligated stewards passed on to consumers (e.g. between CAD 3 and CAD 15 per desktop computer across provinces). • The greatest system costs are payments made to collectors (CAD 59* – CAD 150 per tonne), transporters (CAD 40 – CAD 200 per tonne), and processors (CAD 150 – CAD 700 per tonne). Total costs vary by province from CAD 350 – CAD 870 per tonne to CAD 880 – CAD 1030 per tonne.[2] • Overhead costs (oversight and monitoring, education and enforcement costs) account for 15-25% of overall costs.
Cost coverage	Operational costs range from CAD 1 105 to CAD 1 822 per tonne across provinces
Environmental performance	125 000 tonnes of end-of-life electronic collected in 2012 and amount of WEEE collected increase year after year.
DfE incentive	/
Cost efficiency	• Most programs have achieved financial sustainability. • In British Colombia, EPR program added CAD 110 000 000 to the province's GDP.

* Currency shown in document is in Canadian dollars.

1. Description of EPR set-up

a) Legal context

In 2009, the Canadian Council of Ministers of the Environment (CCME), an intergovernmental forum made up of 14 environmental ministers from federal, provincial and territorial governments, developed the Canada-wide Action Plan (CAP) for EPR. The goal of the CAP is to increase diversion and recycling of municipal solid waste through the harmonization of provincial EPR programs. In particular, twelve principles for electronics stewardship were approved in order to support the jurisdictions of provincial and territorial authorities in the development of WEEE programs. Today, nine out of ten Canadian provinces have WEEE regulations and programs in place. The CCME recognized that EPR may not be appropriate in Canada's remote Northern territories given the unique circumstances and high transportation costs, and is examining a combination of other measures in order to achieve the desired results in those regions.

b) Allocation of responsibilities (distribution of roles, financial flows)

In Canada, the provincial governments are responsible for developing, monitoring and administering regulations for the treatment of WEEE. These governments set performance targets for the designated materials, review and approve industry stewardship plans, monitor and oversee the programs' operations, and provide the compliance and enforcement measures. Producers (manufacturers and first importers) may fund and operate individual WEEE stewardship programs, or establish and register with a non-for-profit producer responsibility organization (PRO aka an industry funding organization

[IFO]) to act as the stewardship agent representing producers. In either case, they have to report designated product sales and fees to the PRO or the government oversight agency on a monthly basis, and to report overall program performance on an annual basis. PROs may act as collectors and/or establish depots for collection. They register and contract with authorized services providers (WEEE processors and recyclers must comply with minimum health and environment criteria and be approved by the Recycler Qualification Office, which operates under the EPRA), and are also in charge of publicly reporting performance, and of information and education campaigns towards consumers. Municipalities may also contract with PROs and participate in the collection of WEEE as a service provider. They further support EPR programs through green procurement policies or through the adoption of secondary policies such as landfill bans.

A key component of the funding model of EPR programs in Canada is the use of environmental handling fees (EHFs), which are regularly reviewed by the Electronic Products Recycling Association (EPRA) or another appointed body[3] where EPR programs exist. These fees, which are paid by obligated stewards and passed on to customers at the point of purchase, are determined on a per unit basis and vary between provinces depending on the costs required to operate the program, as well on other factors such as the product weight and quantity, the composition of products and the presence of orphan products in the market. According to August 2013 data, EHFs levied on WEEE ranged from CAD 3.00 for desktop computers in Ontario to CAD 15in Saskatchewan and Manitoba. EHFs levied on computer printers ranged from CAD 4.80 in Alberta to CAD 10.35 in Ontario.

c) Governance system and sanctions

In Canada, most provincial environment ministries lead the establishment of WEEE EPR regulations, provide program oversight and ensure compliance. In some provinces however, oversights and management is delegated to mandated organisations such as Waste Diversion Ontario (WDO) and Recyc-Québec, which report directly to the provincial ministries of environment and are entirely funded by industry. Regarding sanctions, Québec is the only province to provide financial penalties for producers who fail to meet collection targets (the obligation will become effective in 2018). These penalties, which are up to CAD 10per desktop computer and CAD 15per display device, are expected to encourage producers to join a collective PRO that can facilitate administration. Outside Québec, most other provincial regulations include general sanctions for other non-compliance aspects. For example, Waste Diversion Ontario has the power to remove the mandate from an underperforming PRO.

2. Environmental effectiveness

a) Collection and recycling rates

Together, Canadian WEEE programmes collected over 125 000 tonnes of end-of-life electronics in 2012. Table C.1 shows the amount of material collected in selected provinces.[4] In addition to having the highest overall collection rate both per capita (5.61 kg) and in total (75 702 tonnes), Ontario exceeded its previous year's collection total by 45%. Overall, provinces are improving in performance year after year. The Canadian experience demonstrates that a diversity of collection approaches, consumer outreach strategies, infrastructure facilities and market access can work and co-exist, and that regulating authorities take into account the many factors that can influence performance in their respective jurisdiction.

Table C.1. **Performance indicators for Canadian WEEE programs (data from annual 2012 reporting)**

Indicator	British Columbia	Alberta	Saskatchewan	Ontario	Nova Scotia	Prince Edward Island
Population	4 582 000	4 025 100	1 108 300	13 538 000	940 800	145 200
Tonnes Collected	21 963	17 280	3 080	75 702	4 719	649
Kilograms Collected/Capita	4.8	4.67	2.85	5.61	4.97	4. 44
# of Collection Sites	142	325**	72	444	37	6
Population Awareness (%)	75	80	89	67	79	69
Program Cost/Tonne (CAD)	1 208	1 117**	1 822	1 105	1 269	1 393

* Data is collected from 2012 annual reporting.
** Data represents fiscal year 2011-12.
Source: Séguin, J. (2014), "Promoting Sustainable Materials Management Through Extended Producer Responsibility: Canadian Waste Electrical and Electronic Equipment (WEEE)", Case study prepared for the OECD, *www.oecd.org/env/waste/gfenv-extendedproducerresponsibility-june2014.htm*.

b) Design-for-Environment (DfE)

In Canada, most programs have collection targets but there are no specific incentives in the existing provincial WEEE regulations for producers to address DfE. Moreover, there are no regulatory requirements for electronics to contain a certain percentage of recyclable components. The province of Québec is currently developing a differentiated fee structure to reward DfE initiatives by producers. However, this may add administrative complexity and it is questioned whether those objectives would not be better pursued outside the EPR governance framework as complementary measures. In particular, design changes have appeared to be most responsive to the emergence of Reduction of Hazardous Substances (RoHS) type regulations initiated in the European Union and other targeted toxics reduction initiatives, as opposed to general EPR regulations.

3. Economic efficiency (including competition aspects)

a) Cost efficiency

In 2012, the costs to operate Canada's WEEE collection programs ranged from CAD 1 105 per tonne in Ontario to CAD 1 822 per tonne in Saskatchewan, which includes the overall costs of delivering the programs, including collection, consolidation, transportation, audits, processing, administration, communications, R&D, and management. The greatest system costs are payments made to collectors, ranging from CAD 59 per tonne in Prince Edward Island to CAD 150 per tonne in Ontario; transporters, ranging from CAD 40 per tonne in Prince Edward Island to up to CAD 200 per tonne in Alberta; and processors, ranging from at least CAD 150 per tonne in Ontario to CAD 700 per tonne in Alberta.[5] Generally, overhead costs, which include general administration, oversight and monitoring, education and enforcement costs, are said to account for 15-25% of overall program costs.

While EPR programs are generally established with a financial sustainability objective in mind, not all Canadian programs have achieved this objective at this time.

In 2008, a study was conducted to assess the economic impacts of British Colombia's recycling regulation. It was estimated that total employment generated by electronics recycling in that province was over 123 full-time equivalents in 2007. In addition to job creation, the program added over CAD 110 000 000 to the province's GDP, meaning for that every tonne of material diverted from landfill CAD 4 150 total value was created both by industries directly and indirectly involved in waste management and by induced consumer spending.

Table C.2. **Canada's WEEE collection programmes – Revenues and expenses by program, 2012 (in Canadian dollars)**

	British Columbia	Alberta	Saskatchewan	Manitoba	Ontario	Quebec	Nova Scotia	Prince Edward Island
Revenues								
Environmental Handling Fees (EHFs)	$16,644,653	$12,146,972	$6,410,688	$3,388,486	$58,282,533	$12,974,079	$3,412,346	$446,427
Interest	139,574	10,739					10,057	1,519
Investment Income		7,729,823			177,176			
Other			176,325		150,000			
Total Revenues	**$16,784,227**	**$19,887,534**	**$6,587,013**	**$3,388,486**	**$58,609,709**	**$12,974,079**	**$3,422,403**	**$447,946**
Expenses								
Direct Operating Expenses								
Processing	$8,328,081			$1,113,374		$4,097,806	$1,083,982	$177,237
Handling	3,098,071			349,578		1,030,701	335,579	38,252
Transportation and Storage	2,330,657			333,332		1,702,735	350,187	25,690
Quality Assurance/Recycer Audits	179,087	248,822		60,000		150,000	55,665	8,408
Total Operating Costs	13,935,896	248,822	4,168,486	1,856,284	75,645,372	6,981,242	1,825,412	249,587
Other Expenses								
Consumer Awareness (P&E)	507,249	1,419,756	576,822	397,014	323,557	121,800	100,236	28,293
Research and Development		682,273						
Administration	1,087,040	690,703	532,085	274,858	3,042,365	787,287	311,378	73,811
Government/Professional Fees		1,601,671	222,906	33,491		389,500	0	29,166
Miscellaneous		14,706,654	111,000		2,608,007		66,616	
Total Other Expenses	1,594,289	19,101,057	1,442,813	705,363	8,973,929	1,298,587	411,614	131,271
Total Expenses	**$15,530,185**	**$19,349,879**	**$5,611,299**	**$2,561,647**	**$83,619,301**	**$8,279,829**		
Program Deficit/Surplus	**$1,254,042**	**$537,655**	**$975,714**	**$826,839**	**-$25,009,592**	**$4,694,250**	**$1,185,849**	

Notes: Data for Alberta and Saskatchewan is for fiscal year ended March 31, 2013. All other data is for year ended December 31, 2012.

In Ontario, miscellaneous expenses include $1,044,677 for Waste Diversion Ontario administration and program delivery, and $1,563,330 for program delivery. No breakdown of operating expenses.

In Saskatchewan, miscellaneous expenses refer to effectiveness and efficiency restricted expenses. No breakdown of operating expenses.

In Alberta, environmental handling fees (EHFs) are called advance disposal surcharges (ADSs). Miscellaneous expenses include $14,521,685 for the recycling incentive program and $184,969 for the rural Alberta project and municipal assistance. No breakdown of operating expenses.

It is impossible to determine total expenses for Nova Scotia and Prince Edward Island because the figure reported for program development expenses ($66,616) is combined.

Source: CM Consulting (2013), *The WEEE Report: Waste Electrical and Electronic Equipment Reuse and Recycling in Canada.*

b) Operation deficit plan

In order to avoid operating deficits including because of the presence of orphan products, EPRA has established an Operating Contingency Fund, which is intended to accumulate the equivalent of one year of projected operating costs by transferring all surplus of revenues over expenses.

c) Trade and competition

By assuring and stimulating the collection of a wide range of WEEE, EPR programs have created opportunities to increase recycling and resource recovery. Because WEEE recycling is a very competitive business, processors in Canada have cited concerns about their ability to compete and the need for clearer guidance on controls applied internationally to recycled materials exported from Canada. The regulatory regime that controls the transboundary movement of these products, whether a waste or not, can have an impact on the development of local recycling industries in Canada.

In order to ensure a level playing field that encourages competition and innovation, provincial program operators use a competitive tender process when selecting WEEE processors. This can sometimes pose challenges in a federation like Canada where EPR program requirements have been developed and implemented individually by provincial authorities; which calls for increased harmonization efforts by the CCME. Certain provinces allow producers to organize either collectively or individually to meet their requirements, which should encourage competition and the arrival of new PROs on the market. In practice however, once large PROs have been formed, economies of scale can make it difficult for smaller operators to compete.

4. Key issues and possible reforms

a) Towards harmonization

Despite many shared standards and policies, the independent operation and administration of provincial WEEE programs in the beginning was not ideal. In addition to creating significant policy and operational inefficiencies, it led to reduced service quality for program members and higher costs[6]. Other features such as the visibility or non-visibility of fees in advertised prices may hinder consumer acceptability of different EPR programs. In response to these problems, a new governance structure to streamline electronics stewardship programs across the country was developed by Electronics Product Stewardship Canada (EPSC) in partnership with the Retail Council of Canada (RCC). EPRA is now responsible for managing electronics recycling programs in every province except for Alberta and New Brunswick. This transition has generated multiple benefits for both stewards and consumers, including: streamlined administrative processes and lower administrative costs; lower operational costs; consolidated communication to members to minimize duplicate messaging; lower unit costs for shared services, due to distribution over more programs; and improved service quality.[7] But despite the differing legislative approaches, the Canadian EPR experience remains effective, with increasing amount of WEEE collected year after year, new collection depots established across the country and higher consumer awareness across provinces.

b) Difficult access to remote and rural areas

Another key challenge in Canada is access of rural and remote communities to EPR programs. Even though access has reached at least 92% of the population in several provinces and all consumers pay an environmental handling fee in provinces with an EPR program for WEEE, rural residents may have fewer opportunities to participate in electronic take-back programs. Several factors make operating a WEEE collection and recycling program in rural areas considerably more difficult than in urban locations such as the lack of infrastructure, the high costs associated with collection, and the long distances and resulting high transportation costs, Canada's three northern territories, the Northwest Territories, Yukon, and Nunavut, face the additional challenge of not being able to rely on large and profitable urban programs to support the development of infrastructure in remote areas. Continued discussions and efforts are undertaken in order to establish the necessary partnerships for assisting in creating further opportunities for EPR in northern Canada.

Notes

1. Full source available at: Séguin, J. (2014), "Promoting Sustainable Materials Management Through Extended Producer Responsibility: Canadian Waste Electrical and Electronic Equipment (WEEE)", Case study prepared for the OECD, *www.oecd.org/env/waste/gfenv-extendedproducerresponsibility-june2014.htm*.
2. These values are not found in the Canadian WEEE case study; however, they are published in *The WEEE Report: Waste Electrical and Electronic Equipment Reuse and Recycling in Canada – 2013* by CM Consulting.
3. With the exception of Alberta (represented by the Alberta Recycling Management Authority), the Canada-wide EPRA is the PRO responsible for representing obligated stewards selling electronic products covered by EPR regulations in a given province.

4. It is important to note that each province uses a different approach for determining program performance and may cover different scope of products, which limits performance comparisons.
5. These values are not found in the Canadian WEEE case study; however, they are published in *The WEEE Report: Waste Electrical and Electronic Equipment Reuse and Recycling in Canada – 2013* by CM Consulting.
6. Electronics Product Recycling Association, "Annual Report 2012," *http://eprassociation.ca/ar/en/2012/EPRA2012%20-%20Annual%20Report.pdf*, accessed 20 October 2013.
7. Ibid.

Extended Producer Responsibility
Updated Guidance for Efficient Waste Management

ANNEX D

China's e-waste treatment fund

by
Liu Chunlong, Ministry of Environmental Protection (MEP)[1]

SUMMARY BOX	
Cost allocation	Fund is distributed to the certified recyclers.
Cost coverage	Recycling and disposal of e-waste, information management systems for the recycling and disposal of e-waste and the production and sale of electrical and electronic equipment, relevant information collection and release, management on fund collection. Currently, the charge rate covers between 1/8 and 1/5 of the subsidy allocated to recyclers.
Role of government	• The Ministry of Environmental Protection is the general coordinator of the Chinese Waste of Electric and Electronic Equipment (WEEE) regulation, responsible for developing policies for WEEE collection and treatment, together with the National Development and Reform Commission and the Ministry of Industry and Information Technology. • The recyclers' administrators are responsible for developing and implementing criteria for certification of e-waste recyclers and for monitoring their environmental compliance by verifying the data provided by recyclers aided by the local environmental protection agencies. • The fund is controlled by the central government, in which the Ministry of Finance is the general administrator responsible for co-ordinating collection, utilization and administration of the fund. • The State Administration of Taxation and the General Administration of Customs collect the fees from producers and importers. • The Ministry of Commerce is responsible for administering the management of WEEE collection.
Environmental performance	• 7 678.989 units of e-waste have collected until mid-2013 (Television sets accounted for 93.9% of the total collected e-waste). • 70.45 million units of e-waste have been recycled in 2014 (Television sets accounted for 81.8% of the total collected e-waste).
DfE incentive	MoF develops (with other relevant authorities) regulation that incentivizes producers to improve their products' design, such as using environment-friendly materials. Further incentives are needed here.
Cost efficiency	Charges from the producers and importers are collected by the State Administration of Taxation and General Administration of Customs, respectively. Although the rate is much lower than the subsidy, the charge can cover the costs for the disposal and recycling of the e-waste due to the huge production of electrical and electronic equipment compared with those recycled e-waste.

1. Description of EPR set-up

a) Legal context

The Administrative Measures on Prevention and Control of E-waste Pollution was initiated in 2007 to promote the development of the e-waste recycling industry (e.g. formal e-waste recycling plants were established). From 2009 to 2011 the National Old-for-New Home Appliance Replacement Scheme (HARS) was implemented and gave the opportunity for consumers to buy new electronic appliances with a 10% discount if they sold their e-waste to certified recycling facilities. To replace this programme, an e-waste disposal fund was set up in July 2012 under the Ordinance for Administration of Collection and Disposal of Waste Electronic and Electrical Products. The scheme was jointly implemented by the Ministry of Finance (MoF), the Ministry of Environmental Protection (MEP), the National Development and Reform Commission (NDRC), the Ministry of Industry and Information Technology (MIIT), the General Administration of Customs (GAC) and the State Administration of Taxation (SAT). These bodies published the Measures for the Collection and Administration of the Funds for the Recovery and Disposal of Waste Electronic and Electrical Products. Five common household electronic and electrical appliances are covered by these Measures namely televisions, refrigerators, washing machines, air conditioners and personal computers.

b) Governance and Enforcement

MEP develops certifications for recyclers which include four aspects (differentiated between the eastern, central region (relatively more developed) and the western region): 1) sufficient capacity and infrastructure, including qualified treatment and recycling, workshops and storage; 2) central monitoring systems and facilities to deal with emergencies, including 24-hour surveillance; 3) compliance with environmental management regulations, including waste water discharges, waste gas and noise emissions and, if necessary, solid waste must be sent to certified landfill sites, and 4) sufficient technical support specialised in safety and health, quality control and environmental protection (i.e. each recycler must have at least three technicians). To obtain a certification, a recycler must submit an application to the local environmental agency who will publish the report publically ten working days before receiving the final approval. MoF, MEP, NDRC and MIIT make the final decision and publish the list of recyclers who are eligible to receive the fund subsidy. The national government also encourages producers to set up their own recycling facilities by providing a faster track procedure for the establishment and certification of their recycling operations. The certified recyclers must report the types and quantity of e-waste recycled to the local environmental protection agency on a quarterly basis. They are also required to submit reports including records on the products going in and out of warehouses,- on the recycled e-waste, disassembly operations, disassembled products obtained from e-waste, and the corresponding sale vouchers (invoices) and treatment certification of these final products. The reports are collected and verified by the local environmental protection agency and sent to MEP for final verification before the subsidy is endorsed. Producers, importers and recyclers who receive subsidies can face legal action if they fail to fulfil their reporting obligations.

To ensure high quality collection, proper utilization of the subsidies and avoid cheating, the following measures are employed: 1) MEP has established an on-line administrative information system to monitor production and the sale of electronic and electrical appliances, recycling and disposal of e-waste, where recyclers connect their monitoring systems with the government's main data base (some recyclers have established open on-line websites where they publish their data regularly, e.g. Beijing Hua Xin Green Spring Environmental Co. Ltd; 2) the tax and customs authorities are responsible for monitoring and inspecting the fund to ensure that the fees are effectively collected, 3) the local environmental protection agencies check the data provided by recyclers and comparing it with the data of the on-line administrative information system, and 4) the National Audit Office also supervises collection and the fund. Public supervision is also welcome. All statistics are published online to ensure transparency.

c) Allocation of responsibilities (distribution of roles, financial flows)

MEP is the general coordinator who is responsible for developing policies for WEEE collection and treatment. MoF is the general administrator responsible for coordinating the collection, utilization and administration of the fund. SAT and GAC collect the fees from producers and importers from their agencies across PRC. MEP is also the recyclers' administrator responsible for developing and implementing criteria for certification of e-waste recyclers and monitoring their environmental compliance by verifying the data provided by the recyclers (aided by the local environmental protection agencies). The National Audit Office also provides supervision to ensure the scheme runs smoothly. Producers and importers of electronic and electrical products must pay for each unit produced or imported. Producers must declare and pay into the fund on a quarterly basis (via the tax authority) and importers pay their obligation to the customs authority. Certified recyclers can be eligible for a subsidy if they provide their output figures.

2. Environmental effectiveness

a) Collection and recycling rates

There are currently 106 WEEE recycling facilities. Among the 31 main cities, provinces and autonomous regions excluding Chinese Taipei, Hong Kong, China and Macau, 29 of them have certified recycling companies. Only the Hainan Province has not received any certification, and the Tibet Autonomous Region has no plan to establish recycling facilities. Among the 106 certified companies, 32 are from areas in the eastern part of China, accounting for 50%; 20 are from areas in the central part of China accounting for 31.25%, and only 12 are from areas in the western part of China accounting for 18.75%. In total, 81.25% of the certified recycling companies are from the relatively more developed areas (see Figure D.1).

Figure D.1. **Percentage of certified recycling enterprises in different areas of China**

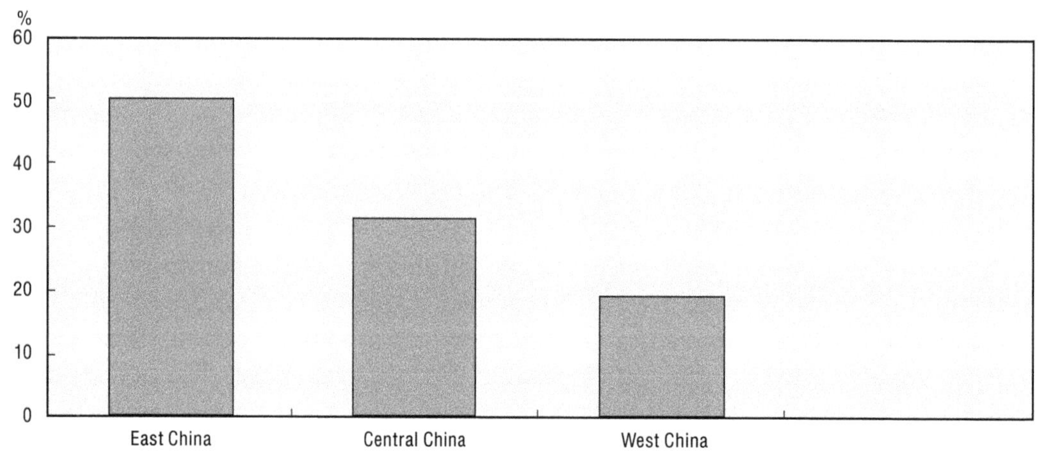

Television sets represent the majority of e-waste (see Figure D.2). A number of reasons explain this factor: 1) PRC households are switching to LCD sets in large quantities; 2) television sets are more easily collected by the certified recyclers than the other four products, and 3) collecting and recycling television sets is more profitable than other e-waste products because the subsidy rate for television sets is higher than for any other subsidized e-waste product.

After on-site inspection, 7 678 989 units were verified until mid-2013 (accounting for 85.1% of the total declared), and the television sets accounted for 93.9% of the total collected e-waste. In 2014, the figures were 81.80% for TV sets, 2.24% for refrigerators, 4.68% for washing machines, 0.16% for air conditioners and 11.1% for computers.

b) DfE

Producers are encouraged to design products in a way to facilitate the comprehensive utilization of the natural resources and innocent treatment. Moreover, they are proposed to use environment-friendly, easily recycled and reused materials to produce their electrical and electronic equipment.

Figure D.2. **Percentage of the five types of e-waste collected and recycled by Beijing Hua Xin from 1 April to 30 June 2013**

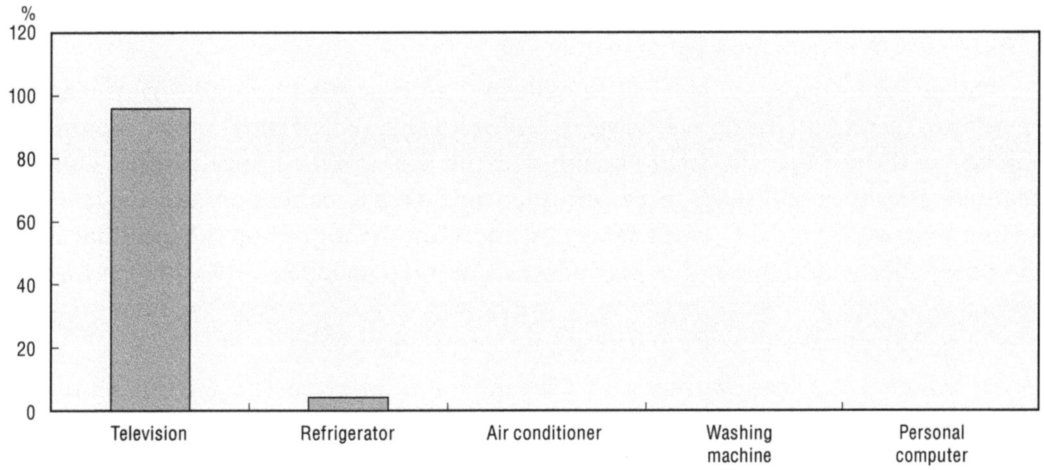

3. Economic efficiency

a) Cost efficiency

The State Administration of Taxation collects the charges from producers of electrical and electronic equipment, and the General Administration of Customs collects charges from the importers or their agents. The fee and subsidy rates were set after several consultations with all the relevant stakeholders, and may be adjusted depending on changes in the costs of collection and disposal of e-waste. The rate is lower than the subsidy in order to avoid any surplus (see Table D.1). The subsidy is based on the basic cost for disposing and recycling the e-waste, and does not include the collection.

Table D.1. **Charged rates and subsidy values**

Products or e-waste\rate	Rate of Charge (CNY/unit)	Rate of subsidy (CNY/unit)
Television	13	85
Refrigerator	12	80
Washing machine	7	35
Air conditioner	7	35
Personal computer	10	85

The fund raised to CNY 854 million in the second half of 2012 since the scheme began, and was expected to increase to CNY 2.8 billion in 2013, when this case study was elaborated. MEP conducts on-site visits in order to facilitate the disbursement of subsidies.

b) Leakages and free riders

Issues to do with free riders and orphan products should be limited because the e-waste subsidies apply to all products entering the waste stream, independent of their type or production date.

4. Key issues and possible reforms

The e-waste fund will require a significant increase in recycling capacity in the future. Ultimately, each province should have at least one certified recycling facility to promote quality and efficiency through competition. By way of comparison, the theoretical e-waste recycling capacity is about 2 kg/capita/a, compared to that of 4 to 16 kg/capita/a in the EU, which is 2-8 times more (for comparison, GDP per capita in China is USD 6 000 compared to USD 40 000 in France, i.e. a factor of only 6.5).

The current quantities of e-waste collected and recycled are still insufficient, as the informal sector profits from e-waste recycling. In fact, the informal sector captures the majority of the e-waste products included in the scheme, except television sets. The incentive structure will need to be reformed to increase collection and recycling for the four remaining products, while taking into account the impact on the informal sector. One possibility would be to give producers more responsibility and to encourage the creation of a PRO. The scope needs to be broadened to include more products – e.g. the sales of mobile phones and cars (batteries) are increasing rapidly.

In March 2013, the Ministry of Commerce published the Measures on the Administration of Circulation of Old Electronic and Electrical Products to regulate the second-hand market of these goods. The China COOP is now involved in collecting and recycling, and setting itself high targets. Their operations are spread across PRC with 150 000 depots and 1 million employees. DfE needs to be further incentivized, on which the MIIT is currently working.

Note

1. Full source available at: Liu, C. (2014), "How does the Chinese E-waste Disposal Fund scheme work?", Case study prepared for the OECD, *www.oecd.org/env/waste/gfenv-extendedproducerresponsibility-june2014.htm*.

Extended Producer Responsibility
Updated Guidance for Efficient Waste Management

ANNEX E

EPR schemes in Colombia

by
Colombian Ministry of Environment and Sustainable Development[1]

SUMMARY BOX	
Cost allocation	No information the National Environmental Licensing Authority (ANLA) does not hold data on costs and the regulations do not determine the financing mechanisms). However, producers should ensure the financing of the schemes through their own contributions
Cost coverage	No information (ANLA does not hold data on costs and the regulations do not determine the financing mechanisms)
Role of government	The Ministry of Environment approves producers' schemes. ANLA assesses and oversees the functioning of the schemes.
Environmental performance	In the case of computers, for example, individual schemes have a 100% compliance rate for the collection target (% of target weight). Collective schemes, on the other hand, have only reached 32,41% of the target. Schemes for batteries and light bulbs have higher compliance rates.
DfE incentive	• restrictions on the use of hazardous substances for fluorescent bulbs • restrictions on the quantity of lead, cadmium and mercury for primary cells. • These restrictions are within the technical regulations for the products issued by the Ministry of mines and energy (light bulbs) and by the Ministry of trade, industry and tourism (batteries)
Cost efficiency	No information

1. Description of EPR set-up

a) Legal context

In Colombia, there are specific regulations on waste collection schemes under the EPR principle, including four resolutions focused on WEEE streams (i.e. batteries, used tyres, bulbs and computers), implemented since 2011. Law 1672 of 2013, issued by the Congress of the Republic, establishes policy guidelines on environmentally responsible management of WEEE in Colombia. The upcoming implementation of the Act may require some adaptation of the obligations and processes stipulated in the resolutions.

b) Allocation of responsibilities (distribution of roles, financial flows)

Producers are financially and organizationally responsible for the proper management of covered products during collection, transport, storage, processing, reuse, recovery and/or final disposal. They assume the costs of separate collection and environmentally responsible management of waste. They are also in charge of designing and paying for any public information campaigns needed to raise awareness of the scheme. Producers have to design and submit waste management and separate collection schemes for approval by the National Environmental Licensing Authority (ANLA), Producers can fulfil their obligations individually or as a group under a collective scheme, either through an agreement between producers (most frequent) or through the establishment of a legal entity (i.e. a PRO). Suppliers and retailers support producers in the running of the schemes by allowing consumers to return used products at no cost and providing a space for containers or receptacle provided by producers. Waste management companies that participate in the schemes must require an environmental licence issued by the regional environmental authorities. Public bidding processes for post-consumer waste management environmental services have been undertaken under the collective schemes. Consumers have to separate covered products from municipal solid waste and are required to return or deposit covered products to collection points.[2] Municipalities inform consumers on their duties, and support the community education and awareness raising programmes run by producers. The overall assessment and oversight of the Schemes is under the responsibility of the National Environmental Licensing Authority (ANLA), a special public entity with administrative and financial autonomy.

c) Governance system and sanctions

ANLA performs initial assessment and annual follow-ups regarding producers' obligations set in separate collection schemes. Sanctions in case of non-compliance are set forth in Act 1333 of 2009, which identifies two types of environmental infractions: i) breaches of prevailing environmental legislation; and ii) damages to the environment. A breach is any case of non-compliance with the obligations imposed on producers, retailers and distributors of covered products, and any acts by consumers (both households and professionals) that do not result in post-consumer waste being deposited at collection points. For minor infractions, the sanctions stipulated in Act 1259 of 2008 on community work and environmental penalties provided for in Act 1333 of 2009 may be imposed. For more serious infringements, preventive measures,[3] daily fines up to USD 1.3 million statutory monthly salary payments or temporary closure of an establishment may be imposed.

2. Environmental effectiveness

a) Collection and recycling rates

The resolutions include two kinds of collection targets: year-on-year increases, and a medium-term (10-12 years) target that must be reached and maintained. For example, in the case of batteries, the year-on-year increase is of 4% until 2016 and 5% from 2017, and the final value to be maintained is of 45% from 2021. Collection rates are generally calculated as a percentage of average product sales in previous years. While collection rates are improving effectively (see Table E.1), the possibilities for recovering the components of post-consumer waste remain limited. Indeed, a large number of waste management companies do not currently have sufficient capability to recover the components from post-consumer waste.

Table E.1. **Post-consumer waste collection, 2012**

Type of scheme	Waste	2012 collection target (tonnes)	Waste collected, 2012 (tonnes)	Compliance rate (% of target weight)	Compliance rate (% of schemes hitting targets)
Individual	Batteries	23.4	22.50	96.15	*
	Bulbs	Not disclosed	Not disclosed	*	*
	Computers	0.693	3.0	100	*
Collective	Batteries	135.08	101.4	75.06	66.6
	Bulbs	243.73	253.014	100	100
	Computers	463.4	150.2	32.41	*
Total general		866.40	530.12	61.18	*

No targets are set for recycling rates. In the specific case of the resolution on the treatment of used computers and peripherals however, a preparation for reuse objective is included that is set permanently at 30% of the total collected year on year. In addition, the requirement in the resolutions on used bulbs and batteries can be interpreted as a 100% recovery from 2016, since it requires that as from January 2016 it would only be possible to manage waste through recycling activities designed to recycle the used products.

b) Design-for-Environment (DfE)

Colombia does not currently use any measurements or estimates of the Separate Collection Schemes' impact on prevention or minimisation of waste, or on design including environmental improvements. However, certain regulations in Colombia provide that the government must encourage producers to increase product quality and environmental friendliness but the instruments needed to fulfil these obligations still need to be put in place. Certain restrictions are also included into technical regulations including restrictions on the use of hazardous substances for fluorescent bulbs (issued by the Ministry of mines and energy) and restrictions on the quantity of lead, cadmium and mercury for primary cells (issued by the Ministry of trade, industry and tourism).

3. Economic efficiency (including competition aspects)

a) Cost efficiency

The Ministry does not regulate financial aspects relating to the collection and use of funds raised by individual or collective schemes. However, producers must give assurances that they finance the schemes through their own contributions. The administrative costs incurred by ANLA for implementing the regulations include the assessment and inspection fees paid to external contractors (producers and PRO) and the costs for performing overhead and legal functions.

b) Leakages and free riders

The first version of the resolutions included a provision requiring importers of covered products to request prior approval from the Ministry of Environment in order to control the issue of free-riders and compliance with EPR requirements. However, this obligation was removed at the request of the Ministry of Trade, Industry and Tourism based on the argument that the Ministry of Environment was not authorized to obtain prior approval for imports. Currently, ANLA uses information from the national trade database (BACEX) in order to identify importers that have to comply with obligations laid down in the EPR resolutions. In order to reinforce monitoring and enforcement of compliance with EPR measures, the Ministry of Trade is required under Act 1672 of 2013 to establish a national register of permanent and occasional producers and retailers. However, ANLA does not have an overview of products put on the market other than imports, and cannot currently verify product declarations. This is expected to be resolved through the establishment of a National Register of Equipment Producers and Retailers foreseen in Act 1672.

c) Trade and competition

Tendering procedures and discrimination concerns are not regulated by the Ministry of Environment. However, transparency requirements and approval of the schemes by ANLA are expected to reduce those concerns.

Over the last five years, the increasing demand for waste management services in Colombia has led a large number of hazardous waste management companies to enter the market and obtain environmental licences. There are currently more than a hundred authorised management companies. An issue with these waste management companies is that certain (especially those working with WEEE that can be more easily reused and recovered) use their own transportation and collection system without partnering with producers under a regulated EPR scheme. Even though these companies may comply with environmental requirements, this undermines the sustainability of EPR schemes and their funding. It also encourages the business of "waste auction", whereby waste management companies' auction off the waste collected by them to separate collection schemes that need the waste to meet their collection targets.

4. Key issues and possible reforms

a) Compliance

EPR needs to be enforced on producers that have not presented schemes, including where possible through monitoring and enforcement actions prior to import or production and measures aimed at raising awareness and prioritising compliant producers in public or private procurement processes. No significant progress has been achieved in imposing sanctions for breaching obligations, especially for producers who have not submitted a separate collection scheme for approval by ANLA. This situation undermines fair competition for companies in the regulated sector and encourages non-compliance. Act 1672 is intended to provide the means through which informal channels can begin actively participating in the schemes' collection mechanisms in an organized manner.

b) Informal sector

As explained above, an issue for the sustainability of EPR schemes in Colombia is that certain waste management companies that comply with environmental requirements do not participate in the regulated EPR schemes. Therefore, work is under way to develop the technical means to designate waste management companies that participate in schemes as preferred suppliers or give them higher scores in tender processes (or even to establish EPR participation as required criteria into green public procurement).

Notes

1. The development of this case study received the support of Christoph Vanderstricht and his colleagues at Ernst&Young. Full source available at: *www.oecd.org/env/waste/gfenv-extendedproducer responsibility-june2014.htm*.
2. Remark: no provisions for advanced disposal or recycling fees or similar financial instruments are currently included in the resolutions.
3. Act 1333 of 2008 indicates that "Preventive measures are immediately executed, are pre-emptive and temporary, take immediate effect, cannot be appealed against, and are applied without prejudice to the fines imposed.

ANNEX F

20 years of EPR in France: Achievements, lessons learned and challenges ahead

This note does not intend to give a full description of French EPR schemes, but rather aims to highlight a number of key design elements for EPR schemes, specific features worth considering, as well as attention points, drawing on the French experience to date. For more detailed descriptions of some of the French EPR schemes, please refer to the case studies set up by the European Commission.

1. Overview of French EPR schemes

- 14 schemes, covering a significant part of household waste

For the first time in 1992, French authorities decided to apply the EPR model to tackle household packaging waste. This was the first large-scale EPR scheme to be designed in France. Twenty years later, it remains the largest French scheme, channeling over EUR 600 million yearly towards packaging waste management.

Since then, many more schemes (14 in total) have become operational, mainly in the 2000s. Some of them stem from EU directives – sometimes EPR is directly required by the directives (WEEE, batteries, ELV), and sometimes France decided to set up EPR schemes where EU directives did not explicitly ask for it (e.g. packaging). There is also a variety of purely domestic schemes – including tyres, graphic papers, and textiles. The latest generation of schemes (furniture, infectious healthcare, dispersed hazardous waste) is just starting operations. While most schemes focus on household waste, some also deal with professional waste (e.g. WEEE, furniture).

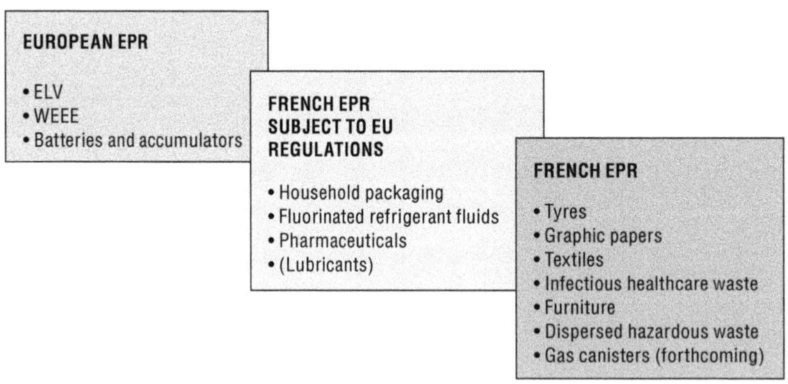

Box F.1. **A recent, innovative EPR scheme for furniture waste**

The furniture scheme is one of the latest-born (2012) of French EPR schemes. Covering both household and professional furniture waste, it is intended to generate over EUR 300 million a year to help develop furniture reuse and recycling, and help create jobs and structure industrial activities around furniture waste management – leading to the creation of many new companies, including in relation to mattress and wood recycling, which were not profitable enough to generate sustainable activities before the scheme. The scheme also strongly promotes furniture reuse, closely involving social economy structures in its organisation model.

- Producers must organise or finance waste management operations

  All schemes are based on making marketers responsible for managing their end-of-life products. Marketers can decide to manage waste in an individual way (which can be an appropriate solution for very vertical distribution systems with robust reverse logistics), but most producers decide to opt for collective organisations – these **collective PROs, called "éco-organisms"** (typically one single PRO for each scheme), are always not-for-profit entities, and can be organised in two general models:

  - In **"organisational" schemes** (e.g. WEEE), PROs are directly in charge of organising waste management operations. To do so, they collect fees from producers and use these funds to contract with waste management operators.

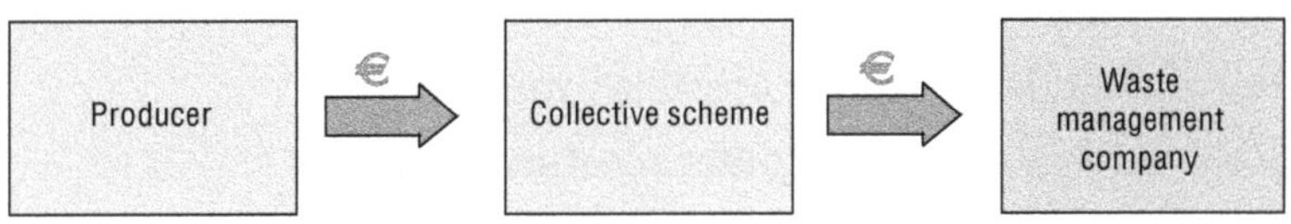

  - In **"financial" schemes** (e.g. packaging, graphic papers), PROs are not directly in charge of waste management operations. They use fees collected from producers to support municipalities, who remain in charge of waste management.

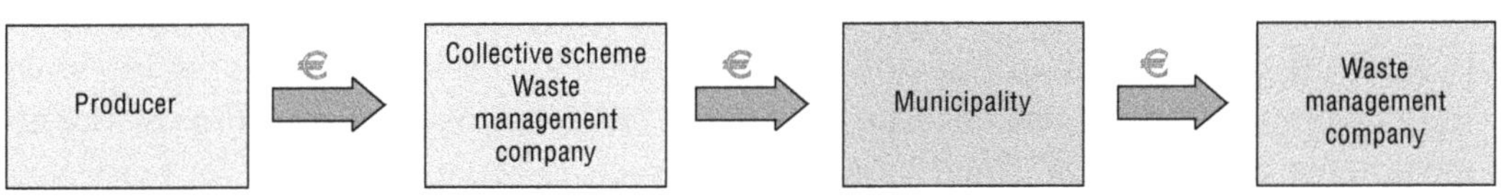

- An inclusive governance model

  All schemes feature an **inclusive governance model associating all stakeholders** – producers, municipalities, waste management operators, environmental NGOs, consumer organisations, public authorities. All stakeholders have the opportunity to participate in the decision-making on the design and ambition of the scheme.

  PROs need to be "approved" by public authorities every six years. All schemes are based on a 6-year cycle, with an in-depth stakeholder consultation process during the last year of the approval period, which leads to the drafting of updated 'terms of reference' for the upcoming period. These terms of reference include performance objectives as well as financial and operational modalities. Once approved, PROs enjoy a good level of flexibility regarding the organisation of their day-to-day operations, provided that they do not deviate from the objectives specified in the 'terms of reference'.

- Significant financial flows mobilised

  All schemes are expected to generate a total of approximately EUR 1.4 billion/yr by 2015, of which around EUR 700 million will be redistributed to municipalities. This accounts for a significant part of the total costs of national household waste management (EUR 9.4 billion/yr).

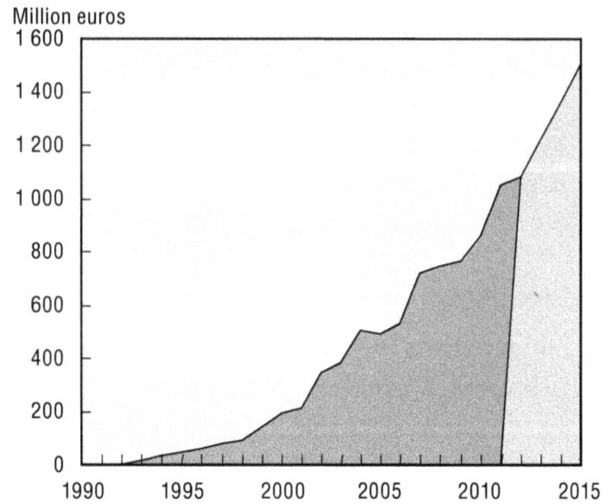

2. Focus on specific design features and attention points

In general, French stakeholders acclaim the EPR model as positive and consider that it contributed to structure large-scale industrial waste management activities, and delivered good performance in terms of collection and recycling in many areas. Over the years, management of the schemes has grown more mature and delivered a number of lessons giving access to a variety of significant design features, some of which are detailed below, alongside with the identified areas for possible progress.

- Producers in the driving seat... but under scrutiny

  One significant element that is at the very root of the EPR system is the fact that producers must be responsible for waste management activities – but in a context where many other stakeholders also play a role at all stages of the waste management process (consumers, municipalities, waste treatment operators...). This raises the question as to **how to strike the right balance between making producers responsible and associating other stakeholders.**

  Over the years, France's position has been to consistently (constantly?) place producers in the driving seat, meaning that not only are producers responsible for paying waste management activities, but also **for settingup PROs themselves and organising the governance of these PROs**. Thus, PROs cannot be seen as 'service providers' for producers – it would be more correct to say that they 'embody' the producers community. In this context, producers are responsible for all the operational decisions taken regarding the way the scheme is run.

  In parallel, **public authorities should pilot the system**, by giving clear orientations at critical moments in the life of the scheme, and other stakeholders should be associated to the governance of the whole scheme, in order to remain informed and to actively provide their guidance and input on the way the scheme is managed. This collective governance is seen as a critical element of success, and all stakeholders consider that the dialogue created and sustained through the scheme per se has a tremendous influence on its overall performance.

Nevertheless, **while all stakeholders must participate in defining general orientations, especially in intense periods of (re-)approval of the schemes, they must not end up micro-managing PROs**. The same applies to public authorities who need to keep their distance in a "command and control" manner. Their role is to ensure that ambitious yet realistic objectives are assigned to PROs, that appropriate indicators are in place (typically separate collection rate, recycling rate, sometimes reuse rate), and to follow up and take sanctions when necessary.

- Performance requires stability and some degree of harmonisation

With four recently-created schemes (2009) still in the process of starting operations, most stakeholders consider that the priority is not to create additional schemes, but rather to **focus on existing schemes to improve their performance and/or address critical design issues.**

With 15 schemes running in parallel, **harmonisation** is also an important task – reproducing good practice from one scheme to the next, finding common solutions is an essential element of credibility for all schemes, as well as for the public authorities managing them.

For instance, recently addressed cross-cutting questions include:

- Addressing potential **free-riders** by proposing and implementing appropriate control operations – a harmonised approach across schemes allows for wider communication and awareness raising.
- Developing a **credible system to control PROs** themselves – for a long time, "un-approving" was the only possibility for public authorities facing a "rogue" PRO, but this proved to be an empty threat as it meant risking to put down the whole waste management system. A more progressive and therefore more credible toolbox of warnings and sanctions across all schemes was put into place.
- **Organising relationships between stakeholders** – for instance, guidelines and safeguards have been developed to ensure that sound competition rules are respected between PROs (often "single buyer" by construction in the waste management market) and their waste management operator clients.
- **Tackling new challenges including internet sales** – where some producers can avoid national regulations through fully dematerialised business models. Dialogue is underway with representatives of online-selling producers, although it remains clear that such models will continue to create difficulties in the future and could benefit from international co-operation.

Box F.2. **The "Triman" label for recyclable products**

In order to improve sorting performance, French law requires that a common label be set up, applicable to all recyclable products covered by an EPR scheme, indicating to consumers that said product needs to be sorted. To that end, the "Triman" logo (see left) has been introduced in 2015.

The benefit of a scheme that applies across product groups is that it makes it possible to communicate more widely to a large audience based on one label, rather than on a variety of labels for each type of product.

- EPR schemes as a driver for "circular economy"

  In times of economic downturn, mobilising political momentum around environmental issues can be a challenge. However, recent high-level stakeholder consultations ("Environmental Conference", September 2013) highlighted the **broad enthusiasm across French stakeholders to develop new production and consumption models focusing on the concept of "circular economy"** – moving away from a "linear" model of extraction, consumption and dumping.

  EPR schemes have an important part to play in developing such "circular" models. They allow for an **inclusive approach covering all stages of a product's lifecycle,** from its (eco-) design to its multiple phases of reuse and recycling.

  They also provide a **very attractive political narrative** by actively linking environmental ambitions and benefits to enhanced economic and social conditions – feeding a virtuous circle of job creation with environmental benefits.

  Indeed, among the economic advantages of EPR schemes experienced, the pre-financing of recycling activities that could otherwise not be self-sufficient stands out. It was noticed that EPR schemes **channel steady financial flows to make recycling chains more sustainable** and less dependent on the great variability of commodity prices. With contracts between PROs and operators established for periods of three to six years, operators are fully in the position to invest in the required infrastructures.

Box F.3. **Creating jobs through WEEE recycling**

In France, over five years the WEEE recycling activities have created at least 30 new plants and more than 3 000 jobs (of which over 1 500 are qualified as "social economy" jobs) directly related to the sorting, depollution and recovery of metals and plastics from WEEE. The biggest of these plants employs 180 people, ensuring treatment of about 50 000 tonnes of WEEE per year, including fine-sorting of plastics.

- Using social economy to catalyse waste management operations

  Several French EPR schemes have historically involved stakeholders from social economy networks, including charities or social economy companies. This was very much the case in the textiles scheme, but also prominent in WEEE or furniture schemes, especially for repairing and reuse activities.

  While social economy cannot be involved at all stages of all schemes, **French regulations will soon require negotiations within each EPR scheme to address to what extent social economy can be involved in their operations** and to introduce facilitating measures to that end.

- Exploring possible ways to promote a full life cycle approach (prevention, eco-design)

  The principle of EPR schemes is to ensure that producers assume responsibility for their end-of-life products. The intention is to drive producers towards internalising end-of-life costs in their business models, as well as moving towards a **full life-cycle perspective** on their products. In particular, robust eco-design at the very beginning of the cycle can be identified as an efficient way of avoiding significant costs when the product is eventually discarded.

To implement this "full life-cycle" approach in practice, a number of concrete tools are proposed:

- **all PROs need to promote prevention actions and eco-design among producers** – some schemes even define quantified prevention targets, e.g. an objective to reduce packaging volumes by 100 000 tonnes over five years for the packaging scheme.
- **all EPR schemes must define "differentiated fees" rewarding pioneers** (producers abiding by eco-design criteria will pay a lesser fee) **or punishing laggards** (greater fees for e.g. products that hinder sorting processes). Stakeholders generally find these "differentiated fees" interesting, especially as they help create dialogue between producers and waste management operators. Some also point out that such criteria would be more efficient if developed at a European or larger, scale;
- **each scheme features specific quantified objectives**, tailored to its specific waste management model (e.g. reuse targets for textiles, standards for downstream recycled materials for packaging).

Box F.4. **Promoting paper recycling with differentiated fees**

The "graphic papers" scheme was re-approved at the end of 2012. Extensive consultations took place on how to promote eco-design, including via the use of differentiated fees. It was finally decided that producers whose papers incorporate more than 50% of recycled materials would benefit from a 10% bonus (i.e. pay 10% less on their fee). This helps encourage producers to undertake concrete eco-design actions.

3. To conclude – EPR as a versatile tool, relevant for future challenges

While EPR schemes cannot be considered a silver bullet to address waste management – especially as they focus on specific products and cannot provide a solution for e.g. residual waste, and remain very oriented towards household waste – seen from a government perspective, they can be a **powerful tool to develop robust waste policies at relatively affordable costs.**

One element that can be appreciated is the versatile nature of this tool. With regular revisions of collective expectations towards each scheme (in our case every six years), it is **possible to channel activities in the right direction and to adjust trajectories in a dynamic manner whenever needed.** This needs, of course, to be well balanced leaving sufficient visibility to economic stakeholders, especially producers, but without putting too much pressure on them.

As new challenges appear, EPR schemes could become part of the solution:

- EPR schemes could contribute to incentivize producers to find operational answers to the need for new prospects regarding **strategic metals/rare earths recycling.**
- Pressure is growing to relocate industries close to where waste is produced. While this cannot be achieved all the time, especially in a globalised context, EPR schemes can be a promising tool to **foster a "principle of proximity" approach** and maximise local waste treatment.
- More generally, EPR schemes can be a robust tool to consistently and methodically **organize the exploitation of "urban mines"** (resource flows in big cities, resources accumulated in infrastructures such as housing and transport) in the future. French EPR

schemes have provided through information gathered from producer registries, extremely valuable knowledge and databases that help to understand and anticipate potential resource flows, predictable waste quantities, nature, and matter contents – the strategic interest of which is self-evident in a long-term perspective. In a globalised market economy, the collective effort organised in a given territory through an EPR approach can and should be a way for this territory to strengthen its capacity to become more resilient and self-sufficient, in a sustainable manner.

Extended Producer Responsibility
Updated Guidance for Efficient Waste Management

ANNEX G

EPR for used rechargeable batteries in Japan

by
Dr Tomohiro Tasaki, National Institute for Environmental Studies, Japan[1]

	SUMMARY BOX
Cost allocation	Producers pay for collection and recycling through membership fees to the producer responsibility organisation (PRO). Fees are proportional to the quantity of batteries produced and sold.
Cost coverage	Producers finance all the operations of the PRO, from collection to recycling of compact rechargeable batteries.
Operational responsibility (downstream)	Collection: Retailers (voluntary; collection box) and the PRO (pick up) Recycling: the PRO (commissioned to private recyclers).
Role of government (MOE and METI)	• ministries in charge of supervision of the scheme • encourage voluntary efforts by producers • issue recommendations and necessary measures for producers • awareness-raising and education activities • could impose fines of up to JPY 500 000 to producers (unlikely).
Environmental performance	Recycling rate targets of between 30% for Lithium batteries and 60% for Sealed nickel-cadmium batteries.
DfE incentive	Producers are encouraged to facilitate disassembly of batteries from waste products but no remarkable (prominent) improvements have been noticed.
Cost efficiency	No information.

1. Description of EPR set-up

a) Legal context

The Act for the Promotion of Effective Utilization of Resources, which was promulgated in June 2000 and came into force in April 2001, is the legal basis for promoting the responsibility of business operators for the recycling of their products, including for rechargeable batteries. It aims at comprehensively promoting the reduction, reuse, and recycling of used products based on a 3R approach. The Act provides flexibility to business operators for the design of the recycling schemes. In the case of compact rechargeable batteries, the Ministerial Ordinance Stipulating the Criteria to be Used by Sealed Rechargeable Batteries Manufacturers was issued in March 2001. The Ordinance stipulates that manufacturers of rechargeable batteries, as well as manufacturers of products using rechargeable batteries should conduct the self-collection of waste sealed batteries by designating self-collection points, installing collection boxes, or taking other measures needed for self-collection, and to organize the recycling of collected batteries. Non-rechargeable batteries and automobile batteries are not covered by the law.

b) Allocation of responsibilities (distribution of roles, financial flows)

Manufacturers and importers are required to collect and recycle waste compact rechargeable batteries, as well as to disclose information about collection and recycling every year. As for manufacturers of products using rechargeable batteries, they must collect used batteries and hand them over to the battery manufacturers. In both cases, manufacturers may ask to be certified by the competent minister, in order to ensure a smooth conduct of operations. To organise and finance the collection and recycling of batteries, most manufacturers register with the Japan Portable Rechargeable Battery Recycling Centre (JBRC), a producer responsibility organization (PRO) that collects the majority of battery types. Another PRO, the Mobile Recycle Network, deals with lithium-ion batteries used in cell-phones, which undertakes collection collectively but the recycling is handled by each communication service provider. For sealed lead acid batteries, most of them are collected by individual battery manufacturers. Retailers that sell compact rechargeable batteries voluntarily register with the JBRC as a co-operation shop and install collection boxes for pick up by the JBRC. There were 21 102 co-operation shops in 2013. For commercial rechargeable batteries, the co-operation businesses also voluntarily register with the JBRC to discard their used batteries to the JBRC. Batteries are sorted by type, which is not the case for retailers. Rechargeable batteries are collected and transported to recyclers by delivery businesses outsourced by the JBRC, who remunerates them with the annual membership fees paid by member manufacturers. Fees are proportional to the quantity of batteries produced and sold. There is no provision in the Act that defines a specific role for municipalities and there is a lack of financial incentive for them to co-operate. However, a small number of them (municipalities) (222) co-operate with retailers for the collection of batteries, which they must insulate and sort by type. The national government (Ministry of Economy, Trade and Industry (METI), and the Ministry of Environment (MOE) are in charge of supervising the scheme and of conducting educational and awareness-raising activities.

c) Governance system and sanctions

The scheme is based on the idea of encouraging voluntary efforts by business operators, and its governance relies essentially on information exchanges between business operators and the government.

Manufacturers of batteries and of products using batteries must disclose information on self-collection and recycling every fiscal year, which is being aggregated and published by the METI and MOE. When necessary, the competent authorities may provide specified resource-recycling business operators with the necessary guidance and advice with regard to self-collection and recycling. When the ministers deem that the self-collection and recycling achievements by a producer (who sells two million pieces of equipment or more) fall far short of the "standards of judgment" (criteria stipulated in the Act), they may recommend necessary measures. When the producer fails to follow the recommendations, the ministers may disclose this shortcoming. If the producer still fails to follow the recommendations, the ministers may order the producer to take necessary measures, after hearing the opinions of the Industrial Structure Council, and may impose a fine up to a maximum of JPY 500 000.

2. Environmental effectiveness

a) Collection and recycling rates

Recycling rates set under the Act (see Table G.1) are defined as the percentage of the total weight of materials recycled from waste compact rechargeable batteries that were turned into a recyclable state, out of the total weight of the waste compact rechargeable batteries collected. Recycling means that waste compact rechargeable batteries are turned into iron, lead, nickel, cobalt, cadmium and other recyclable resources so as to be used. It does not include energy recovery.

As shown in Figure G.1, the targets set by the Act have been achieved and exceeded, with an increasing recycling rate for lithium-ion batteries and constant rates for the other types of batteries.

Table G.1. **Target Recycling rates for waste compact rechargeable batteries in Japan**

	Mandatory target recycling rate (as from entry into force of the Act)
Sealed nickel-cadmium batteries	60%
Sealed nickel-metal hydride batteries	55%
Sealed lead acid batteries	50%
Lithium batteries	30%

Source: Tasaki, T. (2014), "The recycling scheme for compact rechargeable batteries in Japan – under the act on the promotion of effective utilization of resources", Case study prepared for the OECD, *www.oecd.org/env/waste/gfenv-extendedproducerresponsibility-june2014.htm.*

Figure G.1. **Recycling rates of waste compact rechargeable batteries in Japan**

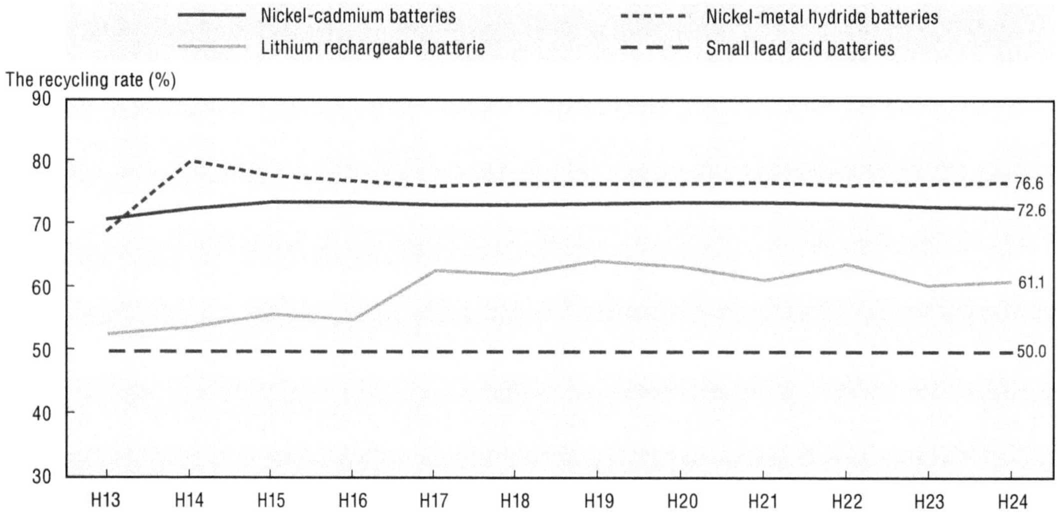

Source: Tasaki, T. (2014), "The recycling scheme for compact rechargeable batteries in Japan – under the act on the promotion of effective utilization of resources", Case study prepared for the OECD, *www.oecd.org/env/waste/gfenv-extendedproducerresponsibility-june2014.htm.*

Collection rates for compact rechargeable batteries are difficult to estimate because they are collected a few years after they have been put on the market. A study carried out in 2001 estimated that 26 % of waste compact batteries were collected in Japan, with low collection rates for nickel-metal hydride batteries and lithium-ion batteries (see Table G.2).

Table G.2. **Estimated Collection rates for waste compact batteries in Japan**

Type of battery		Quantity collected (tonnes/year)	Collection rate (%)
Non-rechargeable batteries (cylindrical)		21 500	30
Non-rechargeable batteries (button shaped)		1	0.1
Rechargeable batteries (compact)	Nickel-cadmium batteries	984	46
	Nickel-metal hydride batteries	205	2
	Lithium-ion batteries	165	6
	Total	1 354	9

Source: Asari, M. et al. (2011), "Current Status of Disposal and Recycling of Small Used Batteries in Japan", *Journal of Material Cycles and Waste Management*, 22 (6), pp. 412-425.

b) Design-for-Environment (DfE)

The scheme encourages product designs that enable easy removal of compact rechargeable batteries from waste products. However, there have been hardly any remarkable improvements in such product designs. According to a survey carried out by the MOE in 2008, about one quarter of consumers remove batteries from waste products, with variable results depending on the product. For example, 42.3% of the respondents reportedly removed batteries from radio-controlled model cars, while only 10% did so for laptop computers, and 13.9% for cell-phones.

3. Economic efficiency (including competition aspects)

a) Cost efficiency

With regard to the costs, the JBRC discloses a balance sheet based on the obligation to report its performance, as stipulated in the Act. However, the information about individual costs is not disclosed and therefore it is not possible to conduct a cost-benefit analysis.

b) Leakages and free riders

As shown in Table G.2, there would be a large amount of uncollected spent batteries, including those not removed from waste products. The free rider issue has been discussed in recycling policies in Japan as well; however, it has not been a topic related to battery collection/recycling so far. All ten battery manufactures are members of the JBRC, whereas only 306 manufacturers of electrical products using rechargeable batteries have joined the JBRC. Thousands of such producers exist in Japan.

c) Trade and competition

No competition-related problems have been pointed out. The Act stipulates that, when a PRO for specified resources-recycled products is to be created, the competent minister may ask for the opinions of the Japan Fair Trade Commission regarding measures for the self-collection and recycling of the products, when necessary.

4. Key issues and possible reforms

One useful improvement to the scheme would be to increase the percentage of batteries that are collected through appropriate paths and adequately sorted by type. According to a survey by the MOE in 2008, 25% of the respondents placed waste batteries in collection boxes at retailers. Currently, there are little incentives for stakeholders other than producers such as municipalities to take part in the collection of waste rechargeable batteries. The quality of collected batteries has, however, increased thanks to steady awareness-raising and guidance activities.

In light of the expected increase in the use of rechargeable batteries for electric vehicles and household photovoltaic power generation systems, sooner or later it will be necessary to examine how the scheme should be developed with a possible option of inclusion of these batteries, from a medium- to long-term perspective.

Note

1. Full source available at: Tasaki, T. (2014), "The recycling scheme for compact rechargeable batteries in Japan – under the act on the promotion of effective utilization of resources", Case study prepared for the OECD, *www.oecd.org/env/waste/gfenv-extendedproducerresponsibility-june2014.htm.*

Extended Producer Responsibility
Updated Guidance for Efficient Waste Management

ANNEX H

Recycling of electronic home appliances in Japan

by
Dr Yasuhiko Hotta, Institute for Global Environmental Strategies,
Atsushi Santo, Institute for Global Environmental Strategies *and*
Dr Tomohiro Tasaki, Institute for Environmental Studies, Japan[1]

SUMMARY BOX	
Cost allocation	Customers pay for the scheme (collection/transportation and recycling fees).
Cost coverage	Negative net income for the private sector (e.g. net incomes of -476 for group A and -375 for group B concerning air conditioners in 2012; and of -20 for group A and -126 for group B concerning CRT TV sets in 2012).
Role of government (MOE and METI)	• supporting activities: R&D, awareness raising activities, technical support • publication of data on annual performance • impose sanctions: fines or imprisonment • reviews of the Act carried out by the joint advisory council.
Environmental performance	• Recycling rates achieved in 2013: 91% for air conditioners, 79% for CRT TV, 89% for LCD and plasma TVs, 80% for refrigerators and freezers, and 88% for washing machines and clothes dryers. • Overall number of recycled units are 174 million units of home appliance in 2001-13. • Avoidance of 50 % potential GHG emissions from using virgin material. • More than 50 % of targeted waste being generated is collected for recycling.
DfE incentive	Publication of guidelines to help manufacturers in DfE Improved communication between recyclers and manufacturers such as training of product designers by dispatching them to recycling facilities.
Cost efficiency	An estimated benefit of JPY 54 billion was achieved through enforcement of the Act.

1. Description of EPR set-up

a) Legal context

The Act for Recycling of Specified Kinds of Home Appliances was enacted in June 1998 and enforced in April 2001. It aims to achieve a reduction in the volume of general waste and sufficient utilization of recycled resources. The Act covers four categories of home appliances: air conditioners; TV sets; electric refrigerators and freezers; and electric washing machines and clothes dryers. Personal computers and small electronic appliances are covered by other legislations.

b) Allocation of responsibilities (distribution of roles, financial flows)

Consumers that dispose of the waste of home appliances are responsible for paying both the collection/transportation fee and the recycling fee as well as for the proper returning of the waste to retailers or to put into municipal collection and treatment route. Business disposer/emitter is out of the scope of this law. Those emitted from business facilities are regulated as industrial wastes. Collection/transportation fees are set by retailers depending on the transportation distance and the size and type of waste products, but many large retailers uniformly charge JPY 540 (including consumers' tax) per unit. Recycling fees are set by manufacturers and are subject to regular reviews by each manufacturer. Retailers are providing a service for (based on a business custom to collect old appliances with a fee (JPY 500 with consumer tax mentioned above: USD 5) when new ones are delivered) collecting the home appliance that they have sold, or to take back the old appliances from customers who buy a new product. Retailers are also responsible of transferring these end-of-life products to designated collection points set up collectively by the responsible manufacturers, or they may contract out these operations to operators certified by the competent authority. Manufacturers and importers have the obligation to set up designated collection points and collect and recycle the home appliance that they produced or imported from these designated collection points. Small and medium-sized manufacturers (defined in the Act according to the category of products) may contract out this responsibility to designated bodies. Municipalities are responsible for managing waste home appliances or WEEE outside the scope of the Act. If they collect waste covered by the Act, they may transfer it to manufacturers or decide to recycle them themselves. The national government (Ministry of Environment (MOE) and Ministry of Economy, Trade and Industry [METI]) are responsible for supporting activities such as promoting research and development, providing information, developing related facilities, providing technical assistance, conducting and implementing environmental education and information dissemination activities. The government also publish data on annual recycling performance.

c) Governance system and sanctions

Retailers and manufacturers who neglect their obligations for collection and recycling as stipulated in this recycling Act, provide false information, or impose unlawful charges are subject to corrective recommendations, corrective orders, or penalties. Monetary penalties range from a fine of JPY 100 000 to JPY 500 000. Individuals who commit illegal dumping are subject up to five years imprisonment or to a fine up to JPY 10 million (or JPY 300 million for corporations) under the Waste Management and Public Cleansing Act.

The Act stipulates that the government must assess the progress of implementation five years after entry into force. For this purpose, a joint advisory council was set up between the MOE and the METI to evaluate processes in, review directions for, and identify issues related to policy implementation, as well as giving advice to the government. Members of the joint advisory council include experts, researchers, representatives of manufacturers associations, consumer groups, local governments, retailers and recyclers. A first evaluation was carried out in 2006-07 that resulted in the publication of a report in 2008. A second review was done in 2013. As a result of this review, the collection target was set at 56% and targets for recycling rates of appliances had also increased.

2. Environmental effectiveness

a) Collection and recycling rates

Recycling targets under the Act refer to the percentage recovered from the WEEE for recycling of the total weight of the components and materials in the WEEE. They exclude thermal recovery. As shown in Table H.1, recycling targets were revised in 2009 except for CRT TVs because the resource value of CRT glass has been declining significantly. Also, in 2015, recycling targets are revised again to promote further recycling of collected items.

Table H.1. **Statutory recycling targets**

	Statutory recycling targets		
	FY 2001-08	FY 2009-14	FY 2015-
Air conditioners	60%~	70%~	80%~
TV sets (CRT)	55%~	55%~	55%~
TV sets (flat screen)	-	50%~	74%~
Refrigerators and freezers	50%~	60%~	70%~
Washing machines	50%~	65%~	82%~

Source: Compiled from the Association for Electric Home Appliances (AEHA), *Annual report on home appliance recycling, for financial year 2004 and* 2012. [in Japanese]; For FY 2015, Press Release of Ministry of the Environment of Japan, March 17, 2015.

Over a period of 13 years since the enforcement of the Act, the number of units received by the designated collection sites has been on the rise, and so has the number of units processed for recycling. Overall numbers of recycled units are 174 million units of home appliances between 2001-13. The recycling rate for each category of home appliances has remained high, although it has varied slightly depending on the category (see Figure H.1). In 2013, the recycling rate was higher than the recycling target as stipulated by the Act for each of all the designated categories of home appliances. More precisely, it stood at 91% for air conditioners, 79% for CRT TV, 89% for LCD and plasma TVs, 80% for refrigerators and freezers, and 88% for washing machines and clothes dryers.

Figure H.1. **Trends in the recycling rates under the Home Appliance Recycling Act**

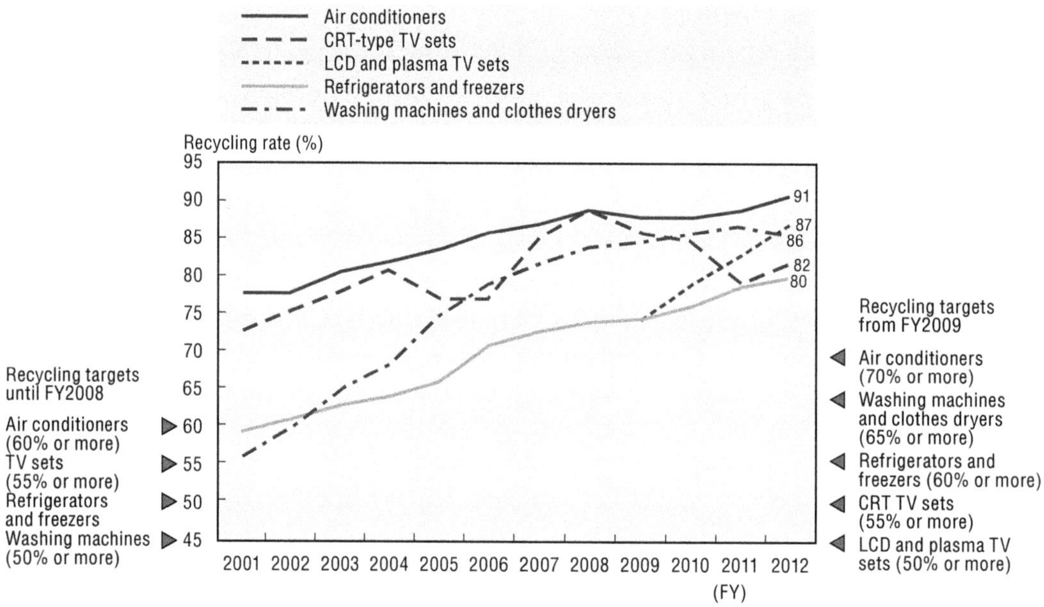

Source: Association for Electric Home Appliances (AEHA), *Annual report on home appliance recycling* for financial year 2012 [in Japanese].

According to a study by the Global Environmental Strategies (IGES), the Act has led to a total recycling volume of 38.4 million m^3 of home appliances over a period of 11 years from 2001 to 2011. It has also estimated that the Act may have resulted in avoiding more than 50 % potential GHG emissions from using virgin materials.

b) Design-for-Environment (DfE)

The Association for Electric Home Appliances' (AEHA) Product Assessment Expert Committee develops guidelines and reports that help manufacturers to work on DfE. To this end, the committee exchanges views with officials at recycling plants to identify specific improvements they want to see in product design. It also conducts a questionnaire survey on waste management procedures at these plants. In 2012, the committee issued the third edition of the "Guidelines on the labelling and recycle symbols on plastic parts of home appliances" to promote product designs that facilitate recycling. Focus is placed on the following aspects: fewer types of plastic materials; fewer parts; product structure designed to facilitate disassembly; and labelling of parts with the type of material, the indication of the positions of screws. In addition, to improve communications between manufacturers and recyclers, some manufacturers second their employees to recyclers as researchers or trainees. Training engineers at recycling facilities would improve communications for design for environment between manufacturers and recyclers. Also, recyclers submit written requests to the manufacturers for a better design in the easy-to-recycle process.

3. Economic efficiency (including competition aspects)

a) Cost efficiency

There is a lack of transparency in the process of setting recycling fees by manufacturers. However, estimation by the joint advisory council indicates that total costs of recycling have decreased, most likely as a result of increased efficiency with technological innovations. Table H.2 shows the results from the cost and revenues estimation. Explanation on Group A and B (two grouping of manufacturers responsible for recycling) is provided in the following section c) on trade and competition.

Table H.2. **Estimated recycling cost per unit under the Home Appliance Recycling Act (Unit:JPY)**

2006	Group	Recycling fee revenues	Revenues from the sale of recovered materials	Total revenues	Administrative expenses	Expenses of secondary logistics	Expenses of designated collection sites	Expenses of home appliance recycling plants	Total expenses
Air conditioners	A	3 500	1 183	4 683	831	469	952	2 466	4 718
	B	3 500	1 366	4 866	831	443	769	3 202	5 244
TV sets	A	2 700	304	3 004	831	305	619	1 678	3 433
	B	2 700	368	3 068	831	288	500	1 989	3 607
Refrigerators and freezers	A	4 600	667	5 267	831	629	1 278	3 815	6 553
	B	4 600	703	5 303	831	594	1 031	4 629	7 085
Washing machines	A	2 400	394	2 794	831	342	695	2 026	3 893
	B	2 400	419	2 819	831	323	561	2 152	3 866

Source: Hotta, Y., A. Santo, and T. Tasaki (2014), "EPR-based Electronic Home Appliance Recycling System under Home Appliance Recycling Act of Japan", Case study prepared for the OECD, *www.oecd.org/env/waste/gfenv-extendedproducerresponsibility-june2014.htm.*

A 2005 analysis suggested that the enforcement of the Act had resulted in a total net benefit of JPY 54 billion. This is due to two parallel tendencies. First, a reduction of JPY 38.1 billion in costs, which comes from a reduction of JPY 44.5 billion in costs borne by the local governments and an increase of JPY 6.3 billion in costs borne by the private sector. Second, an increase of JPY 15.8 billion in benefits which comes from a JPY 1.6 billion reduction for the local governments and an increase of JPY 17.5 billion in benefits or the private sector.

b) Leakages and free riders

As a result of the consultations undertaken in 2006-07 and following a request from stakeholders, the joint council decided to strengthen measures against illegal dumping and improper exports. Indeed, hidden flows outside the recycling chain under the Act are an important concern. It is estimated that almost half of the waste home appliances covered by the Act is not subject to the collection mechanisms stipulated by the Act and about 30% is traded as second-hand goods or metal scarp. In addition, there are concerns that the current system of payment at the time of disposal may encourage illegal dumping by customers. The government is taking measures in order to prevent illegal dumping and export of WEEE as second-hand goods, which may be handled in an environmentally improper manner when exported to developing countries. Such measures include developing clearer criteria for second-hand use in relation to exports, closer partnership with customs authorities, promoting co-operation with trading partners and supporting implementation of the Basel Convention.

c) *Trade and competition*

In order to ensure competition in the home appliance recycling sector, manufacturers are organised in two groups under the Act. Both groups are almost equal in terms of the number of units manufactured and the number of units processed for recycling. Both groups have established their own joint recycling management companies. Designated collection sites accept products from both groups in order to ensure efficiency. The result has been a decline in recycling costs. The MOE-METI joint advisory council are reflecting on how to further reduce fees and ensure they are set in a fair and competitive manner.

4. Key issues and possible reforms

A second review process was completed in 2014 and the joint advisory meeting delivered certain recommendations based on requests made by stakeholders, as follows: setting 56% collection target and increased recycling target rates; greater transparency on the process of setting fees and reduction of the financial burden on consumers; a clearer distinction between the 3R activities; a review of the current license qualifications for collection, transport and recycling plants in order to ensure greater participation, and stricter regulations and measures against the flow of WEEE towards other countries. However, although the possibility of shifting payment at the time of disposal towards a system based on payment at the point of purchase was discussed, the timing of the recycling fee payment was not changed.

Note

1. Full source available at: Hotta, Y., A. Santo, and T. Tasaki (2014), "EPR-based Electronic Home Appliance Recycling System under Home Appliance Recycling Act of Japan", Case study prepared for the OECD, *www.oecd.org/env/waste/gfenv-extendedproducerresponsibility-june2014.htm*.

Extended Producer Responsibility
Updated Guidance for Efficient Waste Management

ANNEX I

The EPR for packaging waste in Japan

by
Dr Hajime Yamakawa, Kyoto Prefectural University, Japan[1]

SUMMARY BOX	
Cost allocation	Size of total expenditure of PRO is equivalent to 18% of total expenditures of municipalities for collection, sorting and management of same packages.
Cost coverage	90% (percentage of PRO's total revenue covered by producers' fees in 2010)
Role of government	• Designation of a PRO. • Approval of rules and regulations on recycling activities of PRO. • Approval of the PRO's business plans and budget for each fiscal year. • Supervision of the PRO' activities including the determination and appropriateness of the unit recycling commissions, and supervision of the PRO's expenditures through reviewing the accounts and business reports. • Authority to conduct an on-site inspection, and give orders if needed. • Authority to issue recommendations and impose financial sanction if a producer fails to fulfil its obligations.
Environmental performance	For PET bottles: recycling increased from 3.5% to 76.7% between 1996 and 2010; and percentage of products dumped at landfills decreased by 72% over that period. (see Figure I.1 and I.2 for other waste products groups)
DfE incentive	Average weight of PET bottles decreased by 7.6% (47 000 tonnes) between 2004 and 2010.
Cost efficiency	• Economic efficiency: net benefits of JPY 144 361 million in 2010. • Cost effectiveness: recycling fees represent 90% of the PRO's revenues and almost cover entirely the recycling operation expenditure incurred after removing impurities.

1. Description of EPR set-up

a) Legal context

The Act on the Promotion of Sorted Collection and Recycling of Containers and Packaging (*"Packaging Recycling Act"*) came into force in December 1995 in order to encourage the collection of "properly-sorted" waste containers and packaging (municipal solid waste) by municipalities and their recycling by business operators.[2] A revision of the Act was carried out in 2006, which added waste reduction as an objective.

b) Allocation of responsibilities (distribution of roles, financial flows)

Under the Act, municipalities have the financial and physical responsibility of the collection of packaging waste discharged by consumers. Producers are financially responsible for recycling the waste of glass bottles, pet bottles, other plastic packaging and paper packaging which are collected and properly sorted by municipalities. Financial responsibility for recycling of paper packaging excludes corrugated cardboard and carton paper. Producers may participate and contribute to a PRO that outsources operations to registered recyclers selected through annual tenders under criteria set by the PRO. The recycling commission paid by individual producers to the PRO is proportional to the quantity of waste packaging that it generates multiplied by the full recycling costs of the product category. The Japan Containers and Packaging Recycling Association is currently the only designed (appointed?) PRO even though the Act allows for multiple corporations. The Japanese government is responsible for raising public awareness and promoting proper sorting through educational activities. In principle, all producers are required to label their products in order to facilitate waste sorting.

c) Governance system and sanctions

The system is supervised by the Japanese government through the Ministry of the Environment (MOE), the Ministry of Economy, Trade and Industry (METI), the Ministry of Agriculture, Forestry and Fisheries (MAFF), the Ministry of Health, Labour and Welfare (MHLW), and the Ministry of Finance (MOF). The national government is responsible for designating PROs upon application, approving the PRO's business plan including its budget, supervising their activities including the calculation of recycling commissions and approving their expenditures and statement of accounts. It also provides advice indirectly to municipalities on their Municipal Sorted Collection Plans. When a producer fails to fulfil its obligations, the government may issue recommendations, order the producer to conduct the recycling, or impose a fine of up to JPY one million. Major revisions of the Act or of regulations and targets are discussed within a committee composed of members of the government and municipalities, representatives of producers' associations, environmental organizations, experts, etc.

2. Environmental effectiveness

a) Collection and recycling rates

The Act does not set targets in terms of collection and recycling rates as the responsibility of producers is limited to recycling all used packaging collected by municipalities. However, the quantity of packaging waste collected from households has reportedly increased for most product categories in the last decade between the introduction of the Act (1995) and 2010 (Figure I.1). Similarly, data shows that the amount of waste ending up in landfills has decreased for each product group over the same period (Figure I.2). Over this time frame, the reduction percentage of products being dumped at landfills was of 39% for glass bottles, 72% for PET bottles, 60 % for other paper packaging and 76% for other plastic packaging. Finally, Table I.1 shows the amount of collected and recycled waste packaging per capita for different product groups, which highlights that the amount collected is in most cases very similar to the quantity being recycled.

Figure I.1. **Percentage of waste packaging collected from households**

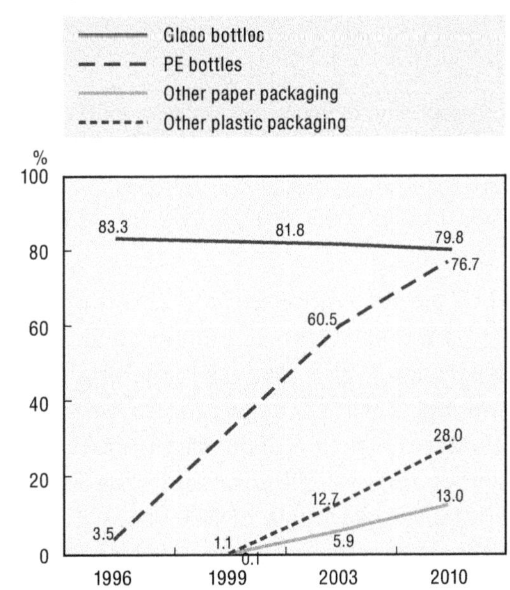

Figure I.2. **Estimates of the quantities of four intems dumped in landfills**

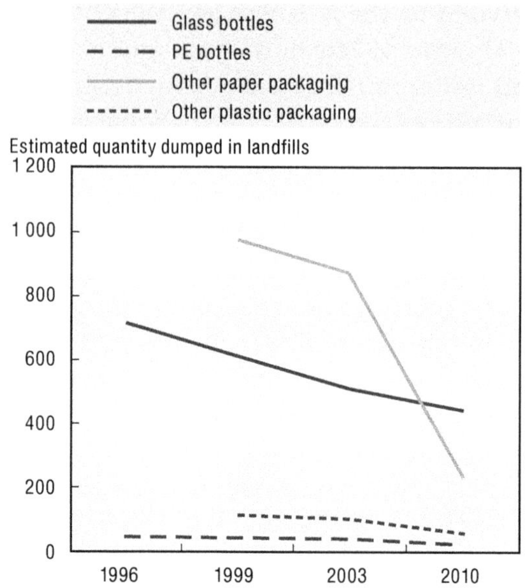

* Collection rate = (Sorted collection by municipalities + Collection by private sector)/Total consumption of packaging by households.

Source: Mitsubishi Research Institute, Inc. (2013), Report on the evaluation and review of the Scheme for the recycling of packages (the Survey for promotion of recycling of packages by the outsourcing expenditure for the research upon environmental programs in FY 2012), *www.meti.go.jp/policy/recycle/main/data/research/h24fy/h2503-yourimri.html* (in Japanese).

Table I.1. **Amount of collected and recycled waste packaging per capita (FY 2010)**

	Sorted collection per capita (kg/person/year)	Recycling per capita (kg/person/year)
Glass bottles	6.2	5.9
Other paper packaging	0.7	0.6
PET bottles	2.3	2.2
Other plastic packaging	5.5	5.2
Steel cans	1.8	1.8
Aluminium cans	1.0	1.0
Corrugated cardboard	4.7	4.7
Paper cartons	0.1	0.1

Note: The amounts of sorted collection and recycling were taken from the Annual Report on Environmental Statistics. The total population was taken from the 2010 Census.

Source: Yamakawa, H. (2014), "The packaging recycling act: the application of EPR to packaging policies in Japan", Case study prepared for the OECD, *www.oecd.org/env/waste/gfenv-extendedproducerresponsibility-june2014.htm*.

b) Design-for-Environment (DfE)

The Act reportedly helped accelerate the reduction of weight for containers and packaging through the introduction of thin and light weight products, products with no aluminium lining, flexible packaging, and changes in materials.[3] In addition, in the context of the 2006 revision of the Act, business associations have developed voluntary plans setting targets to further reduce the weight of packaging and increase recycling rate. These activities led to a reduction of 7.6% (47 000 tonnes) in the average weight of PET bottles designated to be recycled. Other examples of DfE innovations reported include using coloured plastic film labels (easy to remove) instead of coloured PET bottle in order to avoid the sorting of bottles per colour thereby reducing collection costs.[4] Various recycling technologies for waste packaging and material recycling have been developed in Japan in recent years.

c) The waste hierarchy (prevention, re-use, recycling)

Next to increased recycling, the Japanese system also includes provisions to encourage the prevention and re-use objectives in the waste hierarchy. On the *prevention* side, a system for reducing waste packaging was introduced in the 2006 revisions, under which retailers that use particularly large volumes of packaging are required to report their waste prevention strategies to the government that can issue recommendations or impose fines when activities are deemed insufficient. According to the author, the total quantity of packaging items consumed was reduced by 16% between 1996 and 2009; this trend was of 27% for the quantity of containers for liquids and of 21% for other paper and plastic packaging. Surprisingly, the demand for PET bottles greatly increased over this period (by 226%), even though recycling fees were set at higher levels compared to other types of containers. This shows that the incentives and fee structure under the Act were not always effective in reaching the expected results; although the simultaneous shift in the use of lighter containers (demand for glass bottles decreased by 40%) must have contributed to the overall waste prevention in terms of weight. In order to encourage *re-use*, the system also foresees that in the case of specific waste packaging categories with positive value for which more than 90% collection rate is expected, the national government may decide to exempt the category from obligatory recycling.

3. Economic efficiency (including competition aspects)

a) Cost efficiency

The total expenditure of the PRO in 2003 was approximately JPY 40 billion, representing about 18% of the expenditure of municipalities for the four items shown in Table I.2, which amounted to JPY 230 billion in the same year. The total expenditure of municipalities for sorted collection and further sorting of all packaging waste was about JPY 400 billion this year. The high costs paid by municipalities for sorted collection and storage prevents certain municipalities to introduce sorted collection (e.g. only 35% of municipalities in Japan conduct sorted collection for other paper packaging). Table I.2 provides an overview of the PRO's detailed revenue and expenditure in 2010. It can be observed that the overall commissions paid by producers account for 90% of the PRO's total revenue, and almost cover the recycling operations for each product category. Table I.3 highlights the trends of total revenue of the PRO between 2000 and 2013.

The author introduced a cost-benefit analysis of the waste packaging system for the year 2003 and 2010. As highlighted in Table I.4, the net benefits in 2010 were of JPY 144 361 million, compared to net losses of JPY 28 650 million in 2003 at the earlier stages of implementation of the Act. The important increase in benefits over that time frame owes essentially to the declining consumption of virgin resources enabled through recycling and waste prevention activities.

In order to further reduce the overall costs of waste packaging recycling, a *system of contributory commission* was introduced in the 2006 revision of the Act aimed at incentivizing municipalities to provide high quality sorted waste. Under that scheme, when the actual recycling costs fall below the estimated recycling costs thanks to the high quality of sorted packaging waste collected etc., the PRO reimburses 50% of its reduction in payment to the recycler to such municipalities.

Table I.2. **The 2010 Statement of accounts for the PRO (Million JPY) (by Department, after Account Settlement)**

	Settled accounts	Glass bottles	PET bottles	Paper	Plastic
(1) Revenue					
1) Operational revenue					
Recycling commission	38 917	1 692	54	390	36 781
Contributory commission	9 335	0	88	26	9 220
Recycling commissions (from municipalities)	1 008	270	17	18	703
Revenue from sales of recyclables through bidding	4 379	1	4 302	77	0
2) Interest revenue	12	1	1	0	10
Total revenue	53 652	1 964	4 462	512	46 714
(2) Expenditure					
1) Operational expenditure	53 444	1 913	4 410	463	46 658
Outsourcing of recycling					
Recycling commission	38 001	1 733	301	62	35 905
Cost reduction contribution	9 335	0	88	26	9 220
Contribution from sales of recyclables through bidding	3 961	1	3 893	67	0
Labour costs	197	36	45	30	85
Taxes and public charges	776	4	-80	9	842
Other overhead cost *(computer data processing, surveys, equipment, trainings etc.)*	1 086	117	138	247	584
Awareness raising and information provision	87	21	23	21	22
Others	2	0	1	0	0
2) Administration cost	208	51	53	49	56
Total expenditure	53 652	1 964	4 462	512	46 714
Balance	0	0	0	0	0

Source: *www.jcpra.or.jp/english/tabid/603/index.php.*

Table I.3. **Trend of total revenue of the PRO (Settled accounts)**

Year	Million JPY				
	2000	2003	2005	2010	2013
Total	18 445	42 421	50 493	53 652	49 227
Glass bottles	3 265	1 773	1 684	1 964	2 127
PET bottles	6 950	8 435	3 280	4 462	8 043
Paper	1 216	1 010	787	512	539
Plastic	6 952	31 200	44 709	46 714	38 518

Source: Yamakawa, H. (2014), "The packaging recycling act: the application of EPR to packaging policies in Japan", Case study prepared for the OECD, *www.oecd.org/env/waste/gfenv-extendedproducerresponsibility-june2014.htm.*

Leakages under the Japanese system take place especially due to exports of waste PET bottles away from the EPR flow, which reduces the percentage of products contracted with the PRO and undermines regulation on the recycling of PET bottles. Municipalities are not forbidden to sell waste packaging directly to buyers who export them and about 40% of municipalities contract with independent recyclers directly; but recyclers contracting with the PRO are not allowed to export waste packaging received through the PRO. About 30% of municipalities collecting PET bottles separately contract only with independent recyclers directly, and about 10% of them contract with both independent recyclers and the PRO. When a producer acts as a free rider, as explained above, the Japanese government has the authority to issue recommendations and impose financial sanctions on him. As stated above, packages – except the four items – are excluded from the producers' recycling obligations. Packages for business use are also excluded.

Table I.4. **Results of cost-benefit analysis of the Packaging Recycling Act Million JPY**

			FY 2010	FY 2003
Costs	Designated producers	Recycling commission	36 316	39 928
		Contributory commission	9 972	0
		Internal costs (R&D, self-collection, administration, voluntary activities)	12 500	13 615
	Municipalities	Additional costs of sorted collection for waste packaging	174 461	115 427
		Revenue from selling recyclables through bidding	-7 009	
		Cost reduction contribution	-9 972	
		Recycling commission (for micro enterprises)	1 684	2 446
Total cost			217 952	171 416
Benefits	Municipalities	Effect of reduced landfilling of combustible and non-combustible waste (by recycling)	3 175	2 944
		Effect of reduced treatment of combustible and non-combustible waste (by recycling)	104 617	91 711
		Effect of reduced treatment and landfilling of combustible and non-combustible waste (by prevention)	1 541	302
	Society	Effect of reduced consumption of virgin resources (by recycling)	87 071	27 291
		Effect of reduced consumption of virgin resources (by prevention)	165 432	19 372
		Effect of reduced CO_2 emissions from carbon in plastics	476	1 146
Total benefit			362 313	142 766
Benefit – Cost			144 361	-28 650

Source: Yamakawa, H. (2014) "The packaging recycling act: the application of EPR to packaging policies in Japan", Case study prepared for the OECD, *www.oecd.org/env/waste/gfenv-extendedproducerresponsibility-june2014.htm.*

b) Trade and competition

There is no competition at PRO level since there is currently only one PRO operating for waste packaging in Japan; even though the Act allows for multiple ones. On the recycling market, competitive tender procedures are organized by the PRO in order to ensure a level playing field among recyclers. The transparent tender procedure has led to decreasing contract prices over the years. Competition concerns may however arise between recyclers and the PRO as certain recyclers in Japan buy waste from PET bottles and other sellable recyclables directly from municipalities, thereby affecting competition on the recycling market.

4. Key issues and possible reforms

The allocation/assignment of responsibilities were discussed in the 2006 revision of the Act but were finally left unchanged. The issue should be further discussed since the costs paid by municipalities remain much higher than those paid by producers, thereby undermining the effectiveness of the scheme. In 2003, Japanese municipalities paid about JPY 423.5 billion (USD 3.5 billion) for sorted collection, further sorting and storing of waste packages which amount to about 3.4 times the expenditure of producers in the same year (JPY 40 billion). This means that incentives to reduce packaging materials and their recycling cost for producers were lower by USD 3.5 billion than if the producers had paid all costs for recycling of their products' packages.

In addition, in order to compensate the high costs, certain municipalities contract directly with independent recyclers and bypass the PRO route, which may undermine the efficiency of the system.[5]

Two other challenges highlighted are the need to further increase the 2Rs (reduction and reuse) of waste packaging, and the ability to determine the positive and negative environmental effects of changes in material use (e.g. shift from heavy materials such as glass bottles and steel cans to light material such as PET bottles, aluminium cans and paper cartons), which occurred as a result of the Act.

Notes

1. Full source available at: Yamakawa, H. (2014), "The packaging recycling act: the application of EPR to packaging policies in Japan", Case study prepared for the OECD, *www.oecd.org/env/waste/gfenv-extendedproducerresponsibility-june2014.htm*.
2. Designated producers recycling obligation under the Act applies only to properly-sorted waste; therefore, it does not apply in municipalities where no sorted collection is conducted.
3. Munakata, S. and S. Murai (2000), "Influences of the Full Enforcement of Recycling Acts upon Medium and Small Manufacturers of Plastic Products and their Actions Against Them", *JFS monthly report*, 47(12), pp. 12-17 (in Japanese). Yamaguchi and Ishikawa, 2001; Yamaguchi, H., and M. Ishikawa (2001), "Issues and Measures of the Containers and Packaging Recycling Act", *Journal of Resources and Environment*, 37 (12), pp. 50-57 (in Japanese).
4. Hosoda, E. (2004), *Evaluation of EPR Programmes in Japan, in Economic Aspects of Extended Producer Responsibility*, pp. 151-192.
5. Kurita, I (2011), "Independent Disposal of Post-consumer PET Bottles", *Journal of the Japan Society of Waste Management Experts*, 22(1), pp. 61-70.

Extended Producer Responsibility
Updated Guidance for Efficient Waste Management

ANNEX J

EPR in Korea

by

Hyein Heo *and* Mun-hee Jung, Korea Ministry of Environment[1]

SUMMARY BOX	
Cost allocation	
Cost coverage	PROs' costs are fully covered by the recycling fees (does not include collection).
Role of government	• Creates and implements EPR regulation. • KECO (set up by Ministry of Environment) accredits PROs, and monitors and enforces obliged parties (including verifying annual reports and on-sit inspections). • Ministry of Environment has authority to impose fines up to 30% surcharge in case of non-compliance.
Environmental performance	• In 2012, 1 519 thousand tonnes of products and packaging materials were recycled, equal to 62% more than in 2002. • The overall recycling rate increased almost by 103%, while landfill use decreased by 31%.
DfE incentive	Producers are obliged to develop recycling technology, resource efficiency design, and restrict the use of hazardous substances and produce (or import) easier-to-recycle products.
Cost efficiency	• Landfill savings equals KRW 2 888 billion. • KRW 3 055 billion in revenues from selling recycled goods and materials.

1. Description of EPR set-up

a) Legal context

EPR scheme was set up in 2003 for electric products, tyres, lubricant, fluorescent lamps, styrofoam float and packaging. The Act on Resource Recirculation of Electrical Waste and End of Life Vehicles and the act on the Promotion of Saving and Recycling of Resources state that producers and importers are responsible for restricting the use of hazardous substances in their products, collecting and recycling end of life products. Sellers are responsibility for collection.

b) Governance and Enforcement

The Korea Environment Corporation (KECO) monitors compliance by obliging producers and importers to report sales and imports as well the waste collected and recycled (reports are submitted on an online portal and include data on the use of hazardous substances and recyclable components in vehicles, and WEEE). KECO confirms these data and verifies appropriate waste treatment through on-site inspections. PROs are accredited by KECO based on financial stability and potential contribution to the recycling industry's development. If a producer or PRO fails to comply with the obligations the Ministry of Environment imposes a fine up to 30% surcharge (see Table J.1). Producers other than those producing WEEE with a yearly output of less than one billion KRW and importers of less than three hundred billion KRW are exempt.

c) Allocation of responsibilities (distribution of roles, financial flows)

The national government creates and implements EPR regulation. Local governments are responsible for improving collection and recycling as well as encouraging reuse. Private waste collectors also exist, who sign contracts with apartment buildings and sell the collected recyclables to the recycling industry. Producers and importers of EPR products label their goods with information related to disposal and recyclability (see Figure J.1), collect and recycle their end of life products or pay recycling fees to the relevant PRO. Consumers are responsible for separating and disposing their waste.

Table J.1. **2012 Standard fees for recycling and financial sanctions**

Items	Standard fees (KRW)	Financial sanctions imposed (KRW 1 000)
Carton pack	185/kg	30 042
Glass bottle	34/kg	1 301 978
Iron can	87/kg	154 186
Aluminum can	151/kg	56 012
PET bottle	178/kg	450 414
Colored PET bottle	235/kg	7 465
Mixed material PET bottle	360/kg	96 105
EPS	317/kg	127 590
PSP	327/kg	37 850
PVC	981/kg	575 884
Plastic container tray	327/kg	498 049
Film and sheet type plastic materials	467/kg	1 095 986
Lubricant container	327/kg	14 020
Lubricant	20/liter	33 616
Tyre	30/kg	437 872
Fluorescent lamp	143/product	719 909
Styrofoam float	627/kg	1 303
Mercury battery	39.6/g	2 639
Silver oxide battery	35.5/g	68 264
Lithium battery	0.8/g	124 063
Nickel-cadmium battery	0.78/g	29 462
Manganese battery	0.35/g	515 505
Nickel-hydrogen battery	0.16/g	9 661
Television	196/kg	133 123
Refrigerator	197/kg	348 353
Washing machine	137/kg	5 154
A/C	98/kg	4 254
Computer	188/kg	167 436
Audio	225/kg	47 100
Mobile phone	2 649/kg	716 912
Copier	273/kg	29 074
Fax	403/kg	0
Printer	400/kg	32 127

Figure J.1. **Product labelling to ensure correct sorting**

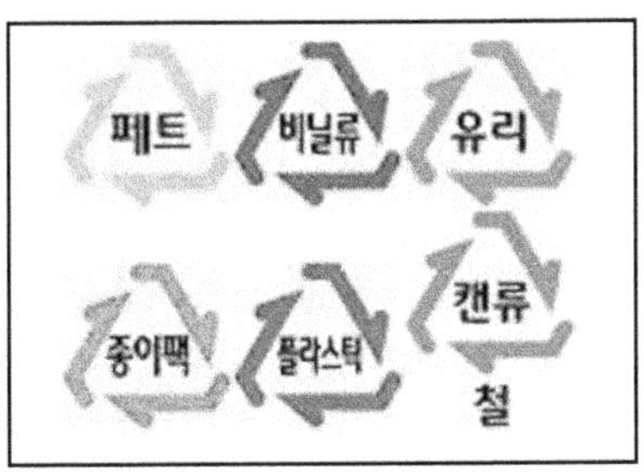

Source: Heo, H. and M.-H. Jung (2014), *Case study for OECD project on extended producer responsibility, Republic of Korea*, Case study prepared for the OECD, *www.oecd.org/env/waste/gfenv-extendedproducerresponsibility-june2014.htm*.

An Advance Disposal Fee (ADF) is charged to producers and importers of goods, materials and containers that are hard to recycle (including pesticides, hazardous chemicals, anti-engeeze solutions, chewing gum, disposable diapers, cigarettes and plastic products not included in the EPR scheme such as PVC pipe, toys and kitchenware). The fees are used for local government subsidies to establish waste treatment facilities. A deposit/refund system has been set up for beverage producers, which equals to around 40% of the cost of manufacturing one bottle. In addition to the EPR schemes, there are complementary policies. These include: a pay as you throw unit price (for non-recyclable waste), separated collection and a recording system for charging bulky waste. The local government can also subsidize or lend funds needed to recycling and waste disposal facilities as well as recyclable materials (e.g. the Special Accounts for Environmental Improvement was set up in 1995 by the Korea Ministry of Environment as a centralised account to fund investments in environmental protection, and KECO provides financial assistance through low interest rate loans to SMEs in the recycling industry seeking technical consulting).

2. Environmental effectiveness

a) Collection and recycling rates

With EPR in effect since 2003, the quantity of recycled products and packaging materials gradually rose to 1 519 thousand tonnes in 2012 – equal to 62% more than in 2002. During the 21st century, the overall recycling rate increased almost by 103%, while landfill use decreased by 31%. The number of obliged companies and recycling facilities rose from 2 747 and 218 in 2003 to 4 567 and 627 in 2012 respectively. The number of producers per recycling business rose from 6.6 in 2003 to 7.3 in 2012.

The Ministry of Environment publishes a yearly recycling rate based on the quantity of products on the market, previous recycling rate and recycling capacity (see Table J.2). Exporting end of life PET bottles for recycling is restricted to less than 20% (currently only at around 0.1%). For other waste products – such as battery, lubricant, fluorescent lamp, and electronics – export is not an approved recycling method.

b) DfE

Producers are obliged to develop recycling technology, resource efficiency design, and restrict the use of hazardous substances and produce (or import) easier-to-recycle products.

3. Economic efficiency

a) Cost efficiency

Savings on landfill expenses amount to KRW 2 888 billion and KRW 3 055 billion has been generated from selling recycled goods and materials. Also, it is estimated that up to 9 769 jobs were created over ten years (see Table J.3). There are currently six PROs, where usually 70-90% of funds collected are used for recycling and 1-5% is allocated for information and public awareness campaigns (funds are generally not used for sorting or collection). PROs' costs are fully covered by the recycling fees whereas KECO is funded by the special account. There is only one PRO per EPR product. PROs are in fact considered public institutions, even though they are private non-profit organisations. Fee levels are set by the PROs, however the standard recycling fees for each EPR product published by the government are taken into account (these standard fees are based on the estimated expenses related to collection and treatment).

4. Key issues and possible reforms

The overall recycling target set for all EPR products equals to only 33% of the capacity of all the recycling facilities combined. Furthermore, the supply chain of recyclable products and materials is unstable and their added value remains low. Thus, there is a plan to launch an online market which will provide demand and supply information of recyclable products

Table J.2. **Increasing target rates**

Items	Long-term Target	2005	2006	2007	2008	2009	2010	2011	2012	2013	2014
Carton pack	**0 360**	0 278	0 266	0 280	0 291	0 305	0 327	0 327	0 341	0 341	0 346
Glass bottle	**0 793**	0 672	0 684	0 708	0 726	0 737	0 751	0 757	0 778	0 760	0 760
Iron can	**0 831**	0 700	0 710	0 720	0 730	0 740	0 756	0 771	0 786	0 786	0 797
Aluminum can	**0 816**	0 700	0 712	0 717	0 730	0 740	0 756	0 771	0 786	0 786	0 791
PET bottle	**0 830**	0 695	0 704	0 717	0 737	0 749	0 764	0 783	0 806	0 806	0 812
EPS	**0 805**	0 613	0 629	0 690	0 748	0 748	0 760	0 772	0 781	0 781	0 790
PSP	**0 423**	0 249	0 280	0 317	0 339	0 339	0 367	0 400	0 423	0 423	0 423
PVC	**0 733**	0 480	0 484	0 556	0 583	0 598	0 600	0 622	0 664	0 644	0 690
Plastic container tray	**0 845**	0 502	0 528	0 585	0 628	0 653	0 702	0 757	0 800	0 800	0 806
Plastic film and sheet	**0 675**	0 368	0 387	0 414	0 433	0 476	0 517	0 562	0 600	0 600	0 603
Lubricant container	**0 790**							0 790	0 790	0 795	0 794
Lubricant	**0 730**	0 687	0 687	0 676	0 677	0 677	0 692	0 708	0 722	0 725	0 726
Tyre	**0 770**	0 718	0 720	0 737	0 748	0 748	0 754	0 758	0 762	0 762	0 765
Fluorescent lamp	**0 394**	0 189	0 207	0 225	0 231	0 240	0 261	0 285	0 300	0 328	0 355
Styrofoam float	**0 277**							0 277	0 277	0 285	0 280
Mercury battery	**0 600**	0 250	0 293	0 380	0 490	0 490	0 600	0 600	0 600	0 600	0 600
Silver oxide battery	**0 560**	0 250	0 250	0 309	0 370	0 390	0 424	0 499	0 560	0 560	0 560
Lithium battery	**0 650**	0 249	0 293	0 380	0 490	0 520	0 577	0 582	0 650	0 650	0 650
Nickel-cadmium battery	**0 400**	0 246	0 246	0 257	0 291	0 310	0 333	0 383	0 400	0 400	0 400
Manganese battery	**0 213**				0 200	0 205	0 236	0 192	0 216	0 216	0 216
Nickel-hydrogen battery	**0 153**				0 250	0 250	0 289	0 136	0 203	0 203	0 203
Television	**0 431**	0 118	0 126	0 133	0 145	0 160	0 190	0 210	0 272	0 564	
Refrigerator	**0 389**	0 165	0 169	0 173	0 189	0 206	0 221	0 250	0 267	0 343	
Washing machine	**0 392**	0 228	0 234	0 242	0 253	0 261	0 274	0 285	0 317	0 330	
A/C	**0 080**	0 015	0 017	0 019	0 021	0 023	0 024	0 024	0 027	0 028	
Computer	**0 260**	0 085	0 094	0 098	0 103	0 111	0 123	0 140	0 153	0 158	
Audio	**0 278**	0 102	0 127	0 131	0 149	0 155	0 170	0 185	0 200	0 268	
Mobile phone	**0 400**	0 119	0 154	0 165	0 180	0 198	0 220	0 230	0 356	0 257	
Copier	**0 280**		0 084	0 094	0 127	0 121	0 134	0 150	0 160	0 234	
Fax	**0 250**		0 084	0 094	0 114	0 119	0 130	0 150	0 156	0 223	
Printer	**0 243**		0 084	0 092	0 112	0 133	0 142	0 150	0 150	0 218	

Source: Heo, H. and M.-H. Jung (2014), *Case study for OECD project on extended producer responsibility, Republic of Korea*, Case study prepared for the OECD, *www.oecd.org/env/waste/gfenv-extendedproducerresponsibility-june2014.htm*.

Table J.3. **Economic benefits of EPR**

	Landfill/Incineration cost saved (A) (billion KRW)	Recycled product's economic value (B) (billion KRW)	Employment effect (Persons)
Total	2 888	3 055	9 769
2003	204	163	685
2004	229	185	776
2005	256	206	866
2006	263	215	904
2007	297	245	1 029
2008	300	372	1 005
2009	398	374	1 010
2010	330	408	1 102
2011	355	442	1 193
2012	355	444	1 199

Source: Heo, H. and M.-H. Jung (2014), *Case study for OECD project on extended producer responsibility, Republic of Korea*, Case study prepared for the OECD, *www.oecd.org/env/waste/gfenv-extendedproducerresponsibility-june2014.htm*.

and materials. The Ministry of Environment has also integrated all packaging PROs into one association to reduce administrative costs. Recycling rates were also be set for WEEE – both annually and for the long-term – as of 2014 (at 3.9 kg per capita). Additionally, there is a plan to expand the WEEE EPR scheme to more products (up to 50 different products, including small sized WEEE). An ELV EPR scheme will be launched in 2015.

Note

1. Full source available at: Heo, H. and M.-H. Jung (2014), *Case study for OECD project on extended producer responsibility, Republic of Korea*, Case study prepared for the OECD, *www.oecd.org/env/waste/gfenv-extendedproducerresponsibility-june2014.htm*.

Extended Producer Responsibility
Updated Guidance for Efficient Waste Management

ANNEX K

Electronics EPR in the United States

by
Scott Cassel, Product Stewardship Institute[1]

SUMMARY BOX	
Cost allocation	• Registration fees paid by manufacturers; usually used to cover states' oversight functions. • In most cases, manufacturers pay a portion or all of the collection and recycling costs based on market share or return share; though some states lack specificity in assigning costs and responsibilities.
Cost coverage	Manufacturers pay partial or total costs: in certain states, manufacturers pay for collection and recycling; in some states local governments share costs; and in some states there's no specificity in now costs are allocated.
Role of government	• State agencies oversee the programs. • 20 states require manufacturers to register and report on sales and collection data. • 11 states requires manufacturers to submit a stewardship plan for review and approval. • 21 states can impose penalties on manufacturers and retailers for non-compliance. • 4 states provide a centralized collection and recycling program.
Environmental performance	• Between 0.3 and 7 pounds per capita of e-waste collected annually
DfE incentive	• Some states include compliance requirements with the EU RoHS Directive. • Some states include green procurement requirements for state agencies. • Some states promote a green electronic equipment database to influence purchasers.
Cost efficiency	• Estimated costs of 19 programs was USD 90 million in 2010. • Cost effectiveness: manufacturers' fee covers a large portion of costs (but cost effectiveness probably undermined as costs are not tracked in most states).

1. Description of EPR set-up

a) Legal context

Since 2003, twenty-five states in the US have passed legislation related to scrap electronics (e-scrap) recycling; of which 23 are considered EPR as per the US definition (i.e. mandatory scheme) and two are considered product stewardship. 19 of the 25 programs were implemented between 2009 and 2011. In addition, sixteen of the 25 states have enacted some form of disposal ban on electronics in order to support the programs. Most e-scrap laws in the US drew concepts from a process convened by the US Environmental Protection Agency (EPA) from 2001 to 2003, the National Electronics Product Stewardship Initiative (NEPSI), which consisted of a series of national stakeholder meetings focused on developing a product stewardship system to increase electronics recycling. These meetings disbanded without an agreement but they did help shape state electronics EPR laws. In contrast, action on e-scrap issues at the federal level has been limited.

b) Governance and Enforcement

State agencies or departments responsible for environmental quality or public health oversee e-scrap stewardship programs and some or all of their costs are covered by manufacturers' registration fees or penalties in most cases. Twenty of the programs require manufacturers of "covered electronic devices" (CEDs) to register with the state as a prerequisite to selling their products, and then to report sales and collection data. Eleven states require manufacturers to submit a stewardship plan to the state environmental agency for review and approval. All states forbid retailers from selling CEDs from manufacturers that are not registered or are otherwise out of compliance with the program. Twenty-one of the 25 state laws include penalty provisions imposed on manufacturers that do not register with the state or that fail to meet financial obligations or collection requirements. State enforcement penalties may also be imposed on retailers that sell a non-registered manufacturer's CEDs.

c) Allocation of responsibilities (distribution of roles, financial flows)

Manufacturers of CEDs are responsible for registering with the state oversight agency. They can organize either individually or collectively through PROs. In order to meet their performance goals, manufacturers can use CEDs generated by other entities (e.g. residents, schools, small businesses) in the state and brought back by consumers to collection points. In addition to registration fees, in some states manufacturers have to assume financial responsibility for collection and/or recycling costs, generally calculated either through the market share or actual sales of CEDs either through the return share (when used, typically for IT manufacturers). In the second case, equipment weight shall be delineated by brand and often by type. Only two states assign municipalities responsibility for ensuring convenient collection locations that will arrange for material pick-up by state-approved consolidators or recyclers (only four states require that recyclers be certified by third-party certification organizations). Those entities then obtain payment from manufacturers to cover the cost of transportation, consolidation, and recycling. Four states offer manufacturers a central collection and recycling program. In these states, a contractor or designated quasi-governmental entity serves as the PRO, managing the recycling infrastructure and in some cases collecting fees from manufacturers to cover the costs. Various entities (manufacturers, municipalities, or retailers) are responsible under the laws for educating consumers about the EPR law.

2. Environmental effectiveness

a) Collection and recycling rates

A recent report concluded that e-scrap recyclers in the US generate up to 3% residue and a small amount of potentially hazardous materials, while the majority of outputs go on to commodity markets.[2] While no hard data exist, most stakeholders report that the environmental performance of the e-scrap recycling sector has improved in recent years. Thirteen of the 25 state laws either set goals in statute or confer authority for setting collection performance goals to the responsible state entity. Because all material collected must be recycled or refurbished for reuse, collection goals are synonymous to recycling goals. Since annual generation of e-scrap is difficult to assess (given varying product life spans and people's tendency to store some materials), per-capita collection rates are typically used to track program performance.

Tables K.1 and K.2 compare state programs based on key policy elements and per-capita collection rates. Table K.1 presents the highest performing states and Table K.2 the lowest

Table K.1. **Program Performance and Program Design in High Performing States, 2012**

State (implementation date)	Pounds per capita 2012*	Program financing	Performance Goals	Convenience Requirements	Product Scope	Disposal Ban	Centralized program	Covered Entities
Vermont (2011)	7.7	Market share	Yes	Geographic	Average	Yes	Yes	Average
Oregon (2009)	6.8	Hybrid	Yes[3]	Geographic	Average	Yes	Yes	Average
Wisconsin (2010)	6.8	Market share	Yes	Rural convenience	Comprehensive	Yes	No	Limited
Minnesota (2007)	6.5	Market share	Yes	Rural convenience	Comprehensive	Yes	No	Limited
Washington (2009)	6.4	Hybrid	No	Geographic	Average	No	Yes	Comprehensive

Source: Product Stewardship Institute (2014), *Electronics EPR: A Case Study of State Programs in the United States*, Case study prepared for the OECD, *www.oecd.org/env/waste/gfenv-extendedproducerresponsibility-june2014.htm*.

Table K.2. **Program performance and program design in low performing States, 2012**

State (implementation date)	Pounds per capita 2012*	Program financing	Performance Goals	Convenience Requirements	Product Scope	Disposal Ban	Centralized Program	Covered Entities
South Carolina (2011)	0.7	Market Share	Yes	Limited	Average	Yes	No	Limited
Oklahoma (2009)	0.6	Funding Obligation Mechanism Not Specified	No	Limited	Limited	No	No	Limited
Virginia (2009)	0.4	Not Specified	No	Limited	Limited	No	No	Limited
Missouri (2010)	0.3	Not Specified	No	Limited	Limited (No TVs)	No	No	Limited

Source: Product Stewardship Institute (2014), *Electronics EPR: A Case Study of State Programs in the United States*, Case study prepared for the OECD, *www.oecd.org/env/waste/gfenv-extendedproducerresponsibility-june2014.htm.*

performing ones. Most of the programs with the highest per-capita collection rate offer a centralized collection and recycling program and set strong convenience standards (i.e. standards that define the extent of collection network required to ensure consumer convenience), which led to the creation of robust infrastructure with year-round financial support regardless of the volumes collected. Clear financial responsibility, whatever the method used (market share, return share or a hybrid system), is also a key indicator of high performers. The relationship of performance goals to program performance is however inconclusive, as while they have clearly been effective in some instances, in other cases they have capped collection levels when manufacturers meet their performance goals or when they are allowed to trade credits. Similarly, the scope of products covered has limited influence on a program's performance. Public education and outreach is an important element impacting performance, but is much more difficult to assess and measure.

b) Design for Environment (DfE)

Though many states implement EPR programs in the hopes of encouraging DfE, it is not necessarily an explicit objective of state legislation. However, seven of the 25 state programs require that manufacturers' products be compliant with the EU Directive on the Restriction of Hazardous Substances (RoHS), or that they notify the state if any of their products exceed these levels. Certain states do also provide for procurement requirements imposed on state agencies to purchase electronic devices containing recycled materials, designed with environmental attributes that consider the product lifecycle, or incorporating DfE elements. Another national program that has significantly impacted the design of electronic equipment is the Electronic Product Environmental Assessment Tool (EPEAT), which serves as an independent environmental rating system that helps purchasers identify greener electronic equipment listed on the database.

3. Economic efficiency

a) Cost efficiency

According to a 2010 report, the estimated total cost (including program and registration fees) for the 19 EPR programs in operation at that time was USD 90 million, including:[4]

- USD 71 million for collection, transportation, and recycling
- USD 14 million for internal compliance costs
- USD 4.5 million for government-incurred administrative costs.

Typically, program costs are evaluated on a cents-per-pound collected basis. They generally do not include education and outreach activities. Costs vary widely among state programs; in 2010 they were reported to be between USD 0.20 and USD 0.30 per pound.

Certain states approve the costs that recyclers charge manufacturers for transportation, consolidation and recycling, which currently range from USD 0.28 to USD 0.34 per pound. Certain products such as cathode ray tube (CRT) glass are costly to handle, while IT equipment can be a revenue generator. As the volume of CRTs goes down over time and the proportional volume of IT equipment increases, program costs are expected to decline.

b) Leakages and free riders

Free-riding problems are minimized in the e-scrap laws thanks to the registration requirements and the penalties imposed on manufacturers or retailers that do not respect the conditions (see 1.b). Some leakages of valuable waste is however noticed in certain states as IT equipment can be resold outside of the EPR programs to brokers or scrap dealers at a high price, leaving the program to manage negative value items such as CRTs.

c) Trade and competition

Since EPR significantly increases the amount of material available for recycling, and some scrap electronics have good value, new recyclers have entered the market in many states and created more competition. The competitive environment for recyclers varies significantly from state to state based on certain program parameters and policy choices. For example, states like Pennsylvania, New York, and Minnesota allow manufacturers flexibility in meeting their recycling obligations, fostering unfettered competition among recyclers. In other states, like Washington, and Vermont, the state or its designated/contracted entity approves the recyclers, thereby managing competition among recyclers through competitive bidding or state approval processes. States with a more competitive environment tend to drive down costs for manufacturers, but may also trim profit margins for recyclers. In addition, it is sometimes difficult for recyclers to increase revenue as equipment that contains high-value materials (laptops and desktop computers) represent a small fraction of the e-scrap stream managed in most state programs, while high-cost CRT TVs and monitors dominate the stream in most states. In order to help recyclers, certain states authorize the landfilling of CRT glass, thereby discouraging investment in CRT recycling.

At external trade level, the export of e-scrap, particularly to under-regulated developing countries, has historically been a contentious issue in the US In 2003, the US International Trade Commission (USITC) issued a report that found that 7% of the sales from US e-scrap recyclers in 2011 went to foreign markets, especially to Mexico, India, Hong Kong, China, China, and other Asia-Pacific countries. However, state e-waste programs are unable to regulate export, as only the US Congress can regulate commerce between countries. The US Congress introduced the Responsible Electronics Recycling Act of 2013 that would make it illegal to send scrap electronics from the US to developing nations. To date, this legislation is still pending.

4. Key issues and possible reforms

The greatest challenge in moving forward is harmonizing the varying approaches used at the state level and expanding them to cover the remaining 45% of the US population. The patchwork of state approaches leads to widely varying program performance, fosters inefficiency, and inflates program costs. The broad disparity in state requirements also drives up manufacturers' internal compliance costs. Until greater harmonization across states occurs, organizations like the Electronics Recycling Co-ordination Clearinghouse reduce administrative overlap between state agencies/departments through inter-state collaboration on key issues and centralized reporting systems.

A national model program would ideally include the following program best practices:

- broad scope of accepted CEDs (which can increase collection efficiency, apportion financial responsibility more fairly, capture more electronic materials, and simplify public messaging)
- comprehensive range of covered entities
- manufacturer funded and led recycling program, including a strong public education and outreach component
- compliance enforcement to ensure a level playing field among manufacturers
- performance requirements with high goals that ensure continuous collection
- strong convenience requirements, and
- minimum standards for processing and recycling e-scrap collected in the program, such as third- party certification programs.

Notes

1. Full source available at: Product Stewardship Institute (2014), *Electronics EPR: A Case Study of State Programs in the United States*, Case study prepared for the OECD, *www.oecd.org/env/waste/gfenv-extendedproducerresponsibility-june2014.htm*.
2. *Used Electronic Products: An Examination of U.S. Exports*, US International Trade Commission, February 2013.
3. Oregon has a program goal for all manufacturers, but only manufacturers who do not participate in the centralized program are penalized if their opt-out program fails to meet a performance goal.
4. Source: National Electronics Recycling Infrastructure Clearinghouse, "Updated Manufacturer E-Waste Compliance Study", March 29, 2009. Internet archive *web.archive.org/web/20130913045004/http://www.ecyclingresource.org/ContentPage.aspx?Pageid=34&ParentID=0*, accessed 4 April 2014.

ORGANISATION FOR ECONOMIC CO-OPERATION AND DEVELOPMENT

The OECD is a unique forum where governments work together to address the economic, social and environmental challenges of globalisation. The OECD is also at the forefront of efforts to understand and to help governments respond to new developments and concerns, such as corporate governance, the information economy and the challenges of an ageing population. The Organisation provides a setting where governments can compare policy experiences, seek answers to common problems, identify good practice and work to co-ordinate domestic and international policies.

The OECD member countries are: Australia, Austria, Belgium, Canada, Chile, the Czech Republic, Denmark, Estonia, Finland, France, Germany, Greece, Hungary, Iceland, Ireland, Israel, Italy, Japan, Korea, Latvia, Luxembourg, Mexico, the Netherlands, New Zealand, Norway, Poland, Portugal, the Slovak Republic, Slovenia, Spain, Sweden, Switzerland, Turkey, the United Kingdom and the United States. The European Union takes part in the work of the OECD.

OECD Publishing disseminates widely the results of the Organisation's statistics gathering and research on economic, social and environmental issues, as well as the conventions, guidelines and standards agreed by its members.

OECD PUBLISHING, 2, rue André-Pascal, 75775 PARIS CEDEX 16
(97 2016 06 1 P) ISBN 978-92-82-25629-3 – 2016

www.ingramcontent.com/pod-product-compliance
Ingram Content Group UK Ltd.
Pitfield, Milton Keynes, MK11 3LW, UK
UKHW051458290926
15471UKWH00008B/1279

9 789264 256293